Lecture Notes in Chemistry

Edited by G. Berthier M.J.S. Dewar H. Fischer
K. Fukui G.G. Hall J. Hinze H.H. Jaffé J. Jortner
W. Kutzelnigg K. Ruedenberg J. Tomasi

46

S.J. Cyvin
I. Gutman

Kekulé Structures in Benzenoid Hydrocarbons

Springer-Verlag Berlin Heidelberg GmbH

Authors

S.J. Cyvin
The University of Trondheim
The Norwegian Institute of Technology
Division of Physical Chemistry
N-7034 Trondheim-NTH, Norway

I. Gutman
Faculty of Science, University of Kragujevac
P.O. Box 60, YU-34000 Kragujevac, Yugoslavia

ISBN 978-3-540-18801-8   ISBN 978-3-662-00892-8 (eBook)
DOI 10.1007/978-3-662-00892-8

© Springer-Verlag Berlin Heidelberg 1988
Originally published by Springer-Verlag Berlin Heidelberg in 1988

2152/3140-543210

PREFACE

This text is an attempt to outline the basic facts concerning Kekulé structures in benzenoid hydrocarbons: their history, applications and especially enumeration. We further point out the numerous and often quite remarkable connections between this topic and various parts of combinatorics and discrete mathematics. Our book is primarily aimed toward organic and theoretical chemists interested in the enumeration of Kekulé structures of conjugated hydrocarbons as well as to scientists working in the field of mathematical and computational chemistry. The book may be of some relevance also to mathematicians wishing to learn about contemporary applications of combinatorics, graph theory and other branches of discrete mathematics.

In 1985, when we decided to prepare these notes for publication, we expected to be able to give a complete account of all known combinatorial formulas for the number of Kekulé structures of benzenoid hydrocarbons. This turned out to be a much more difficult task than we initially realized: only in 1986 some 60 new publications appeared dealing with the enumeration of Kekulé structures in benzenoids and closely related topics. In any event, we believe that we have collected and systematized the essential part of the presently existing results. In addition to this we were delighted to see that the topics to which we have been devoted in the last few years nowadays form a rapidly expanding branch of mathematical chemistry which attracts the attention of a large number of researchers (both chemists and mathematicians).

At first glance one may get the impression that our text comprises sophisticated mathematical considerations, but this is not the case; the level of mathematics required from the reader, matches the skills of normally educated undergraduate chemistry students. Furthermore, all important combinatorial formulas are exemplified and illustrated by pertinent special cases. Thus, organic chemists should be able to use our book without difficulties.

Studies of larger parts of the book in succession, not to mention the whole book, is of interest only for some mathematical chemists, particularly if they actually conduct research in this field. For these readers the book is supposed to be a comprehensive text-book. To a large extent it is self-consistent in the sense that the topics are explained without requiring background references. The topics are partly arranged systematically with respect to the structural characteristics of the benzenoids, and partly progressing from the simplest to more advanced applications of the methods.

In order to make the volume also useful for readers who are not mathematical chemists we have tried to make the results of analyses easily accessible independently of the long path of their derivation. The typical results of this kind are combinatorial formulas for numbers of Kekulé structures (K) for classes of benzenoids and numerical K values. A concrete attempt to increase the usefulness of the book has been made by the arrangement of CHARTS within half of the chapters.

The CHARTS contain collections of K formulas for classes of benzenoids. Each CHART tends to be self-sufficient, giving the necessary definitions for the benzenoid classes in question, supported by figures. The CHARTS also contain full references to the original sources. (All references in the CHARTS are also found in the bibliography of the book.) In effect, the collection of CHARTS may be characterized as a "book within a book".

The designations used for the different benzenoids as members of classes are consistent throughout the book. Nevertheless many of the names are expected to be unfamiliar for most readers, and there are a number of innovations present. Therefore we provide a quick guide to some of the most important benzenoid classes in the following, although only a minute fraction of the classes considered can be covered here. No precise definitions are furnished here, rather the illustrations are intended to give the sufficient preliminary ideas.

October 1987   Sven J. Cyvin

                                                     Ivan Gutman

Quick guide to the definition of some benzenoid classes and their K formulas; K = number of Kekulé structures

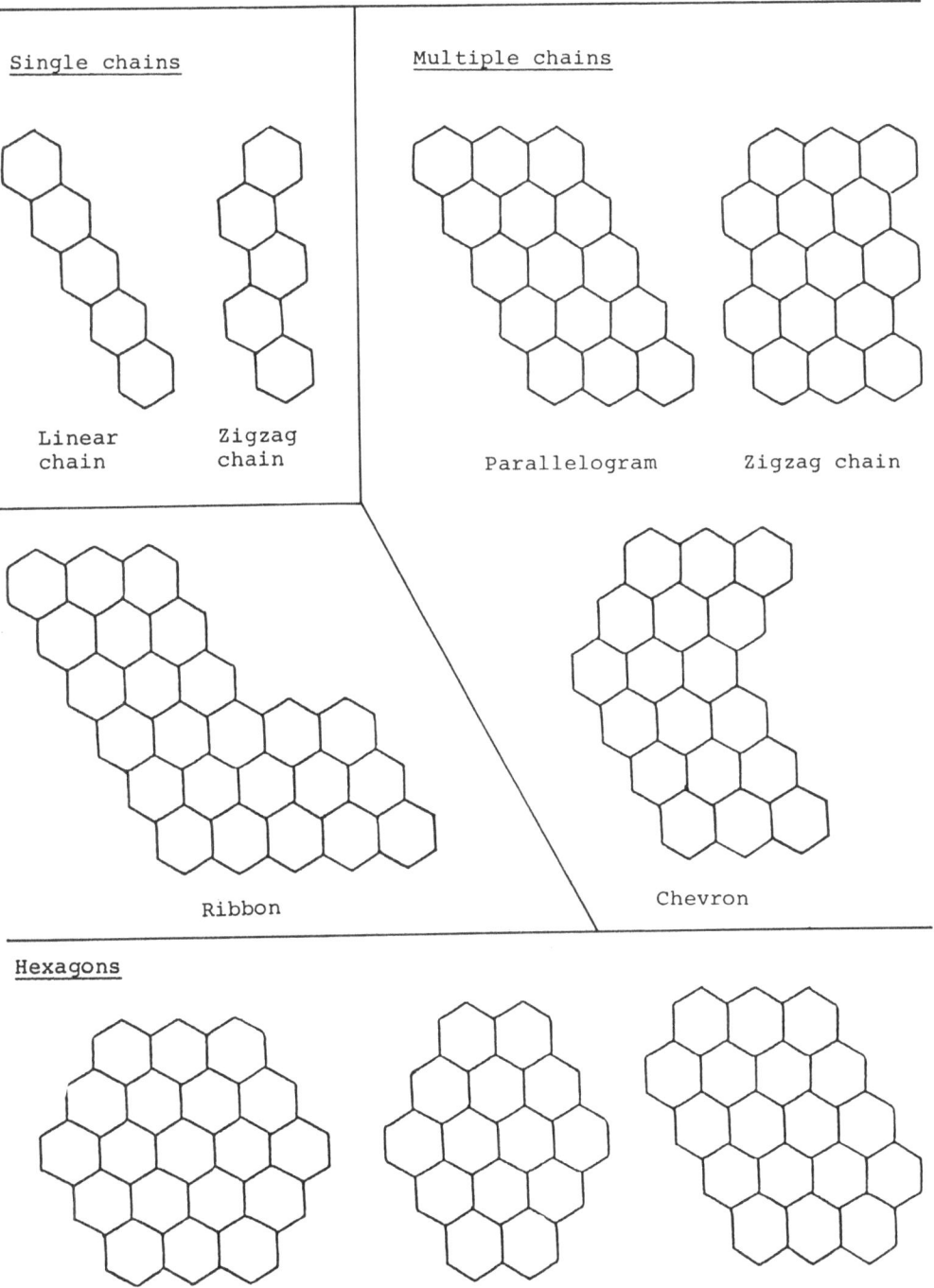

Single chains

Linear chain   Zigzag chain

Multiple chains

Parallelogram   Zigzag chain

Ribbon   Chevron

Hexagons

Rectangles

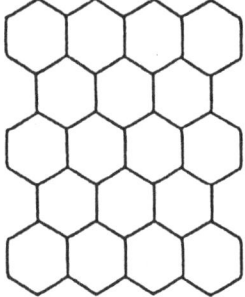 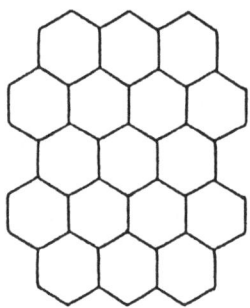

Prolate   Oblate

Pentagons

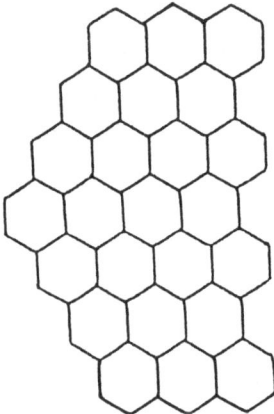 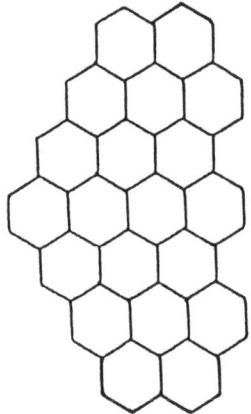

Prolate   Oblate

CONTENTS

LIST OF CHARTS

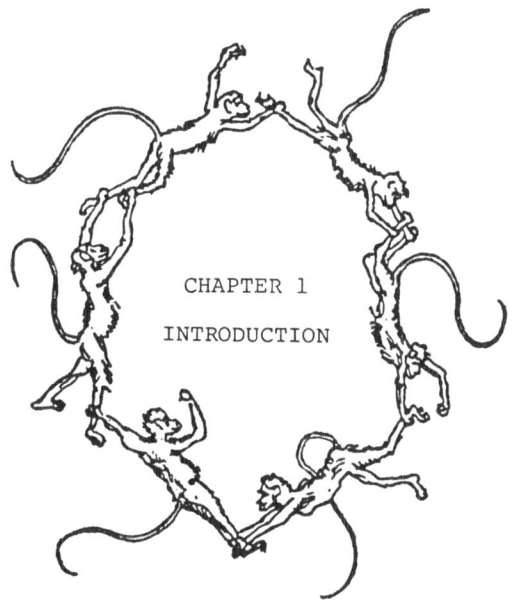

CHAPTER 1

INTRODUCTION

1.1 BENZENOID HYDROCARBONS

Chemists have been faced with benzenoid hydrocarbons and their derivatives
from the earliest days of organic chemistry. These chemical compounds are usually
insensitive, stable over long periods of time, available in large amounts and
cheap. They can be easily purified and characterized and undergo well understood
chemical reactions.

Benzenoid hydrocarbons are ubiquitous substances, produced by incomplete oxi-
dation of wood, coal or petroleum, by frying food etc. They are contained in soot
and smoke. There are strong indications for their existence even in the interstellar
clouds.

Being the starting material for a variety of very important technical and
pharmaceutical products, benzenoid hydrocarbons formed one of the fundaments of the
flourishing development of the organic chemical industries in the second half of
the 19th century.

Several benzenoid hydrocarbons exhibit strong carcinogenic activity.
Especially dangerous is *benzo*[a]*pyrene*, a constituent of tobacco smoke and many
other combustion products. (Alone in USA some 1000 tons of *benzo*[a]*pyrene* are
released annually into the environment by combustion of fossil fuels.)

Around the middle of the last century less than half a dozen of benzenoid

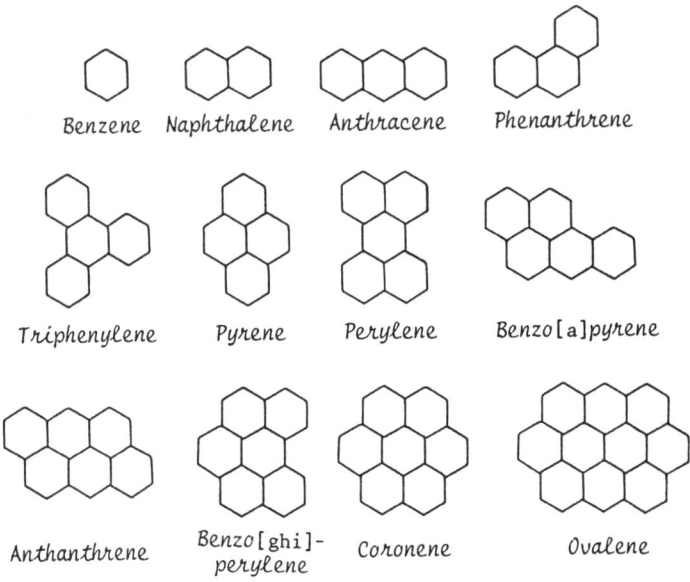

Fig. 1.1. Some benzenoid hydrocarbons.

hydrocarbons were known. In the meantime this family has passed 500 members. For
an exhaustive outline of the chemistry of benzenoid hydrocarbons the classical
monograph of Clar (1964b) should be consulted. Here we wish to get the readers
acquainted with only a few representative benzenoid compounds (Fig. 1). From the
given examples the basic principles of the architecture of benzenoid hydrocarbons
should be clear: they contain fused (condensed) six-membered carbon atom rings.
Hereby it is understood that two rings share exactly two carbon atoms. It can happen that three rings have a carbon atom in common.

As usual, the hydrogen atoms in the formulas (see Fig. 1) are not indicated.
It is tacitly assumed that every carbon atom with two carbon atom neighbours bears a
hydrogen and that no hydrogen is attached to the carbon atoms with three carbon atom
neighbours.

The position of the fourth valence of the carbon atoms (or in modern terms:
the chemical bonding in the molecule) has been a matter of long controversy. The
ambiguities in the valence formulas of the benzenoid molecules puzzled the chemists
almost a whole century. A more or less satisfactory resolution of this problem was
offered by various quantum-mechanical approaches in the thirties and later.

It is not the aim of this book to elaborate or even briefly expose either the
theory of the electronic structure of benzenoid hydrocarbons or its history. These
topics can be found in numerous books on quantum and theoretical chemistry (e.g.
Pauling 1939; Coulson 1952; Streitwieser 1961). Instead we shall remind the reader
of a few facts concerning the history of the so-called Kekulé structural formulas
and then point out the numerous features of chemical theories in which Kekulé
structures or their number play a significant role.

1.2 HISTORICAL REMARKS

Benzene, a liquid of the formula C_6H_6, was first obtained by Faraday in 1825. The discovery of its structural formula by August Kekulé in 1865 is one of the highlights in the theory of chemical structure (Kekulé 1865; 1866).

Kekulé's famous dream (a snake swallowing its own tail) led him to the revolutionary conclusion that the six carbon atoms in *benzene* form a cycle:

Kekulé's early idea was the D_{6h}-type formula I in which, however, the tetravalency of carbon was violated (Kekulé 1866). In order to maintain the usual valency of carbon, the existence of three double bonds in *benzene* had to be postulated, leading to the formula II. This is a Kekulé structural formula (or simply a Kekulé structure) of *benzene*. However, also the formula III obeys all the formal requirements of the theory of chemical structure and thus represents an additional Kekulé structure of *benzene*. Hence, *benzene* has two Kekulé structures.

The fact that the Kekulé structural formula of *benzene* is not unique caused a considerable embarrassment in several generations of chemists and resulted in numerous attempts to resolve the "Benzolproblem". A less serious deliberation along these lines led to the monkey formulas from 1886 (see the vignettes at the start and end of this chapter).

The simplest benzenoid hydrocarbon is *naphthalene*. Its correct chemical formula was first established by Erlenmeyer (1866), only a year after Kekulé's pioneering work on *benzene*. There exist three distinct Kekulé structural formulas for naphthalene, viz.:

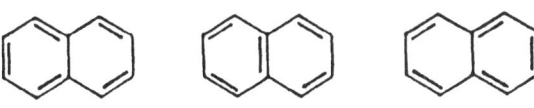

This fact seems, however, not to be mentioned in the literature until 1894. The paper of Marckwald (1894) may be considered as the first publication in the chemical literature which is concerned with the Kekulé structures of a benzenoid hydrocarbon and their enumeration.

The study of Kekulé structures was much promoted by the theory of mesomery, which was put forward in the twenties and was eventually followed by resonance theory and valence bond theory in the thirties. The first publications devoted to elaborate Kekulé structure enumerations seem to be due to Pauling et al. (1935),

Wheland (1935) and Pullman (1946), but still not much was done before the appearance of the classical paper by Gordon and Davison (1952). Even thereafter the progress was not too impressive until some ten years ago, when a rapid proliferation of results occurred and is still in progress. This increase of interest towards Kekulé structures is in a strong causal relationship to the raise of mathematical chemistry (Gutman and Polansky 1986) and the attraction of the attention of numerous scientists to combinatorial issues in organic chemistry.

Although the enumeration of Kekulé structures of benzenoid hydrocarbons has a tradition of almost one century (Marckwald 1894), most of the results in this area were achieved only after 1980. (This fact can be checked by inspection of the list of references.) The aim of the present book is to provide a systematic and, as far as possible, complete survey of these results, along with a considerable amount of original contributions.

1.3 IMPORTANCE OF KEKULÉ STRUCTURES IN THE THEORY OF BENZENOID HYDROCARBONS

1.3.1 *General*

Kekulé structures play (more or less) significant roles in numerous chemical theories, of which resonance theory and valence bond theory are the best known examples (see, e.g. Pauling 1939). The presentation of these theories goes far beyond the ambit of our book. Here we only mention a few theoretical concepts and results which apply to benzenoid hydrocarbons and in which the number of Kekulé structures finds an explicit application. In what follows the number of Kekulé structures of a benzenoid hydrocarbon B will be denoted by $K\{B\}$ or simply by K.

1.3.2 *Total π-Electron Energy*

It has been observed that the total π-electron energy (E) of a benzenoid hydrocarbon, as calculated within the framework of the simple Hückel molecular orbital model, is a linear function of K (Hall 1973). For a hydrocarbon C_nH_s, the following approximate relation could be established (Hall 1981; Gutman and Petrović 1983):

$$E = [0.201n - 0.049s + 0.043K(0.795)^{n-s}] \cdot E(benzene) \qquad (1.1)$$

For some further investigations along the same lines see Cioslowski (1986), Gutman (1986) and the references cited therein.

1.3.3 *Resonance Energies*

A plethora of "resonance energies" and other quantitative measures of "aromaticity" have been proposed in the literature, some of which being explicit functions of K.

In Swinborne-Sheldrake et al. (1975) it was shown that the Dewar resonance

energy, calculated by means of an advanced SCF MO technique, can be well reproduced
by

$$RE = 114.3 \ln K \quad [\text{kJmol}^{-1}] \tag{1.2}$$

This simple expression was eventually applied for the estimation of the resonance
energy of graphite (Stein and Brown 1985).

Randić (1975) defined the benzene-like character of an individual hexagon H in
a benzenoid hydrocarbon B as $2K\{B-H\}/K\{B\}$, where B-H denotes the system obtained by
deleting from B the six carbon atoms belonging to H. The sum of the characters of
all hexagons was proposed as an overall index of aromaticity of B. Later Randić
(1976a; 1980) developed another, somewhat more sophisticated approach to resonance
energy and aromaticity, which goes beyond a simple enumeration of Kekulé structures.

For further relations between resonance energy and K, see Gutman (1981b);
Gutman and Petrović (1981).

1.3.4 *Pauling Bond Order*

Let K_{rs} denote the number of Kekulé structures of a benzenoid hydrocarbon in
which there is a double bond between the carbon atoms r and s. Already in 1935 the
ratio

$$P_{rs} = K_{rs}/K \tag{1.3}$$

was considered as a measure of the order of the chemical bond between the atoms r
and s (Pauling et al. 1935). The quantity P_{rs} is commonly known under the name
"Pauling bond order". The Pauling bond order was successfully correlated with
experimentally determined bond lengths of various benzenoid hydrocarbons (Robertson
and White 1947; Robertson 1948; Herndon 1974a; Herndon and Párkányi 1976; Pauling
1980). The formula

$$d_{rs} = 146.5 - 13.0 P_{rs} \quad [\text{pm}] \tag{1.4}$$

reproduces the interatomic distances (d_{rs}) with an error usually within only 1 pm.

1.3.5 *Miscellaneous Applications*

Various other physico-chemical properties of benzenoid hydrocarbons are found
to be reproducible by means of simple formulas whose one variable is K. These are,
for instance, the first ionization potential (Herndon 1976; Eilfeld and Schmidt
1981), UV/visible absorption band frequencies (Schmidt 1977, Zander 1978), zero-
field splitting parameters (Bräuchle and Voitländer 1982), NMR ortho coupling constants (Herndon 1974a) and rate constants in Diels-Alder reactions of benzenoid
hydrocarbons with maleic anhydride (Zander 1978a; Biermann and Schmidt 1980a; 1980b;
1980c). Kekulé structures have been employed in vibrational force field calculations
(Scherer 1962; Neto et al. 1966; Cyvin BN et al. 1982; Cyvin SJ et al. 1982), and
also in the estimation of heats of atomization (Herndon 1974c). Still more applica-

tions exist; see, e.g. a review by Herndon (1980).

Without going into details we give some more recent references to miscellaneous applications of K numbers: Herndon and Hosoya (1984); Seitz et al. (1985); Klein et al. (1985); Hite et al. (1986); Klein, Alexander, Seitz, Schmalz and Hite (1986); El-Basil (1986b).

Three reviews by Balaban (1982; 1985a; 1985b) dealing with benzenoid hydrocarbons and benzenoid graphs are available.

1.3.6 *Kekulé Structures in Molecular Orbital Theory*

The relation between (Hückel) molecular orbital energy levels and Kekulé structures was first observed by Longuet-Higgins (1950), who found that $K=0$ implies the existence of non-bonding MO's. A more general result along the same line is the discovery that the product of the occupied MO energy levels (in terms of x values) is equal to K (Dewar and Longuet-Higgins 1952). This somewhat unexpected relation between molecular orbital theory and Kekulé structures stimulated a number of subsequent investigations (Ham 1958; Heilbronner 1962; Herndon 1973; 1974b; Cvetković et al. 1974).

1.3.7 *Kekulé Structures in the Aromatic Sextet Theory*

The theory of Clar aromatic sextets[1] (Clar 1972; Gutman 1982a) is another field where the K numbers are found to appear in interesting and non-trivial mathematical relations.

Hosoya and Yamaguchi (1975) introduced the so-called sextet polynomial

$$\sigma(B,x) = \sum_{k=0}^{r} s(B,k) \, x^{k} \tag{1.5}$$

where $s(B,k)$ is the number of ways in which k mutually resonant Clar sextets can be simultaneously drawn in B. By definition $s(B,0) = 1$ for all B, and r is the maximal number of Clar sextets in B. For details of sextet polynomials, see the review of Gutman (1982a). It has been shown (Gutman et al. 1977) that for catacondensed[2] benzenoid hydrocarbons, $\sigma(B,1)$ is equal to the number of Kekulé structures, i.e.

$$\sum_{k=0}^{r} s(B,k) = K\{B\} \tag{1.6}$$

In the case of pericondensed[2] hydrocarbons the above relation (6) is not always obeyed. A complete characterization of benzenoid systems for which $\sigma(B,1) \neq K\{B\}$ was given quite recently (Zhang and Chen 1986b).

---

[1]Defined in the next chapter, Paragraph 2.2.8.
[2]The terms catacondensed and pericondensed are explained in Paragraph 2.2.5.

In the case of non-branched catacondensed benzenoids a further curious identity exists. If B is such a hydrocarbon, then an acyclic graph T_B can be constructed so that $s(B,k)$ is equal to the number of k-matchings of T_B (Gutman 1977). Then
$K\{B\}$ is equal to the Hosoya topological index of T_B (Gutman 1977). The graph T_B has
a number of additional interesting properties (El-Basil 1986a). T_B is sometimes
called a Gutman tree.

For details on k-matchings and the Hosoya topological index, see Hosoya
(1986a); Gutman and Polansky (1986), and references cited therein. Various further
relations in Clar aromatic sextet theory, in which the K numbers are involved, can
be found in Hosoya and Yamaguchi (1975); Ohkami and Hosoya (1983), Gutman et al.
(1984); Zhang et al. (1985).

The vignettes (here and at the beginning of the chapter) are from: Findig FW (1886)
Zur Constitution des Benzols. Berichte der durstigen chemischen Gesellschaft -
Unerhörter Jahrgang: 3535

CHAPTER 2

BENZENOID SYSTEMS: BASIC CONCEPTS

2.1 INTRODUCTION

The main topic of the present book is the enumeration of Kekulé structures in benzenoid hydrocarbons. Since enumeration is essentially a mathematical procedure, it is beneficial to seek for an appropriate mathematical representation of benzenoid hydrocarbons as well as of Kekulé structures. Fortunately both such mathematical objects are already known in combinatorics. These are the benzenoid systems and the perfect matchings.

The present chapter will provide the reader with sufficient information about benzenoid systems and their mathematical properties. In a later part of this chapter the definition of a perfect matching is found together with the explanation of the (obvious) relation between a Kekulé structure and a perfect matching.

A benzenoid system[3] is a combinatorial (or if one prefers: geometrical) object obtained by arranging congruent regular hexagons in a plane so that two hexagons are either disjoint or have a common edge. What is meant under this awkward definition should be immediately clear after a glance on Fig. 1.1 (and subsequent figures). A more precise definition of benzenoid systems is found in the subsequent section.

There is a fairly obvious correspondence between a benzenoid hydrocarbon and a benzenoid system. One example will suffice:

Benzenoid hydrocarbon   Benzenoid system

---

[3] A plethora of names has been proposed in the literature for what we call in this book "benzenoid system". The following list does not pretend to be complete: hexagonal animal, hexanimal, hexagonal system, hexagonal polyomino, polyhex, honeycomb system, fusene (catafusene, perifusene), benzenoid graph, PAH6 (PAH = Polycyclic Aromatic Hydrocarbon).

In order not to elude too far from chemistry, we shall call particular benzenoid systems by the corresponding chemical names. For instance, the benzenoid system of the last example will be referred to as *"phenanthrene."* The counterpart to a six-membered ring is in the following constantly referred to as a "hexagon".

In the following we shall also often abbreviate "benzenoid systems" into "benzenoids".

There exists an extensive literature on mathematical properties of benze-noids. Basic data on this matter as well as further references can be found in Harary (1967); Harary and Harborth (1976); Polansky and Rouvray (1976a; 1976b); Gutman (1974; 1982a; 1983); Trinajstić (1983); Cyvin and Gutman (1986a); Gutman and Polansky (1986).

2.2 DEFINITIONS AND RELATIONS

2.2.1 *Definition of Benzenoid Systems*

Consider a (planar, infinite) lattice of congruent regular hexagons and a cycle Z_B on it. Then the part of the hexagonal lattice which lies in the interior of Z_B and on the cycle Z_B itself, forms a benzenoid system B (Gutman 1982a).

For example, in the below figure the cycle embedded in the hexagonal lattice (heavy line) determines the benzenoid system of *triangulene.*

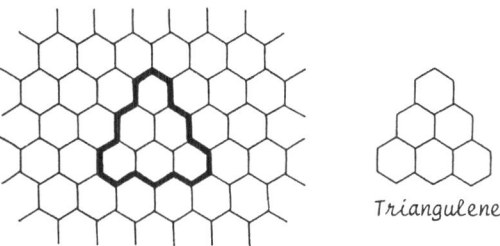

Triangulene

The cycle Z_B is called the perimeter (or sometimes the boundary) of the benze-noid.

2.2.2 *Helicenic and Coronoid Systems*

The definition of a benzenoid system via a cycle on the hexagonal lattice automatically avoids the possibility of overlapping of hexagons or their edges, as well as forming holes of any size. Such systems may still be constructed from con-gruent hexagons, and some members have counterparts in existing hydrocarbons.

An outstanding example of a helicenic system is *hexahelicene* (Newman and Lednicer 1956); cf. Fig. 1. It is realized by a nonplanar structure. Coronoid sys-tems or coronoids[4] are modifications of benzenoid systems. They emerge from a

----

[4]Also called "corona-condensed systems", "coronaphenes", "circulenes", "cycloarenes".

hexagonal lattice by introducing at least two cycles, one embracing the other. Figure 1 shows *kekulene* (Diederich and Staab 1978), a (planar) hydrocarbon corresponding to a coronoid system.

The Kekulé structures and their enumeration have also been studied for helicenic and coronoid systems. These systems are outside the scope of the present book (although some helicenic systems are touched upon in some of the subsequent parts).

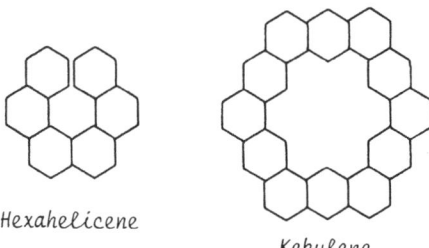

Hexahelicene   *Kekulene*

Fig. 2.1. A helicenic (left) and coronoid hydrocarbon (right).

2.2.3 *Vertices and Edges*

A hexagon contains six vertices, which are connected by six edges: Two adjacent vertices are the endpoints of an edge; they are said to be connected by the pertinent edge. In the very same manner we may speak about vertices and edges of a benzenoid system. Thus, for example, *phenanthrene* (see above) has 14 vertices and 16 edges.

Assume that a benzenoid system is drawn so that some of its edges are vertical. Then we distinguish among the vertices on the perimeter those which point right up or down (Gordon and Davison 1952); they are referred to as peaks and valleys, respectively (Sachs 1984; Gutman and Cyvin 1986). A more precise definition states that a peak lies above its nearest neigbouring vertices while a valley lies below its nearest neighbours. Notice that the identification of peaks and valleys depends on the orientation. *Phenanthrene* in Fig. 1.1, for instance, has two peaks and two valleys. In an alternative orientation (still with some edged vertical; see below) the same system may have one peak and one valley.

2.2.4 *Graphs*

A set of vertices, some of which are connected by edges is called a graph (Harary 1969). Thus a benzenoid system can be viewed as a graph, and then we speak about benzenoid graphs. It is often profitable to consider a benzenoid as a graph.

Benzenoid systems may be represented by different kinds of graphs. Each hexagon of a benzenoid or coronoid system may be represented by a single point or vertex (Smith 1961). If the positions of these vertices are retained in accordance with the centra of the hexagons building up the system in question, the connected vertices are referred to as a dualist graph (also called characteristic graph);[5] Balaban and Harary 1968; Balaban 1982. Figure 2 shows the dualist graphs of *benzene*, *naphthalene, anthracene, phenanthrene, triphenylene, pyrene, perylene, benzo[a]-pyrene, anthanthrene, benzo[ghi]perylene, coronene, ovalene, hexahelicene* and *kekulene*. In the present book dualist graphs will be employed only to a small extent.

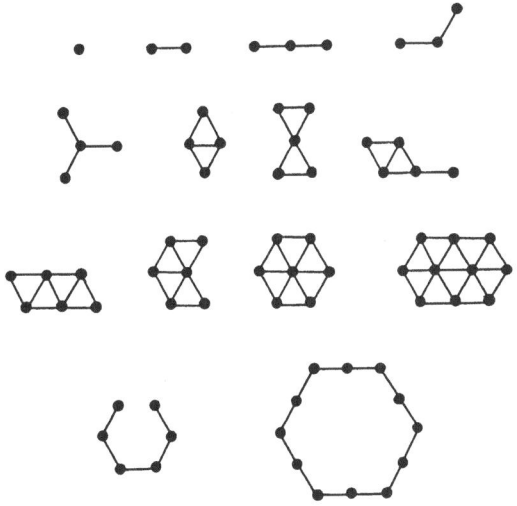

Fig. 2.2. Dualist graphs corresponding to the systems of Figs. 1.1 and 2.1.

2.2.5 *Further Definitions and Relations*

Throughout the present book we use the following notation pertaining to a benzenoid system (B):

h number of hexagons; n number of vertices; m number of edges.

These three quantities are mutually related as

$$h + n = m + 1 \qquad (2.1)$$

The vertices which lie on the perimeter of B are called external. Those vertices (if any) which lie in the interior of the perimeter are internal. Their number is denoted by n_e and n_i, respectively. Then $n_e + n_i = n$. The following relations hold:

---

[5]A "dualist graph" is not a graph in a strict mathematical sense.

$$n + n_i = 4h + 2 \qquad\qquad (2.2)$$

$$m + n_i = 5h + 1 \qquad\qquad (2.3)$$

Let u be a vertex in a graph. The number of vertices which are adjacent to u, is the degree of the vertex u. It is clear that in benzenoid systems only vertices of degree two and three may occur. Their number will be denoted by n_2 and n_3, respectively. Then $n_2 + n_3 = n$. Also:

$$n_2 = 3n - 2m \qquad\qquad (2.4)$$

$$n_3 = 2m - 2n \qquad\qquad (2.5)$$

Here we have defined seven invariants, viz. h, n, m, n_e, n_i, n_2 and n_3. Any pair of them is independent and may be used to express all the rest of them. In addition, Harary and Harborth (1976) have produced inequalities involving h, n and n_i, which are necessary and sufficient for the existence of a benzenoid.

A benzenoid system without internal vertices is said to be catacondensed. Otherwise it is pericondensed. Hence, for catacondensed benzenoids $n_i = 0$ whereas for pericondensed $n_i > 0$. A benzenoid is pericondensed if and only if it contains a vertex which simultaneously belongs to three hexagons.

For instance is *phenanthrene* catacondensed and *triangulene* pericondensed. Further details on different classifications of the benzenoid systems, including the distinction between catacondensed and pericondensed systems will be given in Section 2.3.

2.2.6 *Coloring of Vertices*

The vertices of a benzenoid system may be divided into two groups, which are conventionally called black and white (or starred and unstarred); Coulson and Longuet-Higgins (1948); Longuet-Higgins (1950); Gutman (1974). We require that all vertices adjacent to a black vertex are white and vice versa. Figure 3 shows some examples of the coloring of vertices. Although the assignment of a color to a vertex is arbitrary, the coloring of one vertex determines the colors of all others.

Let the number of black and white vertices be $n^{(b)}$ and $n^{(w)}$, respectively.

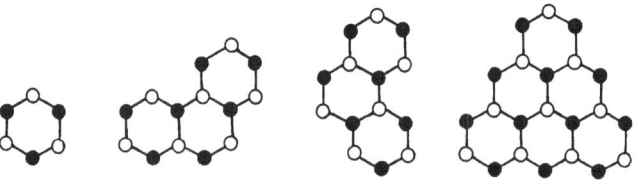

Fig. 2.3. Colored vertices in *benzene*, *phenanthrene* (two orientations) and *triangulene*.

Then $n^{(b)} + n^{(w)} = n$. Correspondingly we designate the numbers of colored external and internal vertices by $n_e^{(b)}$, $n_e^{(w)}$, $n_i^{(b)}$ and $n_i^{(w)}$. Then $n_e^{(b)} + n_e^{(w)} = n_e$ and $n_i^{(b)} + n_i^{(w)} = n_i$. In general

$$n_e^{(b)} = n_e^{(w)} = \frac{1}{2} n_e \qquad (2.6)$$

Let Δ denote the absolute magnitude of the difference between the numbers of black and white vertices, i.e.

$$\Delta = |n^{(b)} - n^{(w)}| \qquad (2.7)$$

In the cases of *phenanthrene* and *triangulene*, for instance (Fig. 3), $\Delta=0$ and $\Delta=2$, respectively.

Introduce the notation n_v and $n_\wedge$ for the number of valleys and peaks, respectively (see Paragraph 2.2.3). Then it has been shown (Cyvin and Gutman 1987a) that Δ also is the absolute magnitude of the difference between the numbers of valleys and peaks or

$$\Delta = |n_v - n_\wedge| \qquad (2.8)$$

Notice that Δ is invariant for a given benzenoid system although the magnitudes of n_v and $n_\wedge$ may depend on the orientation. For *phenanthrene* in the first orientation of Fig. 3, for instance, $n_v = n_\wedge = 2$, while $n_v = n_\wedge = 1$ in the second. In both cases $\Delta=0$.

When $\Delta=0$, then

$$n^{(b)} = n^{(w)}, \qquad n_i^{(b)} = n_i^{(w)}, \qquad n_v = n_\wedge \qquad (\Delta=0) \qquad (2.9)$$

In the remaining of this book when vertices are drawn they are always colored. We will conventionally color all valleys black and consequently all peaks white. As another consequence, the black and white internal vertices will correspond to female and male configurations of edges (Gordon and Davison 1952):

If $\Delta > 0$ we will, as a further convention, draw the benzenoid so that $n^{(b)} > n^{(w)}$ and consequently $n_v > n_\wedge$. All these conventions are fulfilled in Fig. 3.

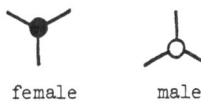

female   male

2.2.7 Perfect Matchings and Kekulé Structures

If e is an edge connecting the vertices u and v, then we say that e covers the vertices u and v. Two edges are said to be independent if they cover different vertices.

A perfect matching[6] of a benzenoid system B is a selection of mutually inde-

---

[6] Mathematicians use "linear factor" and "1-factor" as synonymous for perfect matchings. The name "Kekulé graph" has been proposed in the chemical literature (Trinajstić 1983).

pendent edges of B which cover all vertices of B. Hence if B has n vertices, then its perfect matching contains $n/2$ vertices and n must be even.

As an example we present the three distinct perfect matchings of *naphthalene*. In the below illustration they are compared with the three Kekulé structures of the same system.

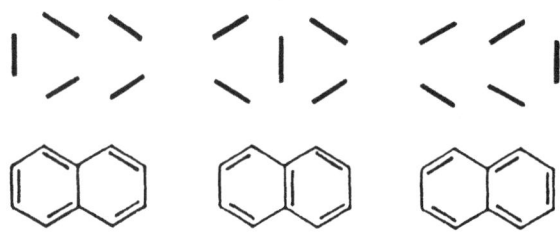

The general one-to-one correspondence between the (mathematical) notion of a perfect matching of a benzenoid system and the (chemical) notion of a Kekulé structural formula of a benzenoid hydrocarbon becomes evident. It is clear in particular that the number of perfect matchings of a benzenoid system is equal to the number of Kekulé structures of the corresponding benzenoid hydrocarbon.

The number of perfect matchings/Kekulé structures in a benzenoid system B will be denoted by $K\{B\}$. Its enumeration is the main topic of the present book.

Consider a benzenoid system B and its perfect matching k. Then some edges of B belong to k and some other edges do not belong to k. For obvious reasons we shall call the edges which belong to k "double bonds" and the edges which do not belong to k "single bonds". Of course the double bond/single bond character of an edge depends on the particular perfect matching considered.

It may, however, happen that a particular edge is single or double in all perfect matchings/Kekulé structures of B. Then we speak about fixed single bonds or fixed double bonds. For instance, the two vertical middle edges of *perylene* in Fig. 1.1 are fixed single bonds. *Zethrene* (Fig. 4) has fixed bonds of both characters.

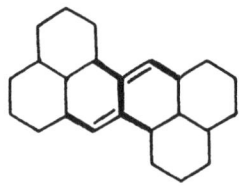

Fig. 2.4. *Zethrene*. Fixed single (heavy) and double bonds are indicated.

2.2.8 *All-Benzenoid Systems*

The Clar aromatic sextets (Clar 1972) occur in varying numbers and constella-
tions in the Kekulé structures. A hexagon is said to be an aromatic sextet when it
has exactly three (alternating) single and three double bonds (as in *benzene*). Pro-
bably the first theory for aromatic sextets was put forward by Armitt and Robinson
(1925), who used the very term "aromatic sextet".

In some benzenoids (see, e.g. Clar and Zander 1958) it is possible to assign
uniquely a constellation of aromatic sextets throughout, so that the remaining
hexagons do not possess additional double bonds. They are referred to as all-benzenoid systems (or all-benzenoids; Polansky and Derflinger (1967).[7] The sextets determine the set of so-called full hexagons, conventionally drawn with inscribed circles,
while the others are referred to as empty. Figure 5 shows a few examples including
the trivial case of *benzene* itself.

Some reports on the topological properties of all-benzenoids are available
(Polansky and Rouvray 1976b; Polansky and Gutman 1980).

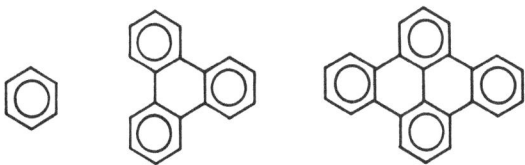

Fig. 2.5. All-benzenoid systems.

2.2.9 *Modes of Hexagons*

We shall classify a hexagon of a benzenoid system according to the number and
position of edges which it shares with the adjacent hexagons. This gives rise to the
total of 13 modes, including the trivial mode L_0 (which occurs only in *benzene*). The
modes are presented in Fig. 6, where also their symbols are indicated.

The complete benzenoid of Fig. 6 possesses all the 12 non-trivial modes. On
the same example we explain the intruding formations of the perimeter called fissure,
bay, cove and fjord.

A hexagon X may be added to the free edge of an L_1, A_2, P_2 or A_3 mode hexagon
of a benzenoid B so that it shares exactly one edge with B; then X acquires the mode
L_1. We will say, by definition, that X is fused to B. In the case of L_1 (for the
hexagon belonging to B) the hexagon X may be fused in two ways so that the L_1 mode
changes into L_2 or A_2. These are the well-known cases of linear and angular annelation, respectively. If the hexagon X is added into a fissure of B so that it

---

[7]Corresponds to "fully benzenoid" in the terminology of Clar (1972).

acquires the P_2 mode, we will say that it is condensed to B. The above defined
terms will also be used in the corresponding sense for a benzenoid X larger than
one hexagon, which may be fused or condensed to B. Then the two units may be described as fused or condensed to each other. Notice also that we use the term "addition" in a general (mathematical) way, referring to any combination of benzenoid
units (including single hexagons). (The term should not be confused with a chemical
addition reaction.)

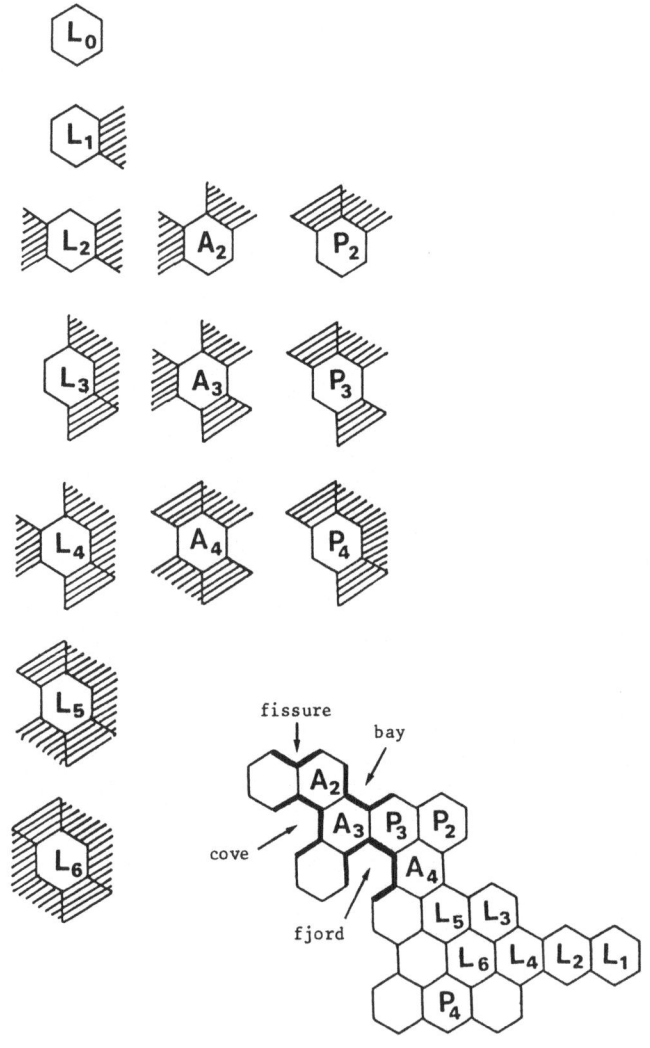

Fig. 2.6. Modes of hexagons in a benzenoid. The subscript of every symbol
indicates the number of edges shared with neighbouring hexagons.

2.3 CLASSIFICATIONS OF BENZENOIDS

2.3.1 *First Classification of Benzenoids*

The concept of modes introduced in the preceding paragraph allows a convenient formulation of a detailed classification of benzenoid systems (Balaban and Harary 1968; Balaban 1969). The fundamental definitions of catacondensed and pericondensed benzenoid systems are also given in Paragraph 2.2.5.

(a) *Catacondensed benzenoids.* Benzenoids of this class possess the modes L_1, L_2, A_2 and A_3. If A_3 is absent, then we speak about unbranched catacondensed systems. Otherwise the systems are branched catacondensed. Examples:

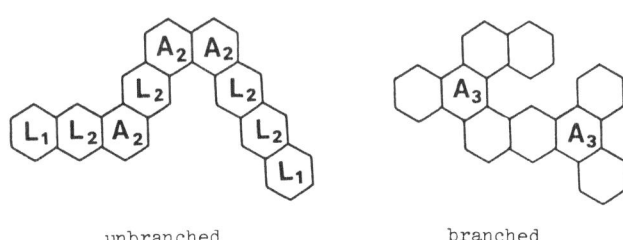

unbranched   branched

In connection with the unbranched catacondensed benzenoids we define the *LA*-sequence as an ordered h-tuple of the symbols L and A (Gutman 1977).[8] The i-th symbol is L if the i-th hexagon is of the mode L_1 or L_2. The i-th symbol is A if the i-th hexagon is an A_2 mode. For instance, the *LA*-sequence of the above unbranched system is *LLALAALLL* or, in abbreviated form $L^2ALA^2L^3$.

(b) *Pericondensed benzenoids.* Benzenoids of this class possess at least one hexagon whose mode **differs** from L_1, L_2, A_2 and A_3. If the L_1, L_2, L_4, A_2, A_3 and P_3 modes are absent, then we have a basic (pericondensed) system (or basic benzenoid). Basic benzenoids possess the modes L_3, L_5, L_6, A_4, P_2 and P_4. A pericondensed benzenoid which is not basic is said to be composite. This means that a composite (pericondensed) system can be cut into two pieces by cutting along only one edge. On the next page we show examples of one basic and three composite benzenoids. A composite (pericondensed) system can be viewed as being composed of two or more units which are basic and occasionally catacondensed; at least one basic must be present.

---

[8]An equivalent of the *LA*-sequence is the ℓ-transform of the three-digit code (Balaban 1977).

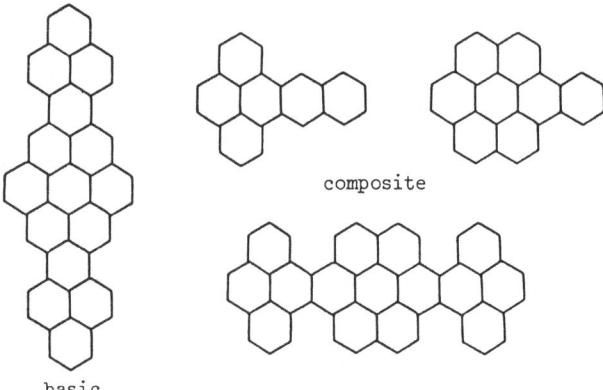

composite

basic

2.3.2 *Second Classification of Benzenoids ("neo")*

From the point of view of the enumeration of Kekulé structures the classification of benzenoid systems into normal (n), essentially disconnected (e) and non-Kekuléan (o) seems to be a rather appropriate one (Cyvin 1986b; Cyvin and Gutman 1986c).

A benzenoid system B for which there exists at least one Kekulé structure is called Kekuléan. Otherwise, i.e. if $K\{B\} = 0$, we say that B is non-Kekuléan.

Kekuléan benzenoids (or Kekuléans) without fixed single or double bonds are by definition normal. Those with fixed single or double bonds are essentially disconnected. For the definition of fixed bonds, see Paragraph 2.2.7.

Among non-Kekuléan benzenoids (or non-Kekuléans) we distinguish between obvious non-Kekuléans for which $\Delta \neq 0$ and concealed non-Kekuléans, for which $\Delta=0$ (Cyvin and Gutman 1987a). The quantity Δ is defined in Paragraph 2.2.6

The essence of such a classification will become clear in the subsequent parts of the book, where also more specific references are included. However, the main bulk of the book deals naturally with Kekuléans, and especially the normal benzenoids among them.

It should be noted that all catacondensed benzenoids are normal (and thus Kekuléan). Only in the case of pericondensed benzenoids we encounter both normal, essentially disconnected and non-Kekuléan systems.

Figure 7 shows all normal benzenoids with $h \leq 6$. Figure 8 follows up with all the essentially disconnected benzenoids for $h \leq 7$. The non-Kekuléan benzenoids for $h \leq 6$ are shown in Fig. 9. They are all obvious non-Kekuléans. Finally the smallest ($h=11$) concealed non-Kekuléans are found in Fig. 10. Since the systematic generation of benzenoids is not a main topic of this book we do not give any special references to it here (but see under the subsequent paragraph 2.3.5).

2.3.3 *Third Classification of Benzenoids (Δ values)*

The benzenoids may be classified explicitly according to their Δ values (cf. Paragraph 2.2.6). A benzenoid with Δ=0 may be Kekuléan (normal or essentially disconnected) or concealed non-Kekuléan. For Δ > 0 the benzenoid is an obvious non-Kekuléan.

2.3.4 *Fourth Classification of Benzenoids (Symmetry)*

A benzenoid may belong to one of eight symmetry groups (Cyvin and Gutman 1986a):

D_{6h}, C_{6h} regular hexagonal; D_{3h}, C_{3h} regular trigonal;
D_{2h} dihedral; C_{2h} centrosymmetrical;
C_{2v} mirror-symmetrical; C_s unsymmetrical.

2.3.5 *Results of Enumeration of Benzenoids*

Benzenoid systems have been enumerated and classified extensively by computer aid. Since this is not a main topic of this book we do not give a complete list of the relevant references here. But one should mention a recent consolidated report by 14 authors (Balaban et al. 1987). It contains a collection of data supplemented by

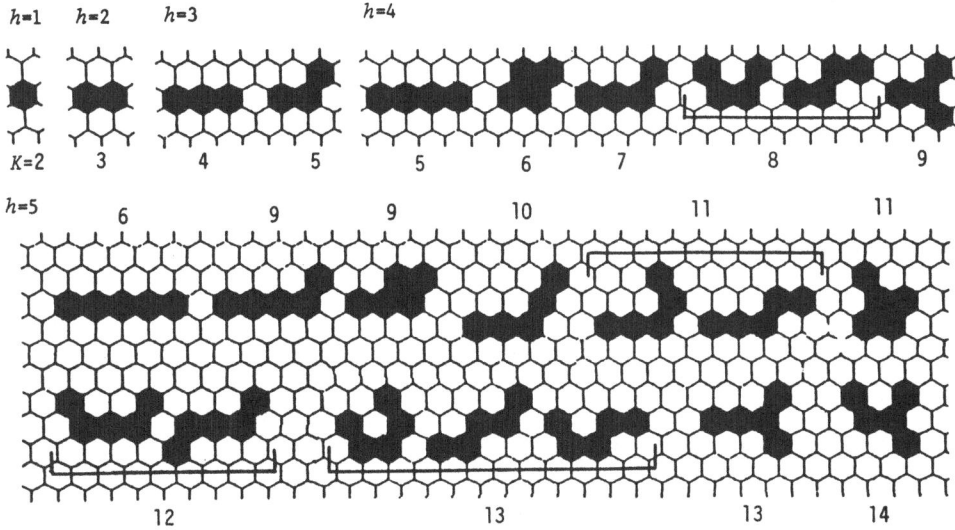

Fig. 2.7. All normal benzenoids with h = 1, 2, 3, 4, 5, and (next page) h = 6. K numbers are indicated.

Fig. 2.7 (continued)

$h=6$

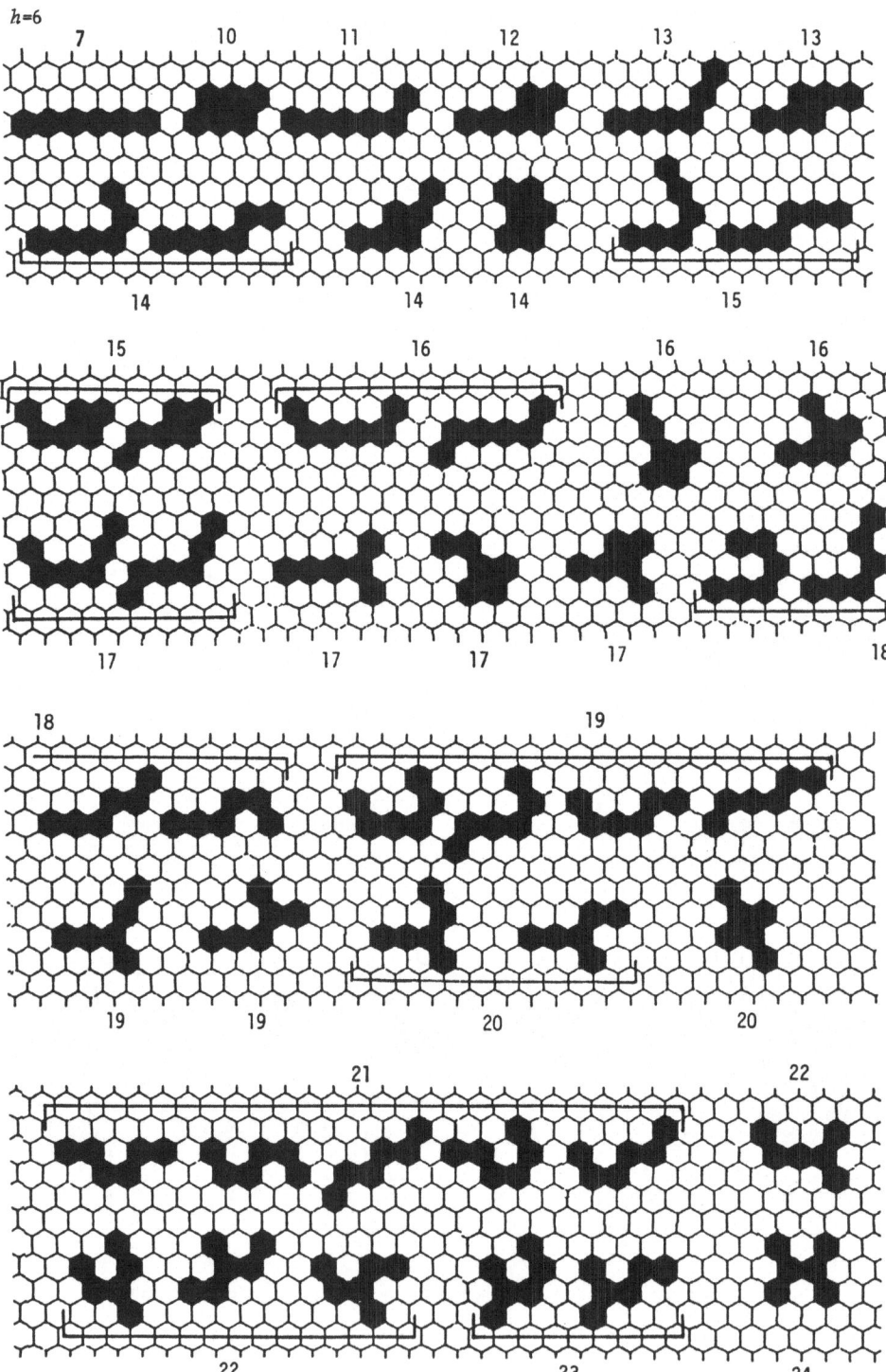

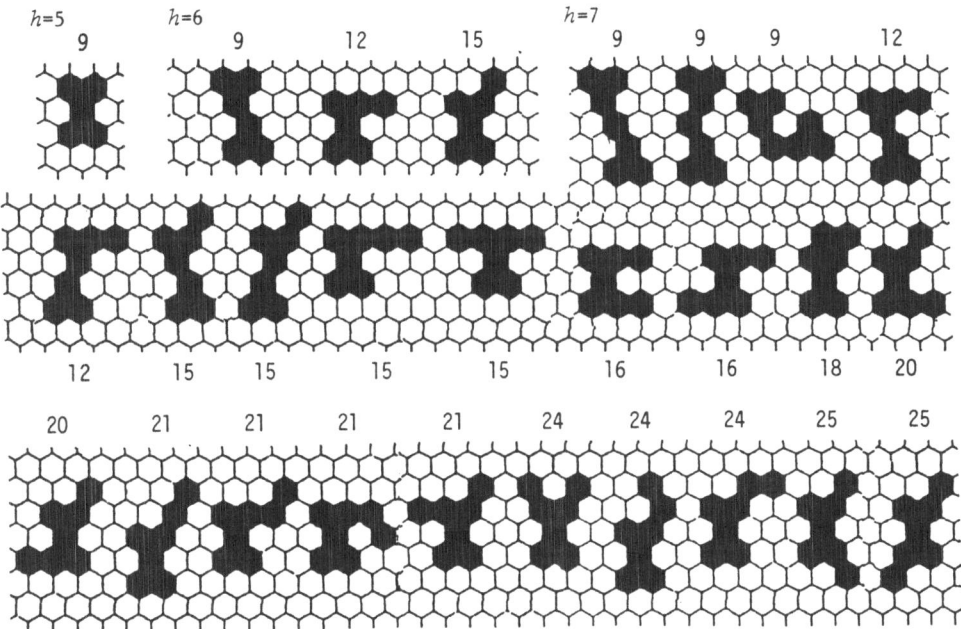

Fig. 2.8. All essentially disconnected benzenoids with h = 5, 6 and 7. K numbers are indicated.

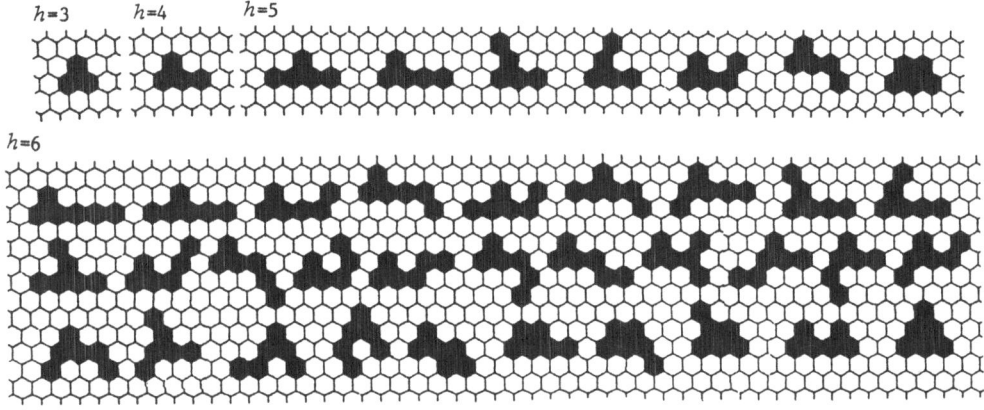

Fig. 2.9. All non-Kekuléan benzenoids with h = 3, 4, 5 and 6. They are obvious non-Kekuléan.

h=11

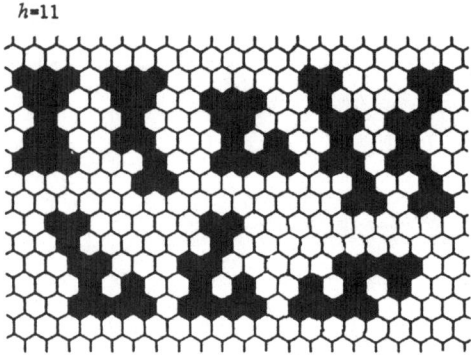

Fig. 2.10. The eight smallest (h = 11) concealed non-Kekuléan benzenoids.

the most recent results and numerous references to previous works. Some of these works are of relevance to our main topic inasmuch as they contain more or less extensive listings of K numbers (Dias 1984; Knop et al. 1985; Brunvoll, Cyvin SJ and Cyvin 1987; Dias 1987).

Table 1 comprises detailed information about the numbers of benzenoids with h up to 10 under different classifications. Most of the data are taken from literature (see Balaban, Brunvoll, Cioslowski, Cyvin, Cyvin, Gutman, He, He, Knop, Kovačević, Müller, Szymanski, Tošić and Trinajstić 1987 and references cited therein), but also some original contributions are present.

Table 2.1. Numbers of benzenoids. The distribution with respect to Δ values and symmetry groups are given.*

h			D_{6h}	C_{6h}	D_{3h}	C_{3h}	D_{2h}	C_{2h}	C_{2v}	C_s	Total
1	Δ=0	n	1	0	0	0	0	0	0	0	1
2	Δ=0	n	0	0	0	0	1	0	0	0	1
3	Δ=0	n	0	0	0	0	1	0	1	0	3
	Δ=1	o	0	0	1	0	0	0	0	0	
4	Δ=0	n	0	0	1	0	2	1	1	1	7
	Δ=1	o	0	0	0	0	0	0	0	1	
5	Δ=0	n	0	0	0	0	1	1	6	6	22
	Δ=0	e	0	0	0	0	1	0	0	0	
	Δ=1	o	0	0	0	0	0	0	3	4	
6	Δ=0	n	0	0	0	0	3	6	10	29	81
	Δ=0	e	0	0	0	0	0	1	0	2	
	Δ=1	o	0	0	0	1	0	0	1	26	
	Δ=2	o	0	0	1	0	0	0	1	0	
7	Δ=0	n	1	0	1	1	3	4	22	135	331
	Δ=0	e	0	0	0	0	0	3	6	14	
	Δ=1	o	0	0	0	0	0	0	10	124	
	Δ=2	o	0	0	0	0	0	0	1	6	
8	Δ=0	n	0	0	0	0	4	28	47	564	1435
	Δ=0	e	0	0	0	0	2	7	2	110	
	Δ=1	o	0	0	0	0	0	0	5	614	
	Δ=2	o	0	0	0	0	0	0	7	45	
9	Δ=0	n	0	0	0	0	5	20	100	2406	6505
	Δ=0	e	0	0	0	0	2	16	29	645	
	Δ=1	o	0	0	0	4	0	0	39	2914	
	Δ=2	o	0	0	1	1	0	0	9	311	
	Δ=3	o	0	0	0	0	0	0	1	2	
10	Δ=0	n	0	0	3	4	10	116	185	10057	30086
	Δ=0	e	0	0	0	0	1	53	31	3647	
	Δ=1	o	0	0	0	0	0	0	20	14004	
	Δ=2	o	0	0	0	0	0	0	38	1878	
	Δ=3	o	0	0	1	1	0	0	0	37	

* n normal; e essentially disconnected; o non-Kekuléan.

CHAPTER 3

KEKULÉ STRUCTURES AND THEIR NUMBERS: GENERAL RESULTS

3.1 INTRODUCTION

In this chapter we outline the basic mathematical results dealing with the
number of Kekulé structures (K) of benzenoid systems. We focus our attention on
generally valid statements, i.e. statements which hold for all benzenoids or for all
Kekuléan benzenoids or (at least) for all catacondensed systems. We are, however,
not going to list here the numerous known relations for K, which apply to more general classes of graphs (namely both benzenoid and non-benzenoid). Exceptions are
Theorem 1 and *Theorem 2* (see below), which happen to be the basis for virtually all
the enumeration techniques employed in the present book.

Some theorems about edges in Kekulé structures are included. An extended definition of isoarithmicity is reported.

Finally some generally valid inequalities involving K numbers are quoted.

3.2 THEOREMS ABOUT K NUMBERS

Theorem 1. Let G be a graph whose edge e connects the vertices u and v. Then

$$K\{G\} = K\{G - e\} + K\{G - u - v\} \tag{3.1}$$

For us this theorem is of special interest when G corresponds to a benzenoid
system, i.e. G = B. The elementary yet fundamental and rather useful result (1) was
certainly known to all early investigators of K (see, for instance, Wheland 1935).
It seems to be first explicitly stated by Randić (1976b). Much of the same can be
stated about the next theorem:

Theorem 2. Let G be a graph where u_1 and u are two adjacent vertices, u_1 being of
the degree one. Then

$$K\{G\} = K\{G - u_1 - u\} \tag{3.2}$$

Here G can not be a benzenoid graph (cf. Paragraph 2.2.5), but benzenoid systems often degenerate to systems with pendent vertices (u_1), for which G is relevant.

An illustration of *Theorem 1* and *Theorem 2* is found in Cyvin and Gutman
(1986a).

Theorem 3. Let G be a graph composed of p disconnected parts: G_1, G_2, ..., G_p. Then

$$K\{G\} = K\{G_1\} \cdot K\{G_2\} \cdot \ \cdots \ \cdot K\{G_p\}$$ (3.3)

This equality is applicable to essentially disconnected benzenoid systems (cf. Paragraph 2.3.2). Otherwise, when dealing with benzenoid systems we are often faced with a degeneration into disconnected subgraphs, which are not benzenoids according to the definitions.

More details concerning the applications of *Theorems 1-3* are found in subsequent sections of this book.

Theorem 4 (Gutman 1981a). Adopting the same notation as in *Theorem 1* we have

$$K\{B - e\}K\{B - u - v\} = \sum_{Z} K\{B - Z\}^2$$ (3.4)

where Z denotes a cycle, and the summation runs over all cycles of B which contain the edge e.

It is instructive to compare *Theorem 1* with *Theorem 4*.

A graph is said to be regular of degree r if all its vertices have the degree equal to r. Let C denote a regular graph of degree two and $p(C)$ the number of its components.

Theorem 5 (Gutman 1984). Let $\{x\}$ symbolize the smallest integer greater than or equal to x. Then for a benzenoid system B,

$$K\{B\} = \left\{ \sum_{C} 2^{p(C)} K\{B - C\} \right\}$$ (3.5)

with the summation going over all regular graphs of degree two, which are (as subgraphs) contained in B.

Let G be a graph whose vertices are labeled by 1, 2, ..., n. Then the adjacency matrix A of G is a square matrix of order n defined via

$$(A)_{uv} = \begin{cases} 1 \text{ if the vertices u and v are adjacent} \\ 0 \text{ otherwise} \end{cases}$$

Theorem 6. If A is the adjacency matrix of a benzenoid system B with n vertices, then

$$\det A = (-1)^{n/2} K\{B\}^2$$ (3.6)

This interesting result was discovered by Dewar and Longuet-Higgins (1952). Its rigorous proof was given somewhat later (Cvetković et al. 1980, pp. 239-243).

According to *Theorem 6* the inverse of the adjacency matrix of a benzenoid system exists only if the system is Kekuléan.

Theorem 7 (Ham 1958, Heilbronner 1962). If B is a Kekuléan benzenoid system, u and v are two adjacent vertices of it, and A is its adjacency matrix, then

$$(A^{-1})_{uv} = \pm K\{B - u - v\}/K\{B\} \tag{3.7}$$

The sign of the right hand side of (7) is a complicated function of the structure of B and the position of u and v; it was determined by Cvetković et al. (1974).

Suppose that the benzenoid system B is drawn in such a manner that some of its edges are vertical. Recall the definition of peaks and valleys in Paragraph 2.2.3. A monotonic path in B is a path connecting a peak with a valley, such that starting at the peak we always go downwards. Two paths are said to be independent if they do not have common vertices. A monotonic path system of B is a collection of independent monotonic paths which involve all the peaks and all the valleys of B.

Theorem 8. $K\{B\}$ is equal to the number of distinct monotonic path systems of B.

The above theorem was discovered by Gordon and Davison (1952) and rigorously proved by Sachs (1984).

If the number of peaks and valleys are equal, we have the following peculiar result.

Theorem 9 (John and Sachs 1985a; John and Rempel 1985). Let p_1, p_2, ..., p_k be the peaks and v_1, v_2, ..., v_k the valleys of B. Define a square matrix W of order k via

$(W)_{ij}$ = number of monotonic paths in B starting at p_i and ending at v_j

Then

$$K\{B\} = |\det W| \tag{3.8}$$

It can be shown (Gutman and Cyvin 1987a) that W_{ij} is equal to the K number of a certain (usually benzenoid) fragment of B.

A benzenoid system B belongs to a symmetry group, say G.

Theorem 10 (Cyvin 1982a; 1983b). Consider the set of Kekulé structures of B as the basis of a representation, Γ_{Kek}, of G. Let χ_{Kek} denote the characters of this representation (see, e.g. Cotton 1971; Gutman and Polansky 1986). When the identity operation is denoted E, then

$$\chi_{Kek}(E) = K\{B\} \tag{3.9}$$

The symmetric structures (Γ_{Kek}, distribtuions into irreducible representations) for a number of benzenoids have been derived (Cyvin 1982a; 1982b; 1983a; 1983b; 1983d; Cyvin and Gutman 1986a).

3.3 VERTICES AND EDGES IN KEKULÉ STRUCTURES

For a Kekuléan benzenoid one has necessarily $\Delta = 0$ (cf. Chapter 2). Then the relations (2.9) are valid, while eqn. (2.6) holds in general.

Let $m_=$ and m_- denote the number of double and single bonds, respectively, in a Kekulé structure. Then $m_= + m_- = m$. When we realize that

$$m_= = \frac{1}{2}(n_2 + n_3) \, , \qquad m_- = \frac{1}{2} n_2 + n_3 \tag{3.10}$$

we obtain easily by means of the relations of Paragraph 2.2.5 (Cyvin and Gutman 1986a):

$$m_= = \frac{1}{2} n \, , \qquad m_- = m - \frac{1}{2} n \tag{3.11}$$

These quantities are invariants, i.e. the same in all Kekulé structures of a benzenoid system.

The following two theorems contain more peculiar statements about some invariants of Kekulé structures.

Suppose again that some edges of the benzenoid system B are drawn vertically. Consider a horizontal line ℓ passing through the center of at least one hexagon of B.

Theorem 11 (Sachs 1984). In all Kekulé structrues of B the horizontal line ℓ intersects an equal number of double bonds.

Theorem 12 (Zhang et al. 1986). All Kekulé structures of B have an equal number of vertical double bonds.

3.4 LOWER AND UPPER BOUNDS OF K

Theorem 13 (Gutman and Cyvin 1988). If B is a normal benzenoid system with h hexagons, then

$$K\{B\} \geq h + 1 \tag{3.12}$$

Equality occurs only in the case of linear single chains (*polyacenes*).

The special case of *Theorem 13* when B is a catacondensed system was proved by Gutman (1982c).

As a consequence of *Theorem 13* it is true that the number of normal benzenoids with a given K is limited. *Benzene* is the unique benzenoid system for which $K=2$. *Naphthalene* and *anthracene* are the unique benzenoids for which $K=3$ and $K=4$, respectively. *Naphthacene* and *phenanthrene* are the only two benzenoids for which $K=5$. *Pentacene* and *pyrene* are the only two benzenoids for which $K=6$. Similarly all benzenoids with $K \leq 8$ have been identified (Gutman and Cyvin 1988). Essentially disconnected benzenoids are encountered when $K \geq 9$, but not for every K value. For every K value when an essentially disconnected benzenoid exists their number is unlimited. All normal benzenoids for $K=9$ and higher K values have also been identified (Cyvin 1986b; Cyvin and Gutman 1986c). This issue (distribution of K numbers) is treated in a later chapter.

Let $K_{max}(h)$ be the maximum number of Kekulé structures of a catacondensed benzenoid system with h hexagons.

Theorem 14 (Chen and Cyvin 1987).

$$K_{max}(h) \leq \begin{cases} 3^{\frac{h}{2}}; & h = 0, 2, 4, 6, \ldots . \\ 2 \cdot 3^{\frac{h-1}{2}}; & h = 1, 3, 5, \ldots . \end{cases} \qquad (3.13)$$

The value for $h=0$ makes sense by virtue of the definition $K_{max}(0) = 1$ pertaining to the trivial case of no hexagons. The inequality (13) improves the upper bound $2^{h-1} + 1$ (Gutman 1982c). A further improvement for $h = 3, 5, 7, \ldots .$ was established, viz. $5 \cdot 3^{\frac{h-3}{2}}$ (Chen and Cyvin 1987).

Another result of this kind, having the form of a recurrence relation, seems to hold (Cyvin 1986b; Chen and Cyvin 1987):

$$K_{max}(h) \leq K_{max}(h-2) + 3K_{max}(h-3); \qquad h \geq 3 \qquad (3.14)$$

With $K_{max}(0) = 1$, $K_{max}(1) = 2$ and $K_{max}(2) = 3$ as initial conditions (14) gives the best results so far. According to a private communication from H. Sachs (1987), the inequality (14) has been proved by P. John.

Theorem 15 (Gutman and Cioslowski 1987). For all benzenoid systems B with n vertices and m edges,

$$K\{B\} \leq \left[\frac{2m}{n} + R\sqrt{\frac{n}{2} - 1} \right]^{1/2} \left[\frac{2m}{n} - \frac{R}{\sqrt{\frac{n}{2} - 1}} \right]^{(n-2)/4} \qquad (3.15)$$

where

$$R = \frac{1}{n} \sqrt{18mn - 12n^2 - 4m^2}$$ (3.16)

Equality in (15) occurs if and only if $n=6$.

3.5 BENZENOIDS WITH EXTREMAL K

Figure 1 shows some benzenoids with minimal and maximal K numbers for given h values. Those with $K = K_{min}$ are simply the linear single chains (*polyacenes*); cf. the first column of drawings. The second column shows the normal pericondensed systems with minimum K, say K_{min}^{np}. The last column pertains to $K = K_{max}$. The maximum is realized for catacondensed systems, which are branched for $h \geq 4$. Every third of them (h = 1, 4, 7, 10) is an all-benzenoid. The normal pericondensed systems with maximum K are depicted in the next-to-last column; they consist of one *pyrene* with catacondensed appendages (for $h > 4$). Within the normal pericondensed systems also the basic benzenoids with minimal and maximal K numbers are shown.

The values of K_{min}^{np} were found to be 6, 9, 10, 14, 15, 20, 21, 27 and 28 for h = 4, 5, 6, ..., 12, respectively (Cyvin BN, Brunvoll, Cyvin and Gutman 1986). The forms up to $h=10$ are found in Fig. 1. Furthermore, it was inferred that

$$K_{min}^{np}(h) = 3h - 6; \qquad h \geq 13$$ (3.17)

and the actual forms were specified in the cited reference. The formula (17) reproduces the K numbers for *aceno*(a)*pyrenes*.

3.6 GENERATION OF NORMAL BENZENOIDS

If a hexagon is added to a normal benzenoid so that it acquires the mode L_1, L_3 or L_5 (see Fig. 2.6), then a new normal benzenoid is generated. The following theorem about K numbers holds under these circumstances.

Theorem 16 (Cyvin and Gutman 1986c). Let B_{h+1} and B_h be two normal benzenoids where B_{h+1} is generated from B_h by adding to it one hexagon. Then

$$K\{B_{h+1}\} > K\{B_h\}$$ (3.18)

3.7 ISOARITHMICITY

It is well known that the K number of an unbranched catacondensed benzenoid is entirely determined by its LA-sequence (see Paragraph 2.3.1), no matter which way the kinks go (Gordon and Davison 1952; Gutman 1983; Balaban and Tomescu 1983). The

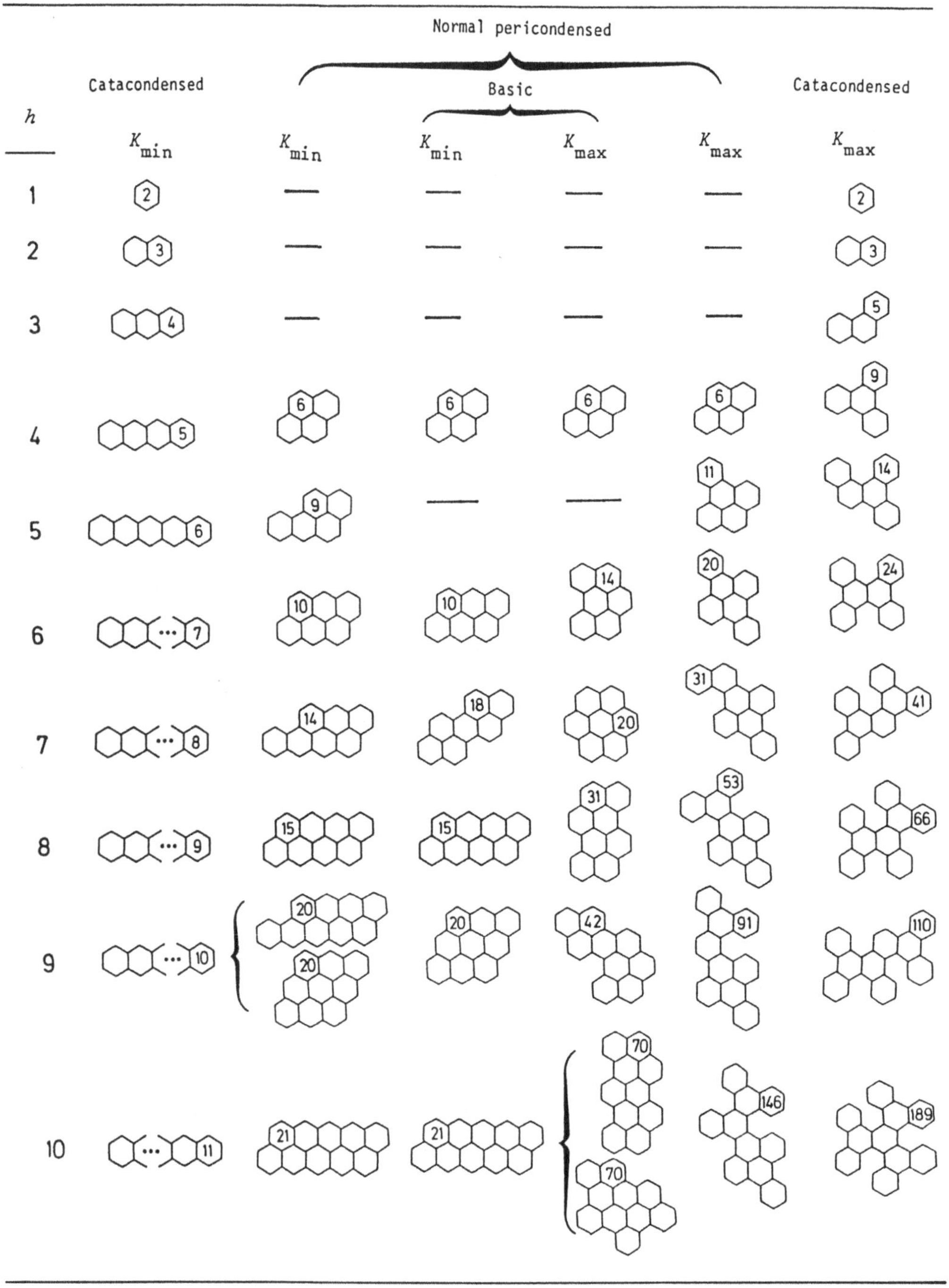

	Catacondensed	Normal pericondensed				Catacondensed
			Basic			
h	K_{min}	K_{min}	K_{min}	K_{max}	K_{max}	K_{max}
1	2	—	—	—	—	2
2	3	—	—	—	—	3
3	4	—	—	—	—	5
4	5	6	6	6	6	9
5	6	9	—	—	11	14
6	7	10	10	14	20	24
7	8	14	18	20	31	41
8	9	15	15	31	53	66
9	10	20, 20	20	42	91	110
10	11	21	21	70, 70	146	189

Fig.3.1. Benzenoids with extremal K. Values of h up to 10. The K numbers are inscribed.

latter authors (Balaban and Tomescu 1983) coined the term isoarithmicity for this phenomenon. As Fig. 2 shows it is also present in branched catacondensed benzenoids and in benzenoids with a catacondensed unit fused to the rest. In Fig. 2.7 the iso-arithmic members are embraced in brackets.

Here we report an extension of the concept of isoarithmicity so that it also becomes applicable to some pericondensed benzenoids without catacondensed appendages. Consider a composite system, say $B_1:B_2$, where two benzenoids B_1 and B_2 are fused to each other. The extended concept of isoarithmicity is based on the following theorem, which holds provided that either B_1 or B_2 or both have an even number of vertices.

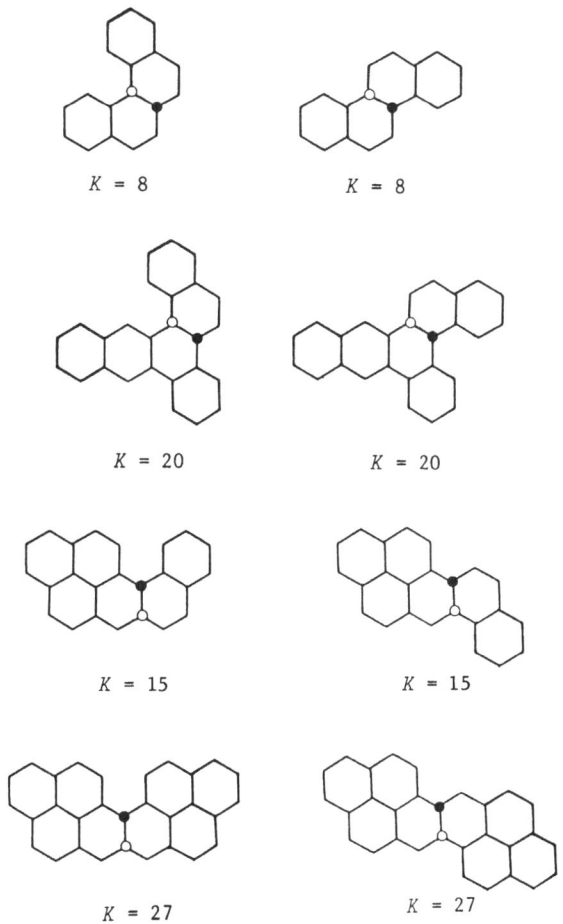

Fig. 3.2. Isoarithmic pairs of benzenoids.

Theorem 17 (Cyvin SJ, Cyvin and Gutman 1987). Produce $(B_1:B_2)^*$ from $B_1:B_2$ by flipping one of the units around the edge of fusion. Then

$$K\{(B_1:B_2)^*\} = K\{B_1:B_2\} \tag{3.19}$$

In Fig. 2 some simple, yet representative examples are shown. The edge of fusion is indicated by (colored) vertices.

CHAPTER 4

INTRODUCTION TO THE ENUMERATION OF KEKULÉ STRUCTURES

4.1 SCHEMATIC SURVEY

The methods which are exploited in this book, are, when it comes to the essence, almost entirely based on *Theorem 1* and *Theorem 2* (Chapter 3). An application of these theorems, or the most useful method of fragmentation (see below), amounts virtually to the same thing.

In an attempt to systematize the topic we have produced the key words:

  (i) empirical methods;
 (ii) algorithm;
(iii) recurrence relations;
 (iv) summation formulas;
  (v) combinatorial formulas.

The issues (ii)-(v) have to do with types of presentation of the derived results.

In addition there are several methods based on principles different from *Theorem 1* and *Theorem 2*. They are mentioned here (Section 4.10) without any detailed treatment.

The examples given in this chapter tend to be the simplest possible ones. More complicated (and less trivial) cases are considered in the subsequent chapters.

4.2 EMPIRICAL METHODS

4.2.1 *Systematic Drawings*

Under point (i) "empirical" one has, first of all, the systematic drawings of all Kekulé structures and counting them at the end. This method is in principle always possible, but needless to say that it very soon becomes impractical for benzenoids of increasing size. Figure 1 shows three simple examples ($K = 4$, 5 and 6). The distinction between *anthracene* and *phenanthrene* ($K = 4$ and 5, respectively) was discussed early; see Pauling (1939). Robertson and White (1947) depicted the six Kekulé forms of *pyrene*. One year later Robertson (1948) presented the forms for *dibenzo*[a,h]*anthracene*, *benzo*[ghi]*perylene* and *coronene*. It has no sense, of course, to attempt to give references to the numerous later works in this direction.

It is more important to point out that it is not necessary to draw all the Kekulé structures in order to determine their number, as is achieved by the method of fragmentation, which plays a crucial role throughout this book.

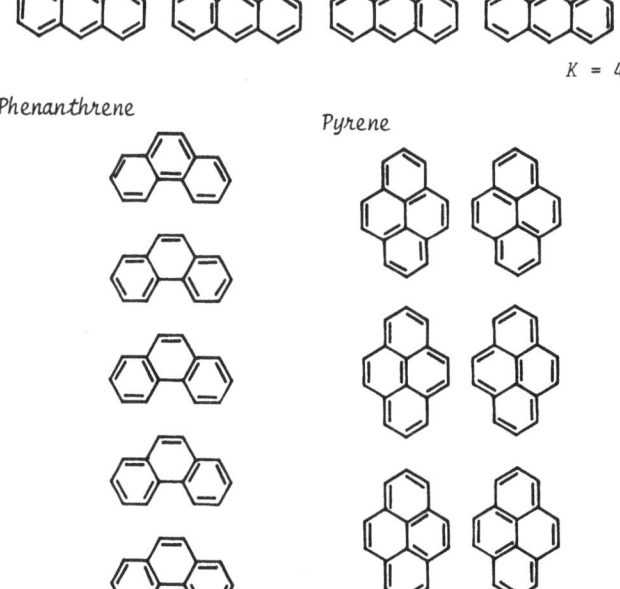

Anthracene

K = 4

Phenanthrene

Pyrene

K = 5

K = 6

Fig. 4.1. Kekulé structures of three benzenoids.

4.2.2 *Method of Fragmentation*

The method of fragmentation is due to Randić (1976b) and summarized in the following.

1. Choose one bond in the benzenoid B and assume it to be double. Delete this bond and all other bonds which are unambiguously determined (single or double) under this assumption. Call the remaining benzenoid $B_=$.

2. Assume the same bond to be single. Delete this bond and all others (if any) which then are unambiguously determined. Call the remaining benzenoid B_-.

3. Then

$$K\{B\} = K\{B_=\} + K\{B_-\} \tag{4.1}$$

It may happen that all bonds become determined during this procedure, as is shown in the example of *pyrene* (Fig. 2). In this case $K\{B_-\} = 1$. It may be interpreted as the number of Kekulé structures for "no hexagons" (see below). Figure 3 shows an application to *coronene*, where the method is applied successively to different bonds.

Figure 4 demonstrates "empirically" that *phenalene* is non-Kekuléan.

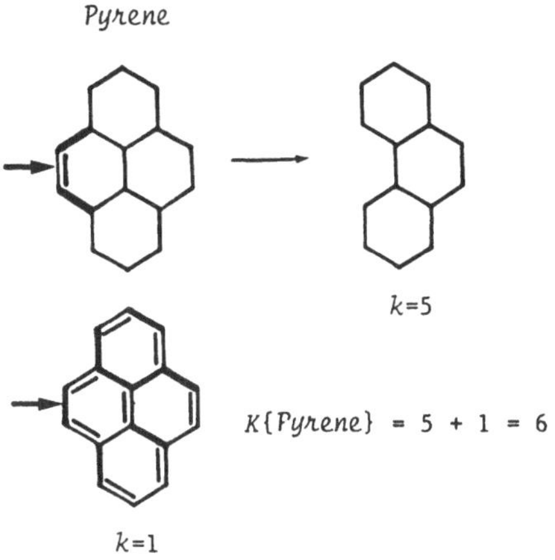

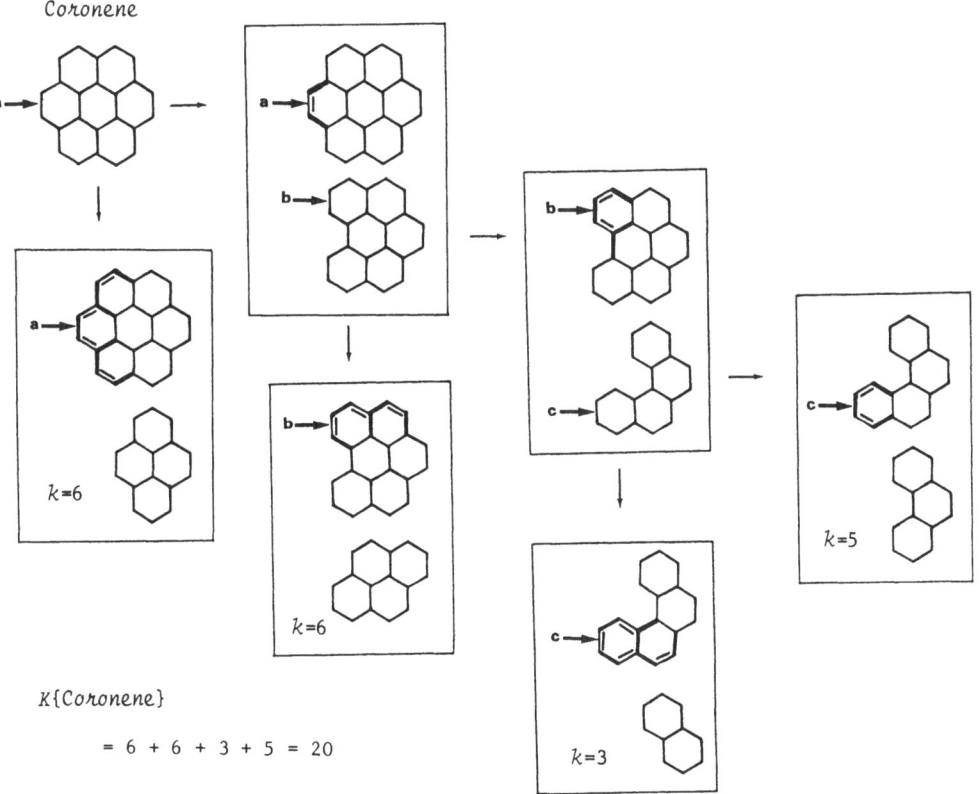

Pyrene

$k=5$

$K\{Pyrene\} = 5 + 1 = 6$

$k=1$

Fig. 4.2. The method of fragmentation applied to *pyrene*.

Coronene

$k=6$

$k=6$

$k=3$

$k=5$

$K\{Coronene\}$

$= 6 + 6 + 3 + 5 = 20$

Fig. 4.3. The method of fragmentation applied to *coronene*.

Phenalene

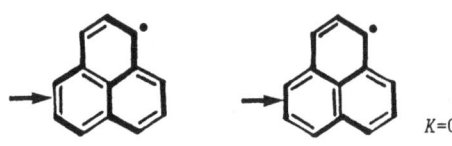

Fig. 4.4. *Phenalene* is non-Kekuléan.

4.2.3 *Degenerate Systems*

The method of fragmentation leads frequently (or always, if carried far enough) to fragments which are not benzenoids and may be considered as degenerate systems. We have already seen a case referred to as "no hexagons". It is necessary to get acquainted with the enumeration of Kekulé structures also in the degenerate cases.

First of all we have the acyclic chains with an odd number of edges (corresponding to *ethylene* and *polyenes*):

This is the typical case of "no hexagons", where $K=1$. In the following we will often tacitly assume that $K=1$ when we refer to "no hexagons" because this case will be encountered so frequently. Care must be taken, however, since non-Kekuléans may degenerate to no hexagons represented by an acyclic chain of an even number of edges:

Such structures have $K=0$.

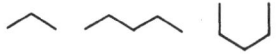

Certain rules have been put forward for acyclic chains attached to a benzenoid (exocyclic bonds; see Trinajstić 1983). When acyclic chains of even number of edges are attached to a benzenoid, they do not change the K number; such exocyclic parts can simply be removed during the enumeration. One acyclic chain with an odd number of edges attached to a Kekuléan makes it non-Kekuléan. Examples:

$K=2$ $K=4$ $K=0$ $K=0$ $K=0$

The results of this kind can easily be found "empirically" by starting with a double bond at the ends of all acyclic chains. For an odd number of edges attached to a non-Kekuléan the situation is more ambiguous.

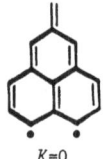

$K=0$

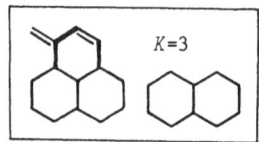

$K=3$

Under this category come also the important cases when several acyclic chains of an odd number of edges each are attached to a benzenoid. For example, two chains, each with an odd number of edges and attached to a Kekuléan benzenoid, can make the system either Kekuléan (occasionally with $K=1$) or non-Kekuléan:

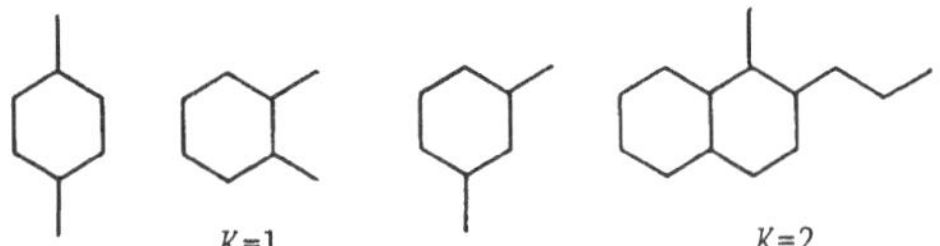

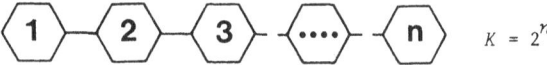

$K=1$ $K=1$ $K=0$ $K=2$

A frequently encountered class of degenerate systems are the disconnected benzenoids. The simplest example is *biphenyl*; cf. Fig. 5 for $n=2$. Here the middle bond (or junction) is a fixed single bond. The two *benzenes* behave independently with respect to their Kekulé structures; hence the K number of the whole system is $2 \cdot 2 = 4$. A simple extension leads to the K formula for *polyphenylenes* in general; see Fig. 5 (Yen 1971; Trinajstić 1983). Two Kekuléan benzenoids form in general a disconnected system when they are joined by an acyclic chain of an odd number of edges. For an even number of edges the system is non-Kekuléan. If non-Kekuléan benzenoids are among the joined units the situation is more complex, but again it can usually be resolved easily "empirically".

$$\langle 1 \rangle - \langle 2 \rangle - \langle 3 \rangle - \langle \cdots \rangle - \langle n \rangle \qquad K = 2^n$$

Fig. 4.5. The class of *polyphenylenes* and its formula for the number of Kekulé structures.

4.2.4 *Modified Method of Fragmentation*

The method of fragmentation (Paragraph 4.2.2) may be modified in many ways, especially by taking advantage of the symmetry properties of Kekulé structures. Here we show a very simple example of a modification, which is applicable whenever the benzenoid B has a vertical plane of symmetry through a peak (or valley). That may occur when B at least has the symmetry of C_{2v}, occasionally D_{2h} or D_{6h} (for non-Kekuléans also D_{3h}).

Assume a double and a single bond at the peak (or valley). This pattern is evidently present in exactly half of the Kekulé structures. Therefore, proceed with this single fragment according to the ordinary fragmentation method, and multiply the result by 2. Figure 6 shows this modified method applied to one of the all-benzenoids (*dibenzo*[e,1]*pyrene*) in Fig. 2.5.

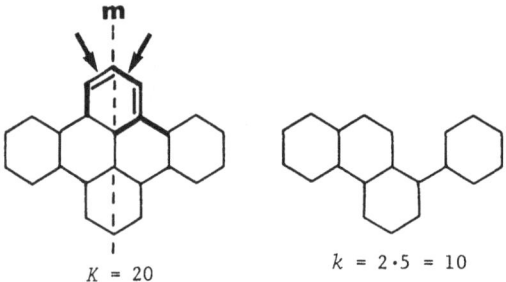

$K = 20$ $k = 2 \cdot 5 = 10$

Fig. 4.6. A modified method of fragmentation applied to a D_{2h} benzenoid with a vertical plane of symmetry, mirror plane m (or two-fold symmetry axis), through a peak
and valley. The total number of Kekulé structures is $K = 2k$.

4.3 COMBINATORIAL FORMULAS, ESPECIALLY FOR THE SINGLE LINEAR CHAIN

Point (i) of Section 4.1 deals with the number of Kekulé structures for a specific benzenoid. The other points are relevant when a class of benzenoids is considered, as is very often of interest. We start at the bottom with point (v), the
combinatorial formulas. Figure 5 shows actually the first combinatorial formula of
K, namely a function of n describing the K numbers for a class of benzenoids depending on an arbitrary value of n. Such a quantity is here referred to as a parameter.
If no special restrictions are imposed, the parameters may have all positive integer
values. Usually also zero is allowed, yielding a degenerate system. In the following
we show the derivation of another very simple one-parameter combinatorial formula
of K.

You can readily deduce the K formula for a single linear chain (*polyacene*) of
n hexagons, say L(n). Figure 1 contains a member of this class, viz. *anthracene* =
L(3). It is practical to include *benzene* itself as belonging to this class (for
$n=1$). When drawing systematically the Kekulé structures you will soon discover that
each of them is characterized by one "vertical" double bond. The number of vertical
bonds is one more than the number of hexagons. Hence the K formula:

$$K\{L(n)\} = n + 1 \qquad (4.2)$$

4.4 RECURRENCE RELATIONS FOR SINGLE LINEAR AND ZIGZAG CHAINS

From eqn. (2) an obvious recurrence relation - see point (iii) of Section 4.1
- follows:

$$K\{L(n)\} = K\{L(n-1)\} + 1 ; \qquad n \geq 1 \qquad (4.3)$$

A less trivial example occurs for the single zigzag chain of n hexagons, say

$A(n)$. Figure 7 shows an application of the method of fragmentation, which leads to

$$K\{A(n)\} = K\{A(n-1)\} + K\{A(n-2)\} ; \qquad n \geq 2 \qquad (4.4)$$

This is a Fibonacci-type recurrence relation. Since $K\{A(0)\} = K\{L(0)\} = 1$ (see below), and $K\{A(1)\} = K\{L(1)\} = 2$ (the case of *benzene*) one may identify the terms of eqn. (4) with the Fibonacci numbers:

$$K\{A(n)\} = F_{n+1} \qquad (4.5)$$

where F_n is the $(n+1)$-th Fibonacci number according to the usual convention in the theory of Kekulé structures. A precise definition reads

$$F_0 = F_1 = 1, \qquad F_{n+1} = F_n + F_{n-1}; \qquad n = 1, 2, 3, \ldots \qquad (4.6)$$

The first Fibonacci numbers are 1, 1, 2, 3, 5, 8, 13, 21, ... A new number is obtained as the sum of the two preceding ones.

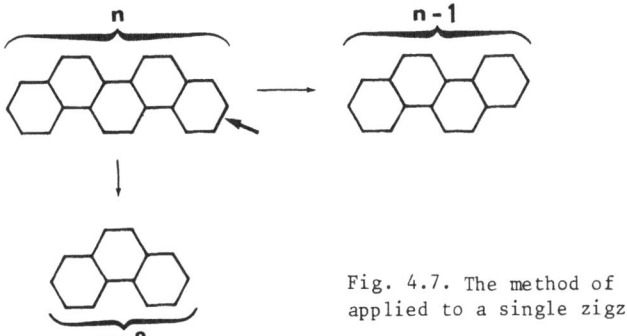

Fig. 4.7. The method of fragmentation applied to a single zigzag chain.

In the above treatment we have already allowed for the possibility of $n=0$, where n is a parameter designating the number of hexagons. It was inferred

$$K\{L(0)\} = K\{A(0)\} = 1 \qquad (4.7)$$

in consistence with eqns. (2) and (5). This is the trivial case of no hexagons (see Paragraph 4.2.3). In Fig. 8 the cases of $L(n) = A(n)$ are identified (Cyvin 1983a).

4.5 SUMMATION FORMULAS FOR SINGLE LINEAR AND ZIGZAG CHAINS

Eqn. (3) along with (7) gives the (rather trivial) summation formula - see point (iv) of Section 4.1 -

L(2) = A(2) $K = 3 = F_3$

L(1) = A(1) $K = 2 = F_2$

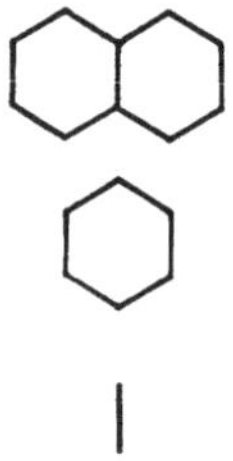

L(0) = A(0) $K = 1 = F_1$

Fig. 4.8. The cases of L(n) and
A(n) for the lowest n values,
where the two types coincide.

$$K\{L(n)\} = K\{L(0)\} + \sum_{i=1}^{n} 1 = \sum_{j=0}^{n} 1 \tag{4.8}$$

In order to pursue this issue further we also put down a summation formula obtained from eqn. (4) along with $K\{A(1)\} = 2$;

$$K\{A(n)\} = 2 + \sum_{i=0}^{n-2} K\{A(i)\} ; \qquad n \geq 2 \tag{4.9}$$

On inserting the Fibonacci numbers into this summation according to eqn. (5) one obtains

$$K\{A(n)\} = 2 + \sum_{i=0}^{n-2} F_{i+1} ; \qquad n \geq 2 \tag{4.10}$$

and furthermore:

$$K\{A(n)\} = 1 + \sum_{j=0}^{n-1} F_j ; \qquad n \geq 1 \tag{4.11}$$

Eqns. (8) and (11) are consistent with (2) and (5), respectively.

4.6 ALGORITHMS FOR SINGLE LINEAR AND ZIGZAG CHAINS

Figure 9 illustrates an algorithm - see point (ii) of Section 4.1 - for the single linear and zigzag chains. A numeral is assigned to each hexagon, in addition

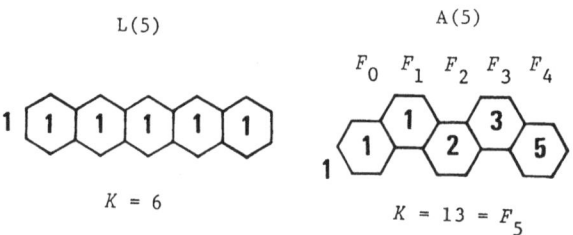

L(5)   A(5)

$K = 6$ $K = 13 = F_5$

Fig. 4.9. Illustration of the algorithm for a single linear (L) and zigzag (A) chain.

to one unity outside the benzenoid. The system of constructing the algorithm is supposed to be evident from the figure. It should also be clear, in view of the algebraic treatment above, that the numerals constructed in this way give the number of Kekulé structures (K) when added together; each of them indicates the increment in K when adding the hexagon in question during a building-up process from the left to the right.

4.7 COMBINATORIAL FORMULA FOR THE SINGLE ZIGZAG CHAIN

Eqn. (5) gives the number of Kekulé structures for a single zigzag chain as a function of n. However, not being an explicit expression in n it still does not satisfy the strict requirements of a combinatorial formula. Such a formula is obtained from Binet's expression for Fibonacci numbers, viz.

$$K\{A(n)\} = \frac{1}{\sqrt{5}}\left[\left(\frac{1 + \sqrt{5}}{2}\right)^{n+2} - \left(\frac{1 - \sqrt{5}}{2}\right)^{n+2}\right] \tag{4.12}$$

This explicit form is obtained by standard mathematical methods (see, e.g. Spiegel 1971) from the recurrence relation (4) and some appropriate initial conditions. One assumes the final formula to be of the form

$$K\{A(n)\} = Aa_1^{\,n} + Ba_2^{\,n} \tag{4.13}$$

where (a_1, a_2) are the solutions of the quadratic equation

$$a^2 = a + 1 \tag{4.14}$$

Here eqn. (14) is to be compared with (4). The solution of (14) is

$$a_1 = \frac{1 + \sqrt{5}}{2}, \qquad a_2 = \frac{1 - \sqrt{5}}{2} \tag{4.15}$$

Hence

$$K\{A(n)\} = A\left(\frac{1 + \sqrt{5}}{2}\right)^n + B\left(\frac{1 - \sqrt{5}}{2}\right)^n \tag{4.16}$$

Here the coefficients A and B are determined by means of the initial conditions $K\{A(0)\} = 1$, $K\{A(1)\} = 2$ (cf. Fig. 8). The result is

$$A = \frac{\sqrt{5} + 3}{2\sqrt{5}}, \qquad B = \frac{\sqrt{5} - 3}{2\sqrt{5}} \tag{4.17}$$

After inserting into (16) that equation indeed reduces to (12).

4.8 TREATMENT OF A PERICONDENSED BENZENOID: THE PARALLELOGRAM

4.8.1 *Introduction*

The derivation and presentation of K numbers will also be exemplified for a class of pericondensed benzenoids in a simple case, viz. the parallelogram. A parallelogram-shaped benzenoid (or simply parallelogram), L(m,n), consists of m condensed rows, each holding n hexagons as in L(n); cf. Fig. 10. For L(m,n) we identify m rows and n columns.

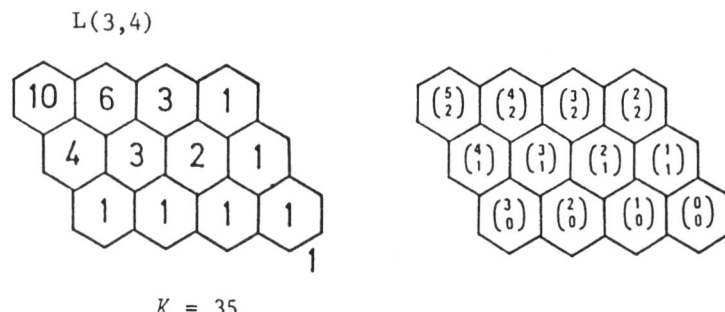

$K = 35$

Fig. 4.10. Example of a parallelogram-shaped benzenoid, L(m,n), with numerals according to the algorithm. Here the benzenoid has 3 rows and 4 columns.

4.8.2 *Algorithm*

Figure 10 includes numerals filled into the hexagons of the parallelogram, along with one unity outside. The numerals inside of the parallelogram constitute a part of Pascal's triangle; hence they may be expressed in terms of binomial coefficients; cf. the right-hand part of Fig. 10. When constructed in this way the number of Kekulé structures (K) is obtained by adding the numerals together. This feature is verified by the subsequent algebraic treatment.

4.8.3 *Auxiliary Benzenoid Class and its Application*

The considered case gives us an occasion to introduce the concept of auxiliary benzenoid classes, which has proved to be very useful in the enumeration techniques for Kekulé structures. Define L(n,m,l) as the parallelogram L(n,m) = L(m,n) augmented by a row of l hexagons, where $0 \le l \le n$; cf. Fig. 11. For the extremal values of l one has by virtue of definition:

$$L(n,m,0) = L(n,m); \qquad L(n,m,n) = L(n,\ m+1)$$

For $m=0$ the system degenerates into L($n,0,l$) = L(l); hence, in accord with eqn. (2),

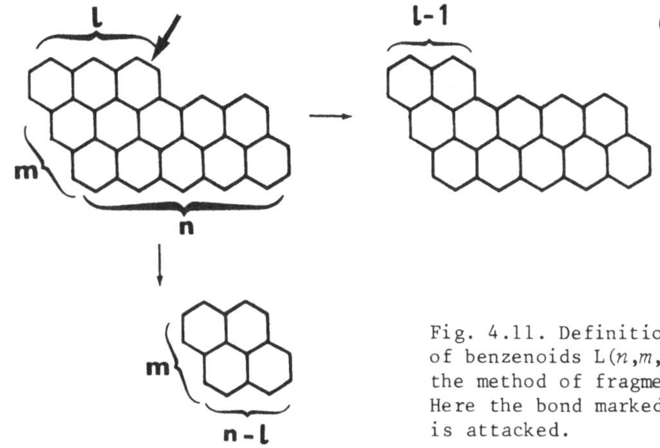

Fig. 4.11. Definition of the auxiliary class of benzenoids $L(n,m,l)$, and the effect of the method of fragmentation applied to it. Here the bond marked with the thick arrow is attacked.

$$K\{L(n,0,l)\} = l + 1 \qquad (4.18)$$

A still more trivial case occurs for $n=0$. Under this condition we have necessarily $l=0$, and

$$K\{L(0,m,0)\} = K\{L(0,m)\} = 1 \qquad (4.19)$$

which is the case of no hexagons (cf. Paragraph 4.2.3). The method of fragmentation (Fig. 11) gives a recurrence formula:

$$K\{L(n,m,l)\} = K\{L(n,m, l-1)\} + K\{L(n-l, m)\}; \qquad l \geq 1 \qquad (4.20)$$

On repeated application of this relation, along with $L(n,m,0) = L(n,m)$, one attains at the summation formula

$$K\{L(n,m,l)\} = \sum_{i=0}^{l} K\{L(n-i, m)\} \qquad (4.21)$$

On inserting $l=n$ one obtains

$$K\{L(n,m,n)\} = K\{L(n, m+1)\} = \sum_{i=0}^{n} K\{L(n-i, m)\} = \sum_{i=0}^{n} K\{L(i,m)\} \qquad (4.22)$$

The usefulness of this relation is most clearly explained by a stepwise application. For $m=0$ it gives, along with eqn. (19)

$$K\{L(n,0,n)\} = K\{L(n,1)\} = \sum_{i=0}^{n} K\{L(i,0)\} = n+1 = \binom{n+1}{1} \qquad (4.23)$$

This is consistent with eqn. (2). As the next step we obtain

$$K\{L(n,2)\} = \sum_{i=0}^{n} K\{L(i,1)\} = \sum_{i=0}^{n} \binom{i+1}{1} = \binom{n+2}{2} \tag{4.24}$$

In general:

$$K\{L(n,m)\} = \sum_{i=0}^{n} \binom{i+m-1}{m-1} = \binom{n+m}{m} \tag{4.25}$$

Hereby we have actually achieved a combinatorial formula for the K number of a parallelogram, $L(n,m)$.

Binomial coefficients are very often encountered in the theory of enumeration of Kekulé structures; for a general mathematical reference, see, e.g. Riordan (1968).

Eqn. (25) may be formulated:

$$K\{L(m,n)\} = \binom{m+n}{n} = \binom{m+n}{m} = \frac{(m+n)!}{m!n!} \tag{4.26}$$

4.8.4 *The Auxiliary Class and the Algorithm Numerals*

Let us turn back to eqns. (20) and (21). By means of eqn. (25) or (26) one obtains

$$K\{L(n,m,l)\} = K\{L(n,m,\, l-1)\} + \binom{n+m-l}{m}\; ; \qquad l \geq 1 \tag{4.27}$$

and

$$K\{L(n,m,l)\} = \sum_{i=0}^{l} \binom{n+m-i}{m} = \binom{n+m+1}{m+1} - \binom{n+m-l}{m+1} \tag{4.28}$$

Now we focus our attention upon eqn. (27). It represents the increment of the K number when l steps one unit. It is obviously the numeral (x) of the algorithm to be placed in the hexagon of the $(m+1)$-th row and l-th column. The equation gives

$$x[i,j] = \binom{n+m-l}{m} = \binom{n+i-j-1}{i-1} \tag{4.29}$$

where we have inserted $i = m+1$ and $l=j$. The labeling of the hexagons in terms of their position with respect to the rows and columns is apparent from Fig. 12. Notice that $i = 1, 2, \ldots, m$, and $j = 1, 2, \ldots, n$. Eqn. (29) is consistent with the algorithm numerals (cf. Figs. 10 and 12).

4.8.5 *Recurrence and Summation Formulas*

From eqn. (29) we obtain the sum of algorithm numerals in the i-th row:

$$\sigma_i = \sum_{j=1}^{n} x[i,j] = \sum_{j=1}^{n} \binom{n+i-j-1}{i-1} = \binom{n+i-1}{i} \tag{4.30}$$

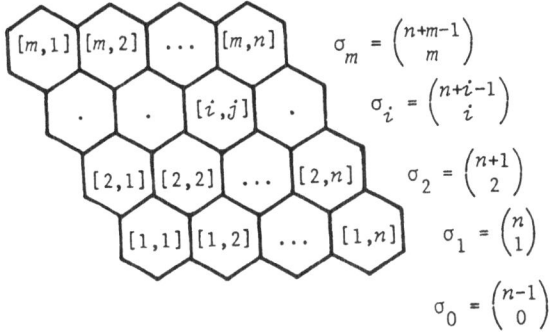

Fig. 4.12. Labeling of the hexagons in a parallelogram. The σ terms are sums of algorithm numerals in each row, supplemented by $\sigma_0 = 1$.

For the sake of completeness we introduce (see Fig. 12)

$$\sigma_0 = \binom{n-1}{0} = 1 \tag{4.31}$$

which represents the algorithm numeral (unity) outside the parallelogram.

It follows

$$K\{L(m,n)\} - K\{L(m-1,\ n)\} = \sigma_m = \binom{n+m-1}{m} = \binom{n+m-1}{n-1}; \quad n \geq 1 \tag{4.32}$$

This is virtually a recurrence relation for the K number of $L(m,n)$. It is consistent with the following recurrence relation put forward by Randić (1976b).

$$K\{L(m,n)\} = K\{L(m-1,\ n)\} + K\{L(m,\ n-1)\} \tag{4.33}$$

A summation formula for $K\{L(m,n)\}$ reads

$$K\{L(m,n)\} = \sum_{i=0}^{m} \binom{n+i-1}{i} = \sum_{i=0}^{m} \binom{n+i-1}{n-1} \tag{4.34}$$

This equation is consistent with eqn. (26). Eqn. (34) may also be formulated

$$K\{L(m,n)\} = \sum_{i=0}^{m} K\{L(i,\ n-1)\} \tag{4.35}$$

Figure 13 shows an illustration of this formula.

4.9 GENERAL REMARKS

A class of benzenoids is defined by one or more parameters, which identify the numbers of hexagons in the different arrangements. Normally the parameters are natu-

ral numbers (positive integers), but allowance is made for zero. Thereby the benzenoid degenerates into a system which still may be meaningful, occasionally the trivial case of no hexagons. If no special restrictions are imposed on a parameter it
may assume anyone of the values 0, 1, 2, ...

An algorithm for K usually covers a wider class of benzenoids than may be
achieved by an algebraic formula.

Among the different algebraic results a combinatorial formula is considered as
an ideal (final) solution. We wish to point out that a final formula, when it depends on more than one parameter, may contain a summation (or product) sign, which
can not be reduced by mathematical manipulations.

The summation formulas we had in mind under point (iv) of Section 4.1 represent in typical cases intermediate steps between a recurrence relation (iii) and a
combinatorial formula (v). Assume in general terms the form of a recurrence relation as

$$K\{B_n\} - K\{B_{n-1}\} = f(n) ; \qquad n \geq 1 \qquad (4.36)$$

Here n is a parameter which characterizes a class of benzenoids, B_n. $f(n)$ is an
algebraic expression containing n (i.e. function of n). Assume furthermore that the
initial condition $K\{B_0\} = f(0)$ holds; in typical cases this equals unity. Then one
obtains the summation formula

$$K\{B_n\} = \sum_{i=0}^{n} f(i) \qquad (4.37)$$

The expression $f(i)$ may often be interpreted as the K number of another class of
benzenoids. It may also be a more complicated expression containing K numbers for
several classes. There are many examples of this kind of formulas in subsequent
parts of the book. A typical feature of these analyses are the extensive manipulations of binomial coefficients (Riordan 1968).

Eqn. (36) represents a two-term recurrence relation. We have also seen (Section 4.4) an example of a three-term recurrence relation. In general such a rela-

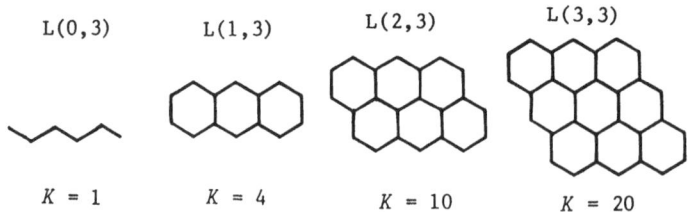

L(0,3)   L(1,3)   L(2,3)   L(3,3)

$K = 1$ $K = 4$ $K = 10$ $K = 20$

Fig. 4.13. Illustration of a summation property of parallelograms. $K\{L(3,4)\}$ for
the parallelogram of Fig. 4.10 is the sum of the four K values displayed here.

tion can be formulated

$$K\{B_n\} = c_1 K\{B_{n-1}\} + c_2 K\{B_{n-2}\}; \qquad n \geq 2 \qquad (4.38)$$

where c_1 and c_2 are independent of n. Standard mathematical methods (Spiegel 1971) lead to explicit (combinatorial) formulas as exponential functions. They are obtained via the solution of a quadratic equation associated with (38). Recurrence relations with more than three terms, in general

$$K\{B_n\} = \sum_{k=1}^{k'} c_k K\{B_{n-k}\}; \qquad n \geq k' \qquad (4.39)$$

lead to higher-order equations: four terms ($k' = 3$) to cubic, five terms ($k' = 4$) to quartic, etc. Depending on the nature of the solution it may be impossible (or impractical) to derive an explicit formula.

Eqn. (39) pertains to one benzenoid class. It may happen that several classes are coupled into recurrence relations. We have in mind benzenoid classes for which the K numbers are linearly dependent so that they each obey the same recurrence relation. Relevant examples are classes which consist of benzenoids with repeated units and their modifications. Benzenoid classes of this category were treated to some extent by Randić (1980), and later by many others (see in particular: Hosoya and Ohkami 1983; Ohkami and Hosoya 1983). A systematic analysis of several classes simultaneously, of which the K numbers are linearly dependent, is due to Gutman (1985b). He employed a method (Spiegel 1971) which we will refer to as the method of coupled recurrence relations.

4.10 OTHER METHODS

4.10.1 *Introduction*

Here we specify a number of methods for determination of K numbers and derivation of K formulas. They are not treated in detail, neither here or elsewhere in the book. However, results which have been obtained by some of these methods are quoted when appropriate.

Klein, Hite, Seitz and Schmalz (1986) have also reviewed the enumeration of Kekulé structures or their equivalents in mathematics and statistical mechanics. They point out: "Often these various methods of solution seem to have been developed (occasionally with some duplications) independently in either organic chemistry or statistical mechanics." A number of references to relevant mathematical methods not cited in the present book are found in the mentioned work of Klein et al. and elsewhere (see, e.g. Hosoya 1986a).

4.10.2 *Application of Coefficients in Hückel Molecular Orbitals*

Herndon (1973) proposed a procedure for determining K from the coefficients of a non-bonding molecular orbital of a radical, obtained by deletion of a carbon atom from the benzenoid molecule considered. Counterexamples have shown (Gutman 1982a) that Herndon's method is not generally applicable.

4.10.3 *Different Combinatorial Methods*

Systematic studies of monotonic independent paths of a benzenoid may be used to enumerate its Kekulé structures in accord with *Theorem 8* (Chapter 3). The method was employed in the classical work of Gordon and Davison (1952).

Reduced Graph and Dualist Graph Models. The reduced graph model is a more modern concept introduced by Džonova-Jerman-Blažič and Trinajstić (1982a; 1982b). In this model each vertical edge is represented by a point, assuming that the benzenoid is oriented so that some of its edges are vertical. The application of the reduced graph model to the enumeration of Kekulé structures is virtually equivalent to the counting of independent path systems. Practical applications have been reported in the two papers cited above and in later works (El-Basil and Trinajstić 1984; El-Basil, Jashari, Knop and Trinajstić 1984; Křivka et al. 1986; Trinajstić and Křivka 1986). By these methods the forms of the Kekulé structures may also be displayed and the aromatic sextets enumerated.

A related, but different method for enumeration and display of Kekulé structures, in addition to the enumeration of aromatic sextets, is referred to as the application of the dualist (graph) model (El-Basil, Křivka and Trinajstić 1984).

Application of the John-Sachs-Rempel Theorem. *Theorem 9* of Chapter 3 has become practically applicable to enumeration of Kekulé structures by virtue of a recent supplement by Gutman and Cyvin (1987a). This is in fact a powerful method, which has been rapidly developed during the last time. In many cases it leads to combinatorial K formulas in terms of determinants. A whole chapter (Chapter 14) is devoted to such determinant formulas. Chapter 14 should also be consulted for additional references, both on theoretical developments and special applications.

4.10.4 *Conjugated Circuits and Kekulé Structures*

Gutman and Randić (1979) have correlated Kekulé structures to conjugated circuits (Randić 1976a). The same idea was employed for catacondensed benzenoids by El-Basil (1982).

4.10.5 *Application of Polynomials*

Trinajstić (1983) has reviewed a number of polynomials in chemical graph theory (see also Knop and Trinajstić 1980), which have different connections with the number of Kekulé structures. One of them is the sextet polynomial; cf. Paragraph 1.3.7.

4.10.6 *Analytical Expressions for the Determinant of the Adjacency Matrix*

Kiang (1980a; 1980b) alias Jiang (1980) has studied the analytical form of the determinant of the adjacency matrix (cf. Chapter 3). With resort to *Theorem 6* he deduced several combinatorial K formulas for classes of benzenoid systems.

4.10.7 *Algorithmic Formula for All-Benzenoids*

Polansky and Gutman (1980) have furnished a general procedure for determining the number of Kekulé structures in an all-benzenoid system, say A. It is manifested in the formula

$$K\{A\} = \sum_{\sigma} 2^{n(F)-f(\sigma)-g(\sigma)} \qquad (4.40)$$

In an all-benzenoid system A there is a unique selection of independent hexagons (called "full") which cover all vertices of A. The number of full hexagons is $n(F)$. Hexagons which are not full are called "empty". A "starring" σ is a set of empty hexagons, such that each hexagon from σ has three double bonds, none of which is shared with a full hexagon. The summation in (40) runs over all starrings. The symbol $f(\sigma)$ designates the number of full hexagons which are adjacent to hexagons of σ; the symbol $g(\sigma)$ stands for the number of hexagons in σ which have three neighbours also belonging to σ.

We will not classify (40) as an explicit (combinatorial) formula since it is not merely composed of mathematical functions of certain parameters. It requires a systematic selection of starrings (σ) and a certain count for each σ; thus it contains an element of algorithm.

4.10.8 *Fully Computerized Method*

The so-called fully computerized method (Cyvin SJ, Cyvin, Bergan 1986; Cyvin 1986a) is actually a polynomial-fitting procedure. In spite of its name it is amenable for analytical treatments for classes of quite large benzenoids; a polynomial of the degree 16 has been determined by this method (Chen 1986b). The method works only when the K formula is known to be a polynomial in one parameter, say n, as often is the case. Also some information about the degree of this polynomial is required. For the sake of exemplification assume that we seek a polynomial of degree 4. Then we write

$$K = P(n) = A + B \binom{n}{1} + C \binom{n}{2} + D \binom{n}{3} + E \binom{n}{4} \qquad (4.41)$$

where the unknown coefficients (A, B, C, D, E) are determined from 5 numerical K values, which must be known. The form (41) makes it possible to find the coefficients successively if the known K values pertain to $n = 0, 1, 2, 3, 4$.

4.10.9 *Transfer-Matrix Method*

A method for studying Kekulé structures and their numbers, referred to as the transfer matrix formulation, is described by Klein, Hite, Seitz and Schmalz (1986). Their paper contains an analytical treatment of the method. It has proved to be a powerful method, which has been used to compute the number of Kekulé structures for *footballene* alias *buckminsterfullerene* (Klein, Schmalz, Hite and Seitz 1986) and other C_{60} carbon cages (Schmalz et al. 1986), all of them non-benzenoid systems. For *footballene* it was reported $K = 12500$ in consistence with Hosoya (1986a).

4.10.10 *Computer Programs*

Džonova-Jerman-Blažič and Trinajstić (1982a) described a computer program for enumeration and generation of Kekulé structures. It is based on the application of the reduced graph model (see Paragraph 4.10.3).

A computer program described by Brown (1983) is based on the determinant of the adjacency matrix; cf. *Theorem 6* (Chapter 3). Basically the same principles have been employed by others (Knop et al. 1985; Brunvoll, Cyvin SJ and Cyvin 1987). These computer methods have also been used for deriving and checking a great number of numerical results in this book.

Different programs have been designed to compute certain polynomials with relevance to K (cf. Paragraph 4.10.5). From these programs the K numbers are obtained as by-products (Hosoya and Ohkami 1983; Ramaraj and Balasubramanian 1985; Hosoya 1986a). A computer program was designed for generating the Kekulé structures of a wide class of graphs (including non-benzenoids), i.e. with all vertices of degree two or three; see Knop et al. (1984).

Finally we mention the computer program of Klein, Hite and Schmalz (1986) employing the transfer-matrix method (cf. Paragraph 4.10.9).

CHAPTER 5
NON-KEKULÉAN AND ESSENTIALLY DISCONNECTED BENZENOID SYSTEMS

5.1 INTRODUCTION

In the present chapter we outline the basic facts about those benzenoids which are not normal. As already explained in Paragraph 2.3.2 these are the non-Kekuléan systems (for which no Kekulé structure can be drawn) and the essentially disconnected systems (which possess fixed single or double bonds).

In Paragraph 2.3.2 the non-Kekuléans have been classified into obvious (for which $\Delta \neq 0$) and concealed (for which $\Delta=0$). Recall that Δ is the difference between the number of black and white vertices (Paragraph 2.2.6); hence Δ is an easily obtainable graph invariant. Consequently, the condition $\Delta \neq 0$ is a trivial, but fully satisfactory structural requirement for a benzenoid system to be obvious non-Kekuléan, and one needs not to elaborate this issue further.

The problem of concealed non-Kekuléans seems to be much less elementary, as it will be documented in the later parts of this chapter.

An essentially disconnected benzenoid possesses (by definition) hexagons in which some of the bonds are fixed. The set of all such hexagons is called the junction of the respective essentially disconnected system. The rest of the system, i.e. when the edges with fixed bonds are deleted, consists of two or more fragments which themselves are normal benzenoids (Gutman and Cyvin 1987b), and which will be called effective units.

The number of Kekulé structures of an essentially disconnected benzenoid is equal to the product of the number of Kekulé structures of its effective units; Gutman and Cyvin 1987b; see also *Theorem 3*, eqn. (3.3).

5.2 INTRODUCTORY EXAMPLES

5.2.1 *Concealed Non-Kekuléan Benzenoids*

The smallest concealed non-Kekuléans have $h=11$. It has been demonstrated (Brunvoll, Cyvin, Cyvin, Gutman, He and He 1987) that there exist exactly eight systems of this category; they are depicted in Fig. 2.10. There are 99 concealed non-Kekuléans with $h=12$ (private communication from W. He and W. He 1987). Some additional examples of non-Kekuléans are found in the subsequent parts of this chapter.

5.2.2 *Essentially Disconnected Benzenoids*

The smallest essentially disconnected benzenoid is *perylene* with $h=5$ (Figs. 1.1 and 2.8). All the essentially disconnected benzenoids with $h = 5$, 6 and 7 are found in Fig. 2.8.

The examples of Fig. 2.8 are all characterized by either (a) two *naphthalenes* as effective units (and therefore $K=9$) and/or the existence of a *perylene* or *zethrene* sub-unit. These conditions are by no means necessary, as shown by the examples of Fig. 1, which are chosen among the $h=8$ systems. Notice especially that the effective unit may be fused to the junction (e.g. the system with $K=18$) and not only condensed to it, as found in the majority of cases.

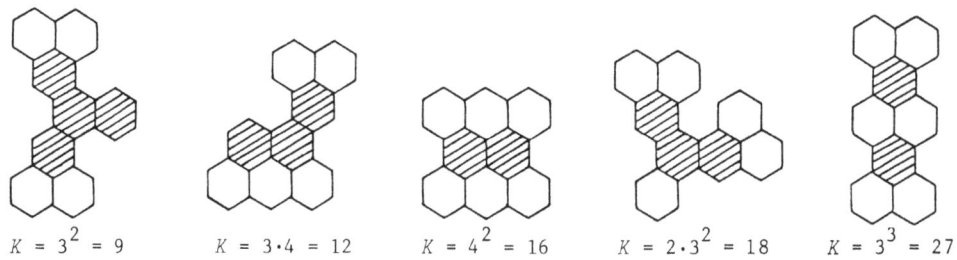

$K = 3^2 = 9$ $\qquad$ $K = 3 \cdot 4 = 12$ $\qquad$ $K = 4^2 = 16$ $\qquad$ $K = 2 \cdot 3^2 = 18$ $\qquad$ $K = 3^3 = 27$

Fig. 5.1. Examples of essentially disconnected benzenoids with $h=8$. The junctions are hatched. K numbers are given.

Following certain rules the junctions may be extended to unlimited sizes (Cyvin and Gutman 1985). Three examples of this kind are given below; all of them have $K=9$.

Other bizarre examples of essentially disconnected benzenoids are found in the subsequent parts of this chapter.

5.3. THE MÜLLER-MÜLLER-RODLOFF RULE

That benzenoids with $K=0$ should be polyradicals and therefore instable and/or highly reactive chemical moietes is a commonly accepted opinion among organic chemists. This statement is sometimes called Müller-Müller-Rodloff's rule because it seems to be first explicitly formulated

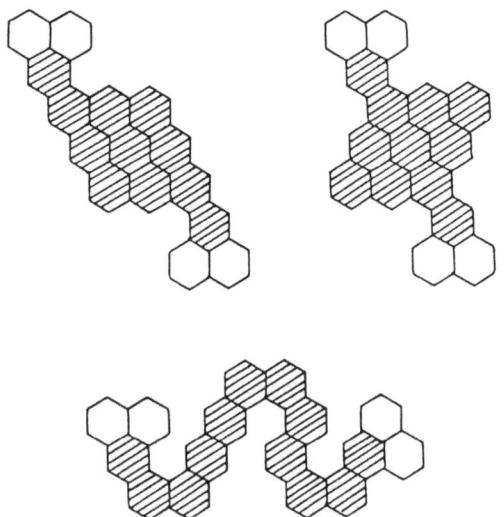

by Müller and Müller-Rodloff (1935). Longuet-Higgins (1950) demonstrated that Müller-Müller-Rodloff's rule follows from molecular orbital theory.

Clar called the chemists' attention to non-Kekuléan benzenoid hydrocarbons (in particular those with $\Delta=2$) and considered them as challenging targets for synthesis (see Clar 1964a; 1972 and the references cited therein). Clar and his coworkers made several synthetic efforts to prepare the following obvious non-Kekuléan benzenoid hydrocarbons, all of which having $\Delta=2$:

None of these attempts was successful; for details, see pp. 108-112 of Clar 1972. At the present moment not a single non-Kekuléan benzenoid hydrocarbon is known to exist.

5.4 CHARACTERIZATION OF CONCEALED NON-KEKULÉAN BENZENOID SYSTEMS

5.4.1 *Introduction*

Finding necessary and sufficient (non-trivial) conditions which a benzenoid system with $\Delta=0$ has to fulfil in order to be (or not to be) Kekuléan was over a long period of time considered as one of the most difficult open problems in this field (Gutman 1982a; 1983; Trinajstić 1983). Recently several important contributions to the problem were published, making it essentially solved (Sachs 1984; Sachs and Zernitz 1984; Zhang, Chen and Guo 1985; Kostochka 1985; He and He 1985; Cyvin and Gutman 1987b).

A plethora of necessary or sufficient (but not necessary *and* sufficient) conditions for a benzenoid system to be Kekuléan were put forward by various authors (Gutman 1974; Balaban 1981; 1982; Sachs 1984; Dias 1985a; 1985b; 1986; Hosoya 1986b). A complete characterization of the structure of Kekuléan benzenoid systems was achieved independently by Zhang, Chen and Guo (1985), Kostochka (1985) and by He and He (1985).

Several algorithms for recognizing and designing concealed non-Kekuléan benzenoids have been developed (Sachs and Zernitz 1984; Hosoya 1986b; Gutman and Cyvin 1986; Cyvin and Gutman 1987a). Counterexamples show that one of these methods (the "peeling algorithm" of Gutman and Cyvin 1986) is not generally applicable (Cyvin and Gutman 1987b).

5.4.2 *First Characterization* (Zhang, Chen and Guo 1985; Kostochka 1985)

The two works report independently the following result which gives necessary and sufficient conditions for a benzenoid to be Kekuléan.

Consider a benzenoid system B and color its vertices as described in Paragraph 2.2.6. The numbers of black and white vertices are denoted by $n^{(b)}(B)$ and $n^{(w)}(B)$, respectively.

An edge-cut of B is a collection e_1, e_2, ..., e_t of edges of B, such that the conditions (a)-(c) are fulfilled.

(a) By deleting the edges e_1, e_2, ..., e_t from B it decomposes into two disconnected parts B' and B".
(b) For each edge e_i; $i = 1, 2, ..., t$, its black end vertex belongs to B' and its white end vertex belongs to B".
(c) For all $i = 1, ..., t-1$ the edges e_i and e_{i+1} belong to the same hexagon of B. Furthermore, e_1 and e_t lie on the perimeter of B.

Theorem 18. A benzenoid system B is Kekuléan if and only if $n^{(b)}(B) = n^{(w)}(B)$ and if for every edge-cut of B,

$$n^{(b)}(B') \geq n^{(w)}(B') \tag{5.1}$$

If for at least one edge-cut of B the relation (1) is violated, then B is a (concealed) non-Kekuléan.

5.4.3 *Second Characterization* (He and He 1985)

Consider a benzenoid system B and its dualist graph D; for details on dualist graphs, see Paragraph 2.2.4.

Any two connected induced subgraphs D' and D" of the dualist graph D is called a cutting of D if the conditions (a)-(c) are fulfilled.

(a) The union of the vertices of D' and D" is the vertex set of D; the union of the edges of D' and D" is the edge set of D.
(b) Both D' and D" possess edges.
(c) Condition (a) is violated if one deletes any vertex from either D' or D".

The number of vertices belonging to both D' and D" is denoted $\nu(D',D")$.

Dualist graphs may possess triangles. We distinguish between upright and upset triangles. A triangle of D is said to be upright if its horizontal edge is at the bottom; a triangle is upset if its horizontal edge is at the top. The numbers of upright and upset triangles of D are denoted by $N_\triangle(D)$ and $N_\triangledown(D)$, respectively.

He and He (1985) arrived at the following result.

Theorem 19. A benzenoid system B is Kekuléan if and only if its dualist graph D has the property that for all its cuttings,

$$\nu(D',D'') \geq \left| N_\triangle(D') - N_\triangledown(D'') \right| \qquad (5.2)$$

The benzenoid system is non-Kekuléan if for at least one cutting of D the relation (2) is violated.

5.5 SEGMENTATION

5.5.1 *Introduction*

Theorem 18 and *Theorem 19* are not easy to use in practice for recognizing the Kekuléan/non-Kekuléan character of a benzenoid system. That is because they imply the inspection of all possible edge-cuts or all possible cuttings, which may be quite incomprehensive for large benzenoids.

In this section we outline some other related conditions which are helpful in deciding whether a benzenoid is Kekuléan or non-Kekuléan and which also can be used for recognizing essentially disconnected systems. Especially for non-Kekuléans the statements are very often sufficient to give the right answer, athough the conditions are not necessary.

5.5.2 *Tracks and Partial Difference Between the Numbers of Valleys and Peaks*

In what follows we assume that the benzenoid system considered is drawn so that some of its edges are vertical. Define an elementary edge-cut as an edge-cut which in addition to the conditions (a)-(c) from Paragraph 5.4.2 also obeys condition (d):

(d) All the edges e_1, e_2, ..., e_t lie in a row (i.e. they are parallel).

The elementary edge-cut defines a segmentation of the benzenoid into the upper and lower segment. Notice that we actually do not separate the benzenoid into two parts.

The number of cut edges in a segmentation is referred to as the number of tracks, t.

Here is an example of an elementary edge-cut (a) and a non-elementary edge-cut (b), both having $t=4$:

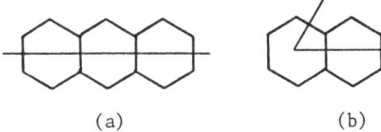

(a)   (b)

Let the numbers of peaks and valleys of the upper segment be denoted by $n_\wedge{}'$ and $n_\vee{}'$, respectively. Correspondingly $n_\wedge{}''$ and $n_\vee{}''$ for the lower segment. Obviously

$$n_\wedge{}' + n_\wedge{}'' = n_\wedge, \qquad n_\vee{}' + n_\vee{}'' = n_\vee \qquad (5.3)$$

If $n_\wedge = n_\vee$ (i.e. $\triangle=0$), then

$$s = n_\wedge{}' - n_\vee{}' = n_\vee{}'' - n_\wedge{}'' \qquad (5.4)$$

is the partial difference between the numbers of valleys and peaks for that particular segmentation.

5.5.3 *Characterizations Based on the Numbers t and s*

In the following statements (Cyvin and Gutman 1987a) the phrase "all segmentations" implies all possible elementary edge-cuts in all possible orientations of the benzenoid system. It is always sufficient to consider three orientations, and this number may be lowered by virtue of symmetry.

1. For Kekuléan benzenoids $s \leq t$ for all segmentations.
2. If $s > t$ in at least one segmentation, the benzenoid is non-Kekuléan.
3. If $s < 0$ in at least one segmentation, the benzenoid is non-Kekuléan.
4. If for a Kekuléan benzenoid $s=t$ in at least one segmentation, the benzenoid is essentially disconnected.
5. If for a Kekuléan benzenoid $s=0$ in at least one segmentation, the benzenoid is essentially disconnected.

Examples. It is clear that most information is gained if the segmentation is executed at the thinnest part of the benzenoid, i.e. a minimum of edges are cut.

Figure 2 shows simple examples, where $s=3$ in all cases, but the number of

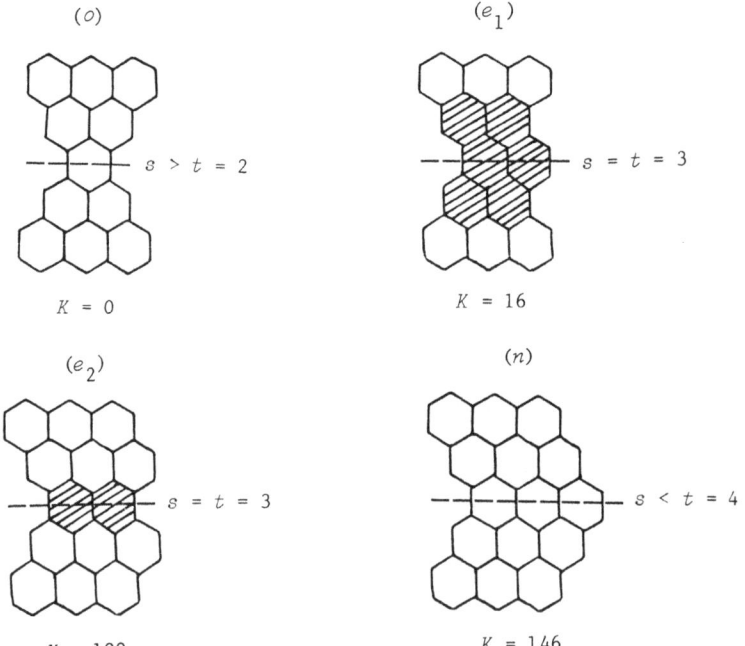

Fig. 5.2. Exemplification of the segmentation (I). See the text for explanations.

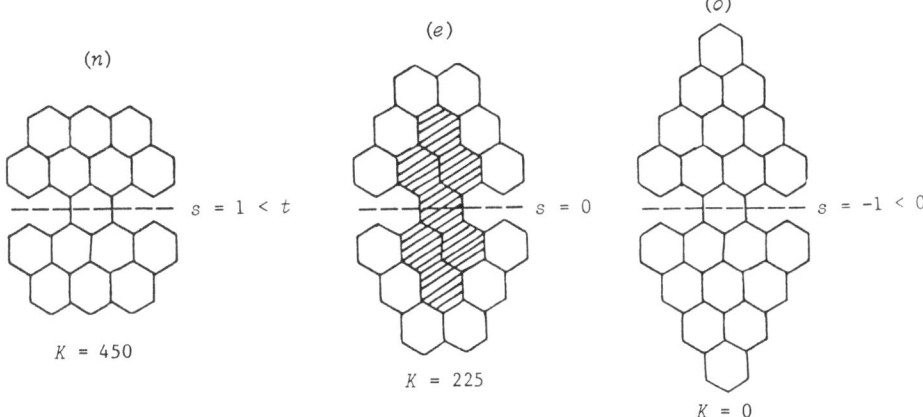

Fig. 5.3. Exemplification of the segmentation (II). See the text for explanations.

tracks differ. It can immediately be inferred (point 2) that the first system (o) is non-Kekuléan. The systems (e_1) and (e_2) must be essentially disconnected if they are Kekuléan (point 4). A closer inspection reveals that this indeed is the case. (Junctions of essentially disconnected benzenoids are hatched as in Fig. 1.) For the sake of completeness also a normal (Kekuléan) benzenoid (n) is included. It does not violate point 1.

Figure 3 includes an example with a negative s number. Here $t=2$ in all cases, while the s number varies. It may immediately be inferred (point 3) that the right-hand system (o) is non-Kekuléan. The two other systems are: (n) normal with $s < t$; (e) essentially disconnected with $s=0$.

The present technique was purposely chosen as simple as possible in order to be applicable in practice. In consequence, there are subtle cases where this segmentation technique fails to be decisive. Below we show an essentially disconnected benzenoid (e) for which neither $s=t$ (point 4) or $s=0$ (point 5) is realized for any segmentation (based on elementary edge-cuts). For the non-Kekuléan benzenoid (o) neither $s > t$ (point 2) or $s < 0$ (point 3) is realized.

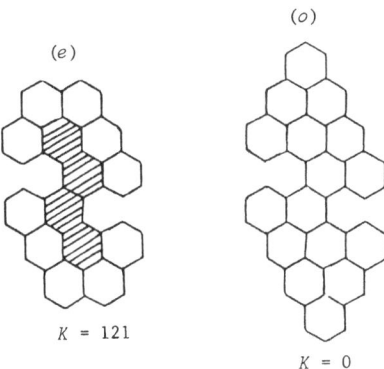

Perhaps more typical examples are those of Fig. 4. The essentially disconnected
benzenoid (e) has $s=t$ in one orientation and $s=0$ in an other; the non-Kekuléan benzenoid has $s > t$ in one orientation and $s < 0$ in an other.

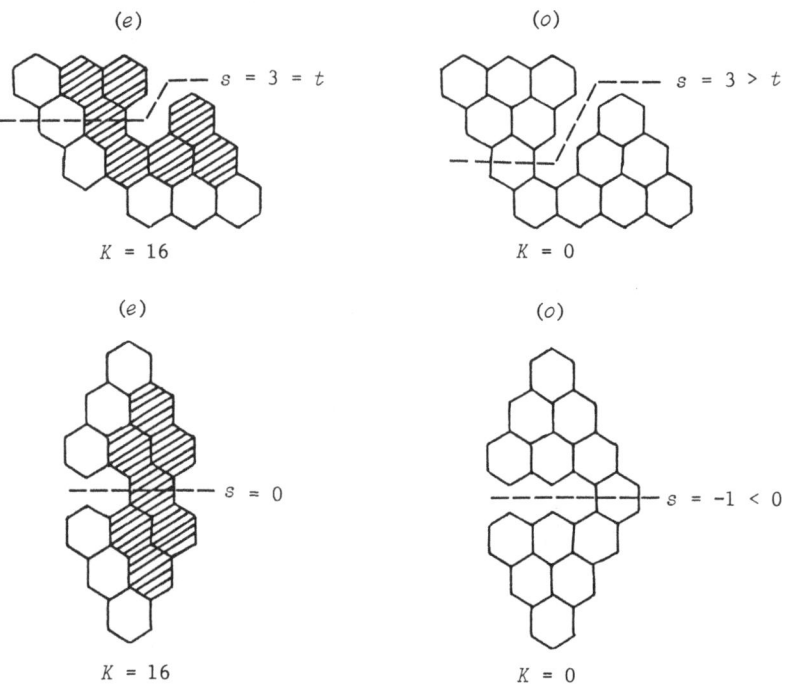

Fig. 5.4. Exemplification of the segmentation (III). See the text for explanations.

6.1 PREVIOUS WORK

The single linear and zigzag chains are treated in Chapter 4. It is reasonable to give Gordon and Davison (1952) credit for eqn. (4.2) therein, although it has certainly been discovered independently many times; cf. the simple considerations of Section 4.3 (Cyvin and Gutman 1986a). The identification of the number of Kekulé structures (K) for a single zigzag chain with the Fibonacci numbers was also mentioned already by Gordon and Davison (1952). The explicit form corresponding to Binet's formula was first given by Yen (1971) and independently by Cvetković and Gutman (1974). Cyvin (1983a) re-derived the connection between Fibonacci numbers and the number of Kekulé structures for single zigzag chains, and supplemented the treatment by group-theoretical considerations of symmetry. A treatise on three connections between Fibonacci numbers and Kekulé structures is due to Balaban and Tomescu (1984); see also Hosoya (1973).

The classical paper of Gordon and Davison (1952) contains a general algorithm for the enumeration of Kekulé structures of catacondensed benzenoids (single chains), branched or unbranched. Cyvin (1983c) gave an alternative derivation for the case of unbranched chains. This case was revisited by Cyvin and Gutman (1986b), who produced a useful modification of the Gordon and Davison algorithm.

Balaban and Tomescu (1983) elaborated a system for producing explicit algebraic formulas for the K number of an arbitrary catacondensed benzenoid. This system had to be somewhat complicated as a consequence of the general nature of the problem.

Supplementary references are cited in the subsequent sections; they contain formulas for more specialized classes of the benzenoids in question.

In conclusion it may be stated that, in a sense, the enumeration problem for Kekulé structures of catacondensed benzenoids is completely solved.

6.2 SINGLE UNBRANCHED CHAIN

6.2.1 *Introductory Remarks Including some Helicenic Systems*

A single chain may have a number of kinks. Their distribution is characterized by the *LA*-sequence (Chapter 2). We will speak about segments of a chain as the linear parts between the kinks. They are defined so that every hexagon of an A mode belongs to two neighbouring segments. For the number of Kekulé structures it is immaterial which way the kinks go (isoarithmicity; see Chapter 3).

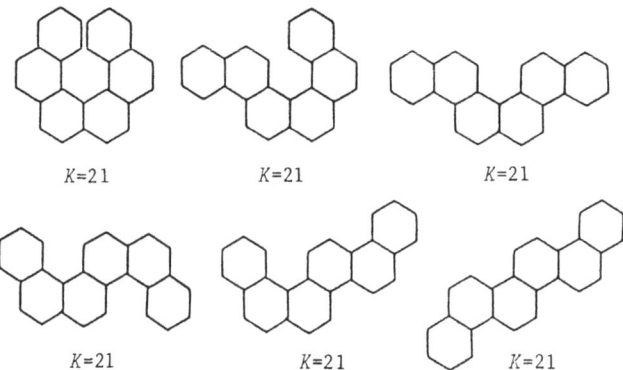

Fig. 6.1. The six isoarithmic systems from *hexahelicene* to *fulminene*.

Figure 1 shows the six isoarithmic members of the LA^4L chain including the helicenic system (*hexahelicene*; cf. also Fig. 2.1). Such systems, which consist of single chains with all segments of the length of two hexagons, will presently be referred to as all-kinked.

In order to show something of the diversity of the forms of helicenic systems we have included the examples of $h=9$ (Brunvoll, Cyvin SJ, Cyvin BN 1987) in Fig. 2.

6.2.2 *Algorithm*

The algorithms for linear and zigzag chains (cf. Chapter 3) are special cases of a more general algorithm for a single chain with an arbitrary distribution of kinks. We adher to the version of the algorithm of Cyvin and Gutman (1986). It has been commented on by Zivković and Trinajstić (1987).

1. Start from one (arbitrary) side of the chain, put down an external unity and the same numeral into every hexagon until the first kink.

2. Right after each kink the numeral is the sum of numerals for the preceding linear segment.

3. Continue with the same numeral until the next kink.

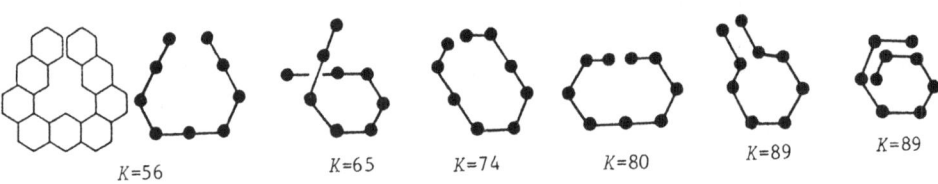

Fig. 6.2. The six existing mirror-symmetrical (C_{2v}) helicenic systems of a single chain with 9 hexagons. Two of them are isoarithmic. Dualist graphs are employed.

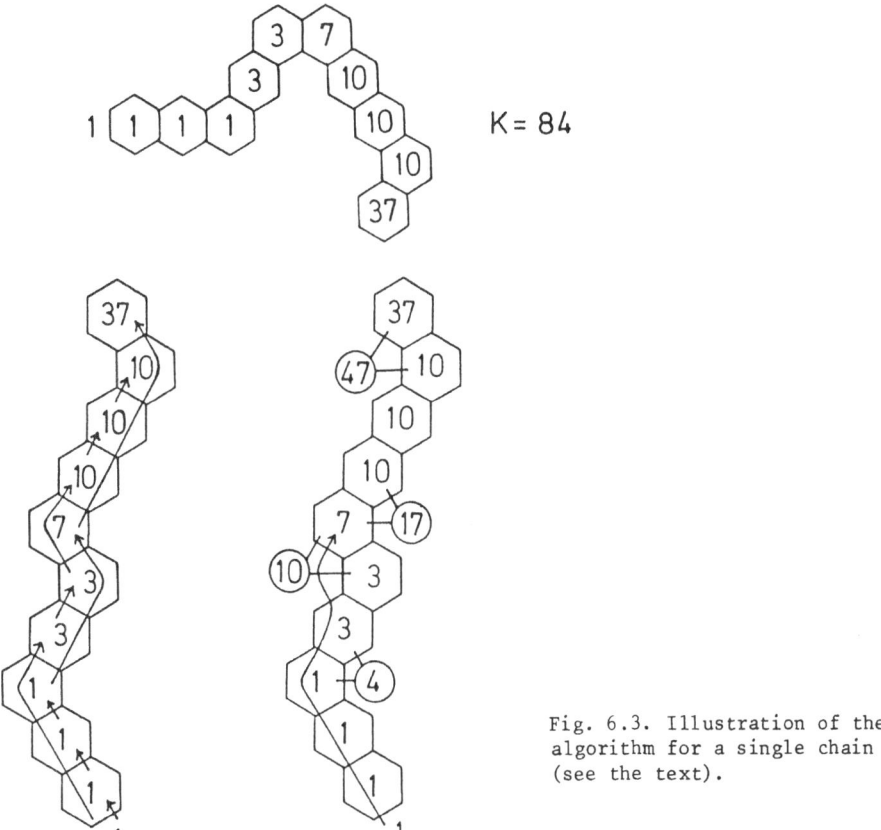

Fig. 6.3. Illustration of the algorithm for a single chain (see the text).

K = 84

4. The procedure suggests a building-up process of the benzenoid. When the numerals are added from the start (the external unity) they give the K number for any benzenoid during the building-up. Accordingly, the K number for the whole benzenoid is obtained by adding all the numerals.

Figure 3 shows a single chain with algorithm numerals filled in. The two benzenoids of the figure are isoarithmic; they have the same LA-sequence, viz. $L^2ALA^2L^2AL$. The bottom-left part of Fig. 3 indicates by arrows how the algorithm numerals are found according to the above rules.

The bottom-right part of Fig. 3 shows a useful alternative for deducing the first numeral after a kink: Delete the preceding (neighbour) numeral, add otherwise all numerals from the start, including the external unity.

As a consequence of these algorithm rules one has also: Take the sum of two neighbouring different numerals, say $x + y$, where $x < y$. That gives the K number for the benzenoid up to the hexagon with x (inclusive). This property is exemplified by the encircled K numbers in Fig. 3.

6.2.3 *Single Linear and Zigzag Chains: Combinatorial Formulas*

Single Linear Chain. This class is designated L(n) and consists of *benzene* followed by the *polyacenes*: *naphthalene*, *anthracene*, *naphthacene* (*tetracene*), *pentacene*, *hexacene*, etc. For the sake of completeness eqn. (4.2) is repeated in CHART I.

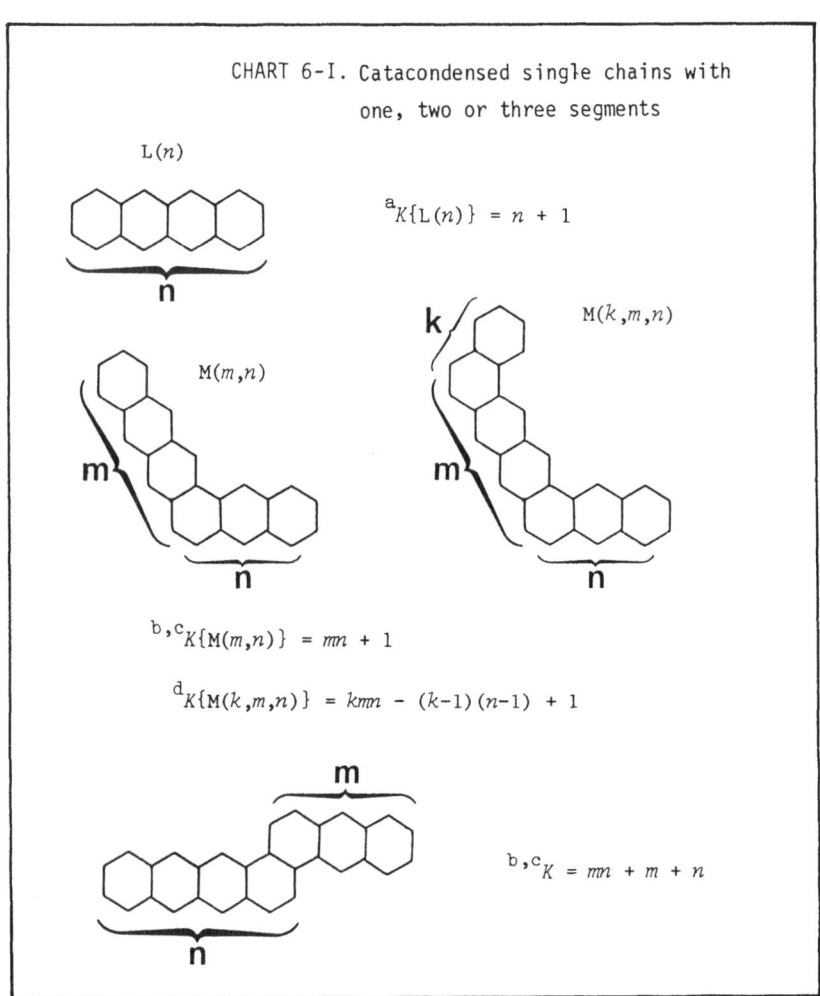

CHART 6-I. Catacondensed single chains with one, two or three segments

L(n)

$$^{a}K\{L(n)\} = n + 1$$

M(m,n)

M(k,m,n)

$$^{b,c}K\{M(m,n)\} = mn + 1$$

$$^{d}K\{M(k,m,n)\} = kmn - (k-1)(n-1) + 1$$

$$^{b,c}K = mn + m + n$$

[a]Gordon M, Davison WHT (1952). J Chem Phys 20: 428

[b]Biermann D, Schmidt W (1980). Israel J Chem 20: 312

[c]Jiang Y (1980). Sci Sinica 23: 847

[d]Balaban AT, Tomescu I (1983). Match 14: 155

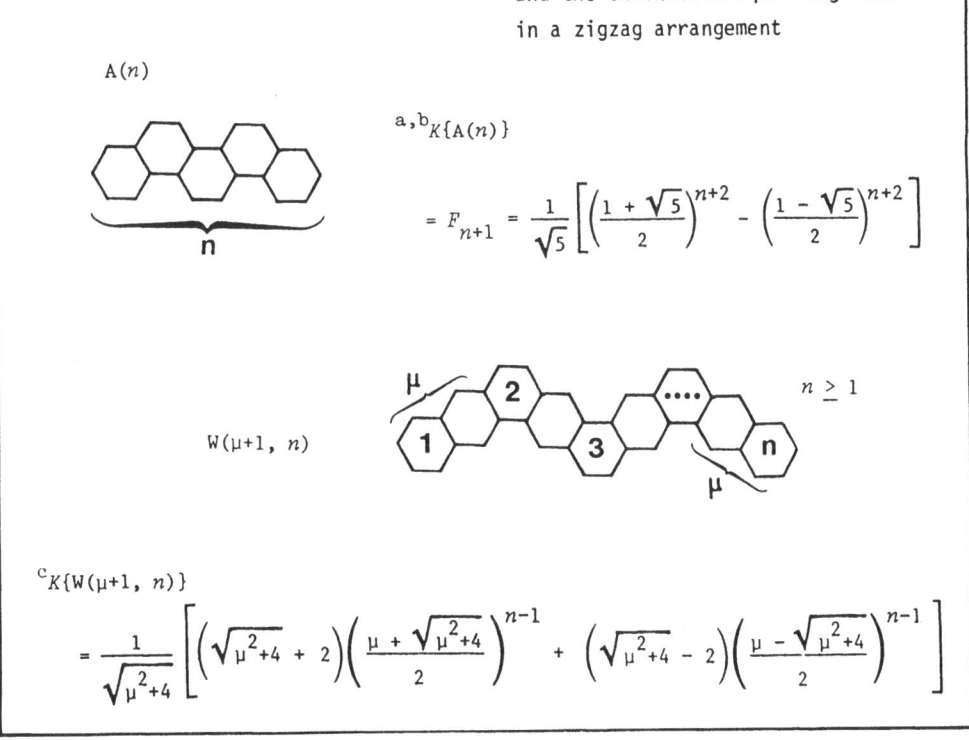

CHART 6-II. Catacondensed single zigzag chain, and the chain with equal segments in a zigzag arrangement

$A(n)$

$^{a,b}K\{A(n)\}$

$$= F_{n+1} = \frac{1}{\sqrt{5}}\left[\left(\frac{1+\sqrt{5}}{2}\right)^{n+2} - \left(\frac{1-\sqrt{5}}{2}\right)^{n+2}\right]$$

$W(\mu+1, n)$

$n \geq 1$

$^{c}K\{W(\mu+1, n)\}$

$$= \frac{1}{\sqrt{\mu^2+4}}\left[\left(\sqrt{\mu^2+4} + 2\right)\left(\frac{\mu+\sqrt{\mu^2+4}}{2}\right)^{n-1} + \left(\sqrt{\mu^2+4} - 2\right)\left(\frac{\mu-\sqrt{\mu^2+4}}{2}\right)^{n-1}\right]$$

[a] Gordon M, Davison WHT (1952). J Chem Phys 20: 428

[b] Yen TF (1971). Theor Chim Acta 20: 399

[c] Balaban AT, Tomescu, I (1985). Match 17: 91

Single Zigzag Chain. This class, $A(n)$, consists of *benzene, naphthalene, phenanthrene, chrysene, picene, fulminene,* etc. Eqns. (4.5),(4.12) are reproduced in CHART II.

Yen (1971) derived an alternative expression for $K\{A(n)\}$ involving a summation of binomial coefficients. A simpler formula of a similar kind emerges from the relation of Lucas for Fibonacci numbers (see Hosoya 1973):

$$K\{A(n)\} = \sum_{i=0}^{\left[\frac{n+1}{2}\right]} \binom{n+1-i}{i} \tag{6.1}$$

Here $\left[\frac{n+1}{2}\right]$ equals $\frac{n}{2}$ for $n = 0, 2, 4, \ldots$, and $\frac{n+1}{2}$ for $n = 1, 3, 5, \ldots$. It is by the way not important to specify this upper limit of summation because the terms automatically vanish thereafter.

Now we are able to derive a mathematical property of Fibonacci numbers "chemically", i.e. by the enumeration of Kekulé structures for a zigzag chain. Apply the method of fragmentation in contrast to the case of Fig. 4.7, to an edge somewhere in the middle of the zigzag chain; see Fig. 4. The fragments are disconnected benzenoids and yield

$$K\{A(n)\} = K\{A(k-1)\}K\{A(n-k)\} + K\{A(k-2)\}K\{A(n-k-1)\}; \qquad 2 \leq k \leq n \qquad (6.2)$$

and consequently

$$F_{n+1} = F_k \, F_{n-k+1} + F_{k-1} \, F_{n-k} ; \qquad 1 \leq k \leq n \qquad (6.3)$$

Especially useful are the cases corresponding to a fragmentation as near as possible to the mid-point of the zigzag chain. One arrives at:

$$K\{A(2p-1)\} = F_{2p} = F_p^2 + F_{p-1}^2 , \quad K\{A(2p)\} = F_{2p+1} = F_p(F_{p+1} + F_{p-1}); \quad p \geq 1 \qquad (6.4)$$

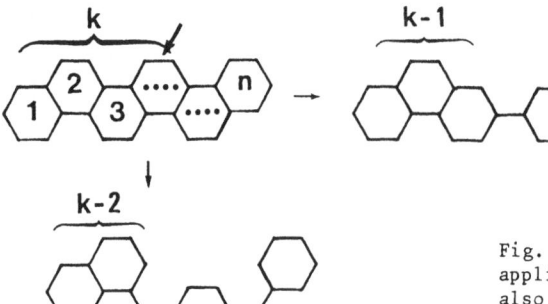

Fig. 6.4. The method of fragmentation applied to a single zigzag chain; see also Fig. 4.7.

6.2.4 *Other Single Chains*

Single Chain with Two or Three Segments. CHART I includes a single chain with (a) one kink and consequently two segments, and with (b) two kinks and three segments. The notation is explained by

(a) $M(m,n) = M(L^{m-1}AL^{n-1})$,   (b) $M(k,m,n) = M(L^{k-1}AL^{m-2}AL^{n-1})$,

where the *LA*-sequences are indicated. M (without subscript) designates a single chain. The present notation implies: $M(m,n) = M(1,m,n)$. For $k=0$ one has $M(0,m,n) = L(n-1)$. The trivial case of no hexagons is realized by $M(0,n) = M(m,0)$.

The bottom system of CHART I is actually a three-segment chain: $M(m,2,n)$.

The works of Jiang (1980) and of Eilfeld and Schmidt (1981) contain several results for K numbers of special single chains. A work of Randić (1980) should also

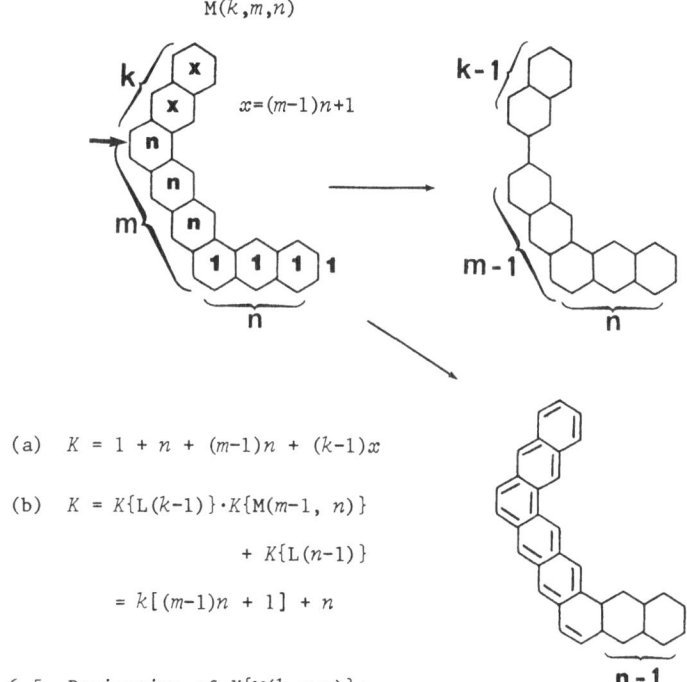

(a) $K = 1 + n + (m-1)n + (k-1)x$

(b) $K = K\{L(k-1)\} \cdot K\{M(m-1, \ n)\}$

$+ K\{L(n-1)\}$

$= k[(m-1)n + 1] + n$

Fig. 6.5. Derivation of $K\{M(k,m,n)\}$:
(a) Addition of algorithm numerals.
(b) The method of fragmentation, where the edge indicated by a thick arrow
is attacked.

be cited in this connection. *Polyphenes* are special two-segment chains; they were
treated by Cyvin (1982b).

Many of the K formulas (cf. CHART I and CHART II) may conveniently be derived
by means of the addition of algorithm numerals in their general form. One example is
sufficient for illustration. Figure 5 shows $M(k,m,n)$ with algorithm numerals in
their general form. The figure also indicates an alternative derivation by means of
the method of fragmentation. One fragment consists of two disconnected units, viz.
$L(k-1)$ and $M(m-1, \ n)$. The other fragment reduces to $L(n-1)$.

*Benzenoids with Equal Segments in a Zigzag Arrangement and Related
Classes.* Balaban and Tomescu (1984; 1985) have studied extensively the single-
chain systems

$$W(m,n) = L(L^{m-2}A)^{n-2}L^{m-1} \quad ;$$

cf. CHART II. It is a catacondensed unbranched single chain with $n-2$ kinks ($n \geq 2$)
and $n-1$ linear segments of equal length, viz. $m-1$ hexagons (for $m > 1$). For $n=2$ the
system coincides with the single linear chain: $W(m,2) = L(m)$. For $m=2$ it coincides
with the single zigzag chain: $W(2,n) = A(n)$. The validity of the formulas (see

below) is extended to the parameter values $n=1$ and $m=1$ when we define these degenerate cases as consisting of one hexagon:

$$W(m,1) = L(1) , \qquad W(1,n) = L(1)$$

Consider two auxiliary classes in addition to $W(m,n)$, viz. $W'(m,n)$ and $W''(m,n)$ as defined in Fig. 6. They emerge by taking away one hexagon from $W(m,n)$ at one end or both ends, respectively. Introduce the notation:

$$\mu = m - 1 \tag{6.5}$$

$$K\{W(\mu+1, \ n)\} = W_{\mu}(n) \tag{6.6}$$

and the corresponding abbreviations for W' and W''.

The method of fragmentation when applied to the edges of $W(m,n)$ marked by a and b in Fig. 6, yields

$$W_{\mu}(n) = W_{\mu}{}'(n) + W_{\mu}{}'(n-1) \tag{6.7}$$

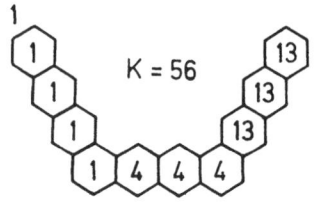

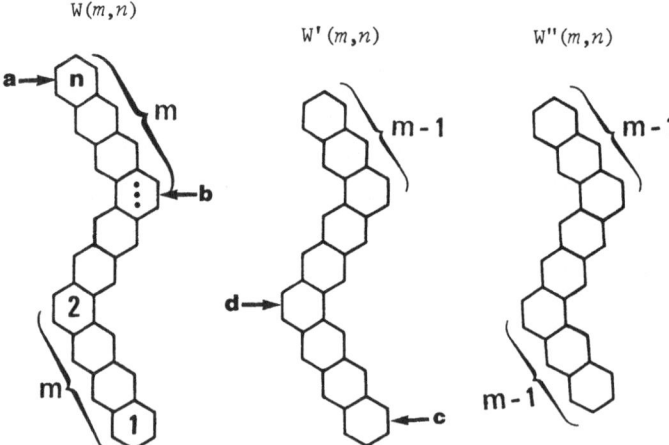

Fig. 6.6. Three related classes of catacondensed single-chain benzenoids. The top drawing (supplied with algorithm numerals) is isoarithmic with $W(4,4)$.

and

$$W_\mu(n) = (\mu+1)W_\mu'(n-1) + W_\mu'(n-2) \qquad (6.8)$$

This is a set of linearly coupled recurrence relations (Spiegel 1971). We wish to deduce the recurrence relation for the W quantities exclusively. Eq. (8) is equivalent to

$$W_\mu(n+1) = (\mu+1)W_\mu'(n) + W_\mu'(n-1) \qquad (6.9)$$

By subtraction of (8) from (9) $W_\mu'(n-1)$ is eliminated, and one obtains

$$\mu W_\mu'(n) = W_\mu(n+1) - W_\mu(n) \qquad (6.10)$$

This equation is equivalent to

$$\mu W_\mu'(n-1) = W_\mu(n) - W_\mu(n-1) \qquad (6.11)$$

On inserting from (10) and (11) into (7) it is finally obtained

$$\mu W_\mu(n) = W_\mu(n+1) - W_\mu(n-1) \qquad (6.12)$$

or the desired recurrence formula (Balaban and Tomescu 1984) in a final form:

$$W_\mu(n) = \mu W_\mu(n-1) + W_\mu(n-2); \qquad n > 2 \qquad (6.13)$$

The quantities $W_\mu(n)$ are polynomials in μ (or m). They are listed in Table 1 up to $n=10$. Balaban and Tomescu (1983; 1984; 1985) gave six of these polynomials and indicated how their general form can be deduced. Notice that all these polynomials (irrespective of n) give $K=2$ for $m=1$ ($\mu=0$). For $m=2$ ($\mu=1$) the Fibonacci numbers from $F_2 = 2$ are reproduced.

By virtue of the linear dependence between the W_μ and W_μ' quantities exactly the same form as (13) is valid as a recurrence relation for W_μ' This is verified most easily by equating the right-hand sides of (7) and (8).

The quantities W_μ'' are also linearly coupled to W_μ (and W_μ'). The clue to these linear dependencies is obtained again from the fragmentation method; by attacking the edges c and d in Fig. 6 one obtains

$$W_\mu'(n) = W_\mu''(n) + W_\mu''(n-1) \qquad (6.14)$$

$$W_\mu'(n) = (\mu+1)W_\mu''(n-1) + W_\mu''(n-2) \qquad (6.15)$$

The quantities W_μ'' also obey the recurrence relation of the form (13).

By suitable manipulations of the formula apparatus (7)-(15) we can express the W_μ' and W_μ'' quantities linearly in terms of W_μ (considered as pertaining to the

Table 6.1. K formulas for $W(m,n) = W(\mu+1, n)$.

n	Polynomial in m	Polynomial in $\mu = m-1$
1	2	2
2	$m+1$	$\mu+2$
3	m^2+1	$\mu^2+2\mu+2$
4	$m(m^2-m+2)$	$(\mu+1)(\mu^2+\mu+2)$
5	$m^4-2m^3+4m^2-2m+1$	$\mu^4+2\mu^3+4\mu^2+4\mu+2$
6	$(m^2-m+1)(m^3-2m^2+4m-1)$	$(\mu^2+\mu+1)(\mu^3+\mu^2+3\mu+2)$
7	$m^6-4m^5+11m^4-16m^3+16m^2-8m+2$	$\mu^6+2\mu^5+6\mu^4+8\mu^3+9\mu^2+6\mu+2$
8	$m^7-5m^6+16m^5-30m^4+39m^3-31m^2-8m+2$	$\mu^7+2\mu^6+7\mu^5+10\mu^4+14\mu^3+12\mu^2+7\mu+2$
9	$m^8-6m^7+22m^6-50m^5+80m^4-86m^3+62m^2-26m+5$	$\mu^8+2\mu^7+8\mu^6+12\mu^5+20\mu^4+20\mu^3+16\mu^2+8\mu+2$
10	$m^9-7m^8+29m^7-77m^6+146m^5-196m^4+187m^3$ $-119m^2+46m-8$	$\mu^9+2\mu^8+9\mu^7+14\mu^6+27\mu^5+30\mu^4+30\mu^3+20\mu^2$ $+9\mu+2$

main class). The result may be written

$$W_{\mu}{}'(n) = \frac{\mu-1}{\mu} W_{\mu}(n) + \frac{1}{\mu} W_{\mu}(n-1) \tag{6.16}$$

$$W_{\mu}{}''(n) = \left(\frac{\mu-1}{\mu}\right)^2 W_{\mu}(n) + \frac{2(\mu-1)}{\mu^2} W_{\mu}(n-1) + \frac{1}{\mu^2} W_{\mu}(n-2) \tag{6.17}$$

By standard mathematical methods (cf. Chapter 4) the explicit K formulas for the classes in question may be derived. The result for the main class, $W(m,n)$, was furnished by Balaban and Tomescu (1985) and is entered into CHART II. It reduces to Binet's formula for F_{n+1} when $\mu=1$. The result for $W''(m,n)$ is in fact somewhat simpler (Bergan et al. 1987):

$$K\{W''(\mu+1, n)\} = W_{\mu}{}''(n) = \frac{1}{\sqrt{\mu^2+4}} \left[\left(\frac{\mu + \sqrt{\mu^2+4}}{2}\right)^n - \left(\frac{\mu - \sqrt{\mu^2+4}}{2}\right)^n \right] \tag{6.18}$$

This equation reduces to Binet's formula for F_{n-1} when $\mu=1$.

6.3 BRANCHED CHAIN

6.3.1 *Systems with One Branching Hexagon*

Assume three branches meeting at a hexagon, which has the mode A_3. If the branches are linear chains the systems are called *starphenes* and have been included in several works (Biermann and Schmidt 1980a; Jiang 1980; Balaban and Tomescu 1983; Cyvin 1983d). CHART III gives the K formula for *starphenes* along with related systems, where the branches also are allowed to be all-kinked.

Below we give an algorithm for a catacondensed benzenoid with one branching hexagon and arbitrary *LA*-sequences of the three branches.

CHART 6-III. Three linear or all-kinked chains meeting at one branching hexagon (*starphenes* and related systems).

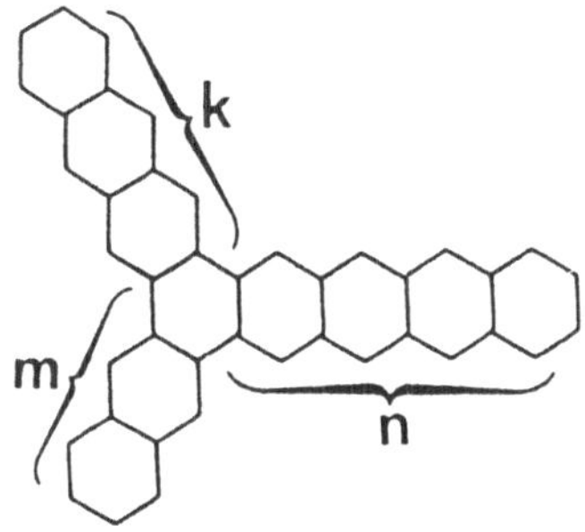

[a,b] $K = kmn + 1$

$$K = kmF_n + F_{n-1}$$

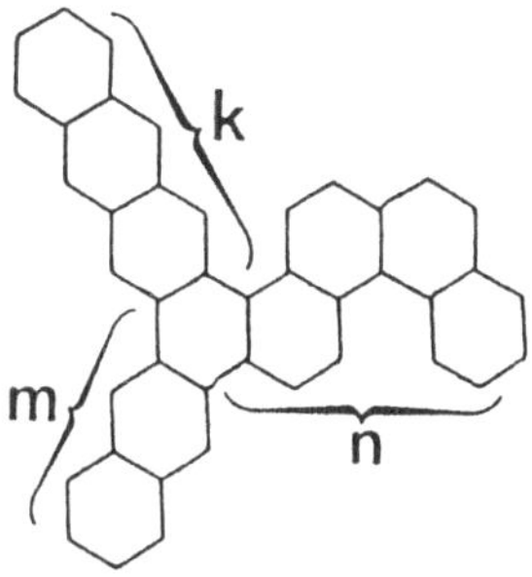

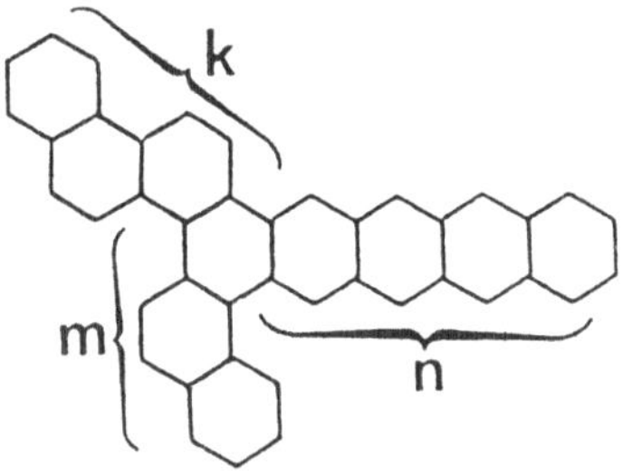

$$K = F_k F_m n + F_{k-1} F_{m-1}$$

$$K = F_k F_m F_n + F_{k-1} F_{m-1} F_{n-1}$$

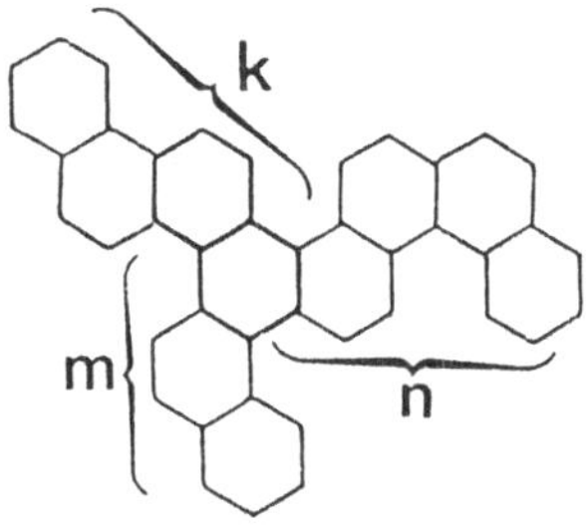

[a] Biermann D, Schmidt W (1980). Israel J Chem 20: 312
[b] Jiang Y (1980). Sci Sinica 23: 847

The total K number is obtained as the sum of K numbers for two disconnected benzenoids generated in the following way. (1) Delete the branching hexagon only. (2) Delete the branching hexagon, and continue deleting hexagons along the branches until the first kink on each of them.

The algorithm is illustrated in Fig. 7.

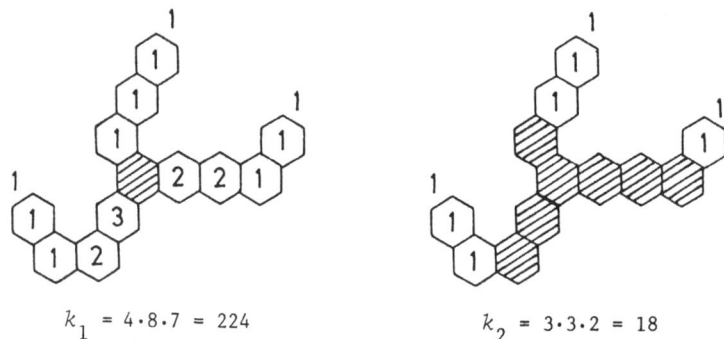

$$k_1 = 4 \cdot 8 \cdot 7 = 224 \qquad k_2 = 3 \cdot 3 \cdot 2 = 18$$

Fig. 6.7. Illustration of the algorithm for three single chains meeting at a hexagon. The total K number is $k_1 + k_2 = 242$.

6.3.2 *Systems with Several Branching Hexagons*

Figure 8 shows the system with two branching hexagons and only linear branches. The K formula is given in the figure.

Several works with relevance to Kekulé structures have appeared where *tetrabenzacenes* are included (Schmidt 1977; Zander 1978; Randić 1980; Biermann and Schmidt 1980a; 1980b; Eilfeld and Schmidt 1981; Bräuchle and Voitländer 1982). The expression for K reported by Eilfeld and Schmidt (1981), viz. $16n' + 24$, emerges from the formula of Fig. 8 on inserting $m_1 = m_2 = m_3 = m_4 = 2$, $n = n' + 2$. For $m_i = 1$ a system is obtained where no hexagon is fused to the appropriate edge.

Otherwise we shall refrain from reporting explicit formulas for additional

$$K = m_1 m_2 m_3 m_4 (n-1) + m_1 m_2 + m_3 m_4$$

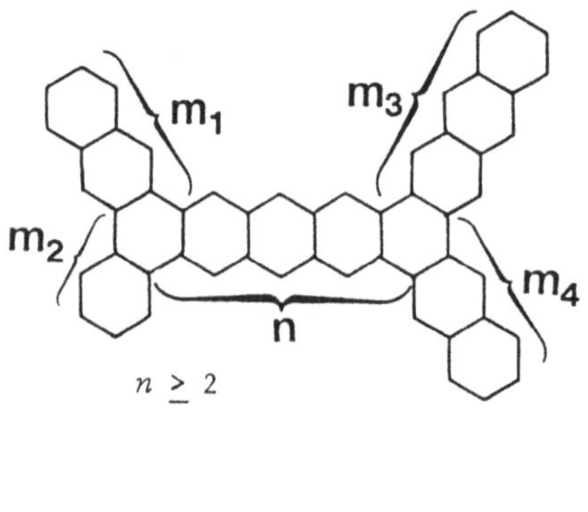

Fig. 6.8. A system with linear branches and two branching hexagons. For the depicted example $K = 306$.

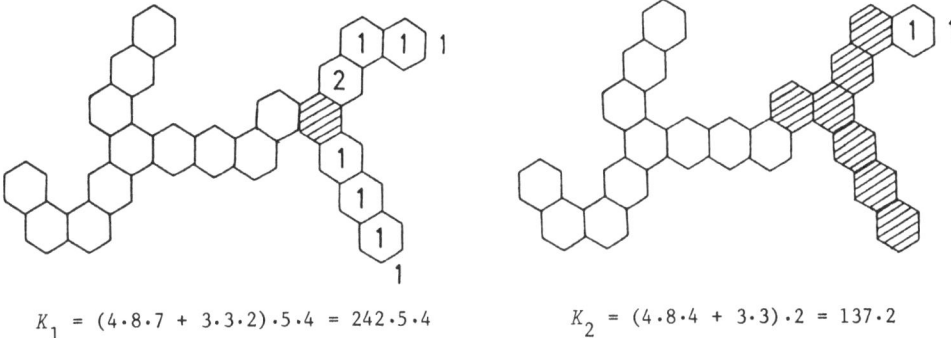

$$K_1 = (4 \cdot 8 \cdot 7 + 3 \cdot 3 \cdot 2) \cdot 5 \cdot 4 = 242 \cdot 5 \cdot 4 \qquad\qquad K_2 = (4 \cdot 8 \cdot 4 + 3 \cdot 3) \cdot 2 = 137 \cdot 2$$

Fig. 6.9. Illustration of the algorithm adapted for chains with two branching hexagons. The total number of Kekulé structures is $K_1 + K_2 = 5114$ (see also Fig. 6.7).

special cases, but only point out that the algorithm of the preceding paragraph may be adapted to systems with more than one branching hexagon. Figure 9 shows an extension of the system of Fig. 7 and indicates the application of the algorithm adapted to the system with two branching hexagons.

6.4 CATACONDENSED LADDER

6.4.1 *Introduction*

In this section we introduce a class which represents typical benzenoid systems with repeated units. The single linear and zigzag chains (Paragraph 6.2.3; CHARTS I and II) may be interpreted as (trivial) benzenoid systems with repeated units. A less trivial example is the benzenoid with equal segments in a zigzag arrangement (CHART II). The benzenoids to be treated here, which are referred to as catacondensed ladders and denoted $\Xi(m,n)$, belong to the branched chains (for $n \geq 3$); cf. CHART IV.

The class of ladders is used here to illustrate in some detail the derivation of explicit K formulas from a three-term recurrence relation. The method was touched upon in the simple case of a single zigzag chain (Section 4.7). The quoted results for the somewhat more complex cases of CHART II and eqn. (18) are obtainable by the same methods.

Classes of benzenoids with repeated units will be a recurring theme of this book.

6.4.2 *The Case of m=2*

Consider a class of ladders as shown in Fig. 10 and designated $\Xi(2,n)$. This is a special case of $\Xi(m,n)$ to be treated in the subsequent paragraph. Following

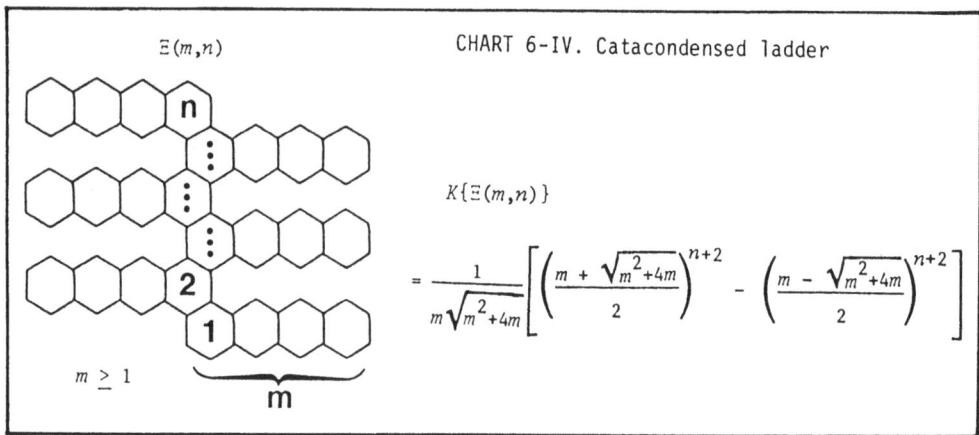

Randić (1980) we apply his fragmentation method trying to produce fragments which are (lower) members of the same class. This is easily achieved by attacking the edge marked with the thick arrow in Fig. 10. Introduce the notation

$$K\{\Xi(m,n)\} = \Xi_m(n) \tag{6.19}$$

for the number of Kekulé structures. According to Fig. 10 we obtain readily the recurrence relation

$$\Xi_2(n) = 2\Xi_2(n-1) + 2\Xi_2(n-2) ; \qquad n \geq 2 \tag{6.20}$$

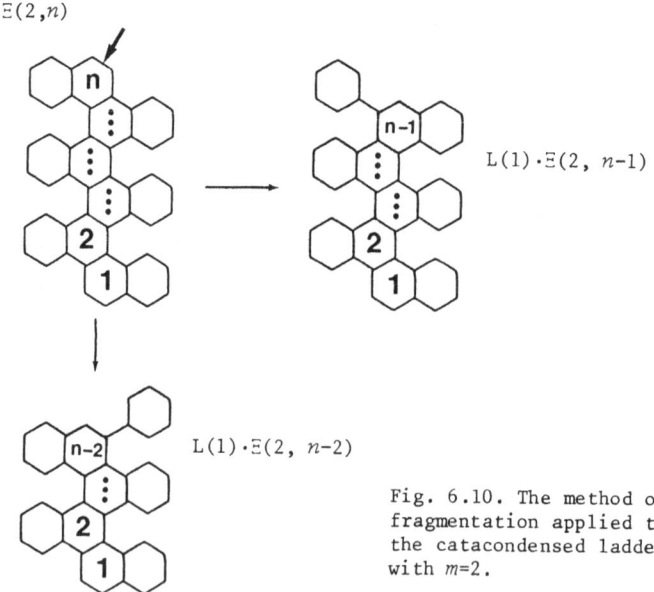

Fig. 6.10. The method of fragmentation applied to the catacondensed ladder with $m=2$.

Assume a particular solution for $\Xi_2(m)$ to be of the exponential form a^n. Hence from eqn. (20)

$$a^n = 2a^{n-1} + 2a^{n-2} \qquad (6.21)$$

and

$$a^2 - 2a - 2 = 0 \qquad (6.22)$$

with the solution

$$a_1 = 1 + \sqrt{3}, \qquad a_2 = 1 - \sqrt{3} \qquad (6.23)$$

The general solution of a formula for $\Xi_2(m)$ which satisfies the recurrence relation (21) is

$$\Xi_2(n) = Aa_1^{n} + Ba_2^{n} \qquad (6.24)$$

The undetermined coefficients (A, B) are obtained from the initial conditions. Here $\Xi_2(1) = 3$ and $\Xi_2(2) = 8$, pertaining to *naphthalene* and *chrysene*, respectively; cf. Table 2. The analysis is somewhat facilitated on introducing $\Xi_2(0) = 1$ as one of the initial conditions. It corresponds to the trivial case of no hexagons. The value is deduced without doubt as consistent with (20) on inserting $n=2$ into this equation. Now the known values for $n=0$ and $n=1$ inserted into eqn. (24), together with the numerical values (23), yield

$$A + B = 1 \qquad (6.25)$$

$$A(1 + \sqrt{3}) + B(1 - \sqrt{3}) = 3 \qquad (6.26)$$

Table 6.2. Numerical values of $K\{\Xi(2,n)\} = \Xi_2(n)$.

n	$\Xi_2(n)$
0	1
1	3
2	8
3	22
4	60
5	164
6	448
7	1224
8	3344
9	9136
10	24960

The set of linear equations (25), (26) has the solution

$$A = \frac{1}{2\sqrt{3}} (\sqrt{3} + 2), \qquad B = \frac{1}{2\sqrt{3}} (\sqrt{3} - 2) \qquad (6.27)$$

Inserting into (24) from (23) and (27) gives

$$\Xi_2(n) = \frac{1}{2\sqrt{3}} \left[(\sqrt{3} + 2)(1 + \sqrt{3})^n + (\sqrt{3} - 2)(1 - \sqrt{3})^n \right] \qquad (6.28)$$

which is the explicit formula for $\Xi_2(n)$. It can be simplified on taking advantage of the relations $(1 + \sqrt{3})^2 = 2(\sqrt{3} + 2)$ and $(1 - \sqrt{3})^2 = -2(\sqrt{3} - 2)$. Hence

$$\Xi_2(n) = \frac{1}{4\sqrt{3}} \left[(1 + \sqrt{3})^{n+2} - (1 - \sqrt{3})^{n+2} \right] \qquad (6.29)$$

In eqn. (25) we have taken advantage of the trivial case with $n=0$. The procedure can be stretched to regions of n values below the indicated limit ($n \geq 2$) of eqn. (20) if we take into account what will be referred to as nominal K values. The relation (20) is amenable for successive extrapolations to $\Xi_2(-1)$, $\Xi_2(-2)$, $\Xi_2(-3)$, ... etc. on inserting $n = 1, 0, -1, \ldots$, respectively. It is obtained for the first two values:

$$\Xi_2(-1) = \frac{1}{2} , \qquad \Xi_2(-2) = 0 \qquad (6.30)$$

We can no longer visualize any corresponding benzenoid systems, neither genuine or degenerate. Therefore these fictious "numbers of Kekulé structures" are referred to as nominal. Nevertheless they fit into our equations and can be used to deduce the explicit formula for $\Xi_2(n)$ as shown in the following. Set

$$\Xi_2(n) = A'(1 + \sqrt{3})^{n+2} + B'(1 - \sqrt{3})^{n+2} \qquad (6.31)$$

Now the initial conditions (30) yield

$$A'(1 + \sqrt{3}) + B'(1 - \sqrt{3}) = \frac{1}{2} \qquad (6.32)$$

$$A' + B' = 0 \qquad (6.33)$$

The solution is

$$A' = \frac{1}{4\sqrt{3}} , \qquad B' = -\frac{1}{4\sqrt{3}} \qquad (6.34)$$

which on inserting into (31) gives the explicit formula (29) directly in its simplified form.

6.4.3 General Case

The computation may be carried through with an arbitrary parameter $m \geq 1$ for the class defined in CHART IV. The recurrence relation in this general case reads

$$\Xi_m(n) = m\,\Xi_m(n-1) + m\,\Xi_m(n-2); \qquad n \geq 2 \qquad (6.35)$$

where again the restriction $n \geq 2$ is released if nominal values are taken into account.

The quantities $\Xi_m(n)$ are polynomials in m:

$$\Xi_m(0) = 1$$
$$\Xi_m(1) = m+1$$
$$\Xi_m(2) = m(m+2)$$
$$\Xi_m(3) = m(m^2+3m+1)$$
$$\Xi_m(4) = m^2(m+1)(m+3)$$

$$\Xi_m(5) = m^2(m^3+5m^2+6m+1)$$

$$\Xi_m(6) = m^3(m+2)(m^2+4m+2)$$

$$\Xi_m(7) = m^3(m+1)(m^3+6m^2+9m+1)$$

$$\Xi_m(8) = m^4(m^4+8m^3+21m^2+20m+5)$$

$$\Xi_m(9) = m^4(m^5+9m^4+28m^3+35m^2+15m+1)$$

$$\Xi_m(10) = m^5(m+1)(m+2)(m+3)(m^2+4m+1)$$

The explicit formula for $\Xi_m(n)$ is entered into CHART IV.

The catacondensed ladder reduces to the single zigzag chain for $m=1$; $\Xi(1,n)$ = $A(n)$. Indeed also the explicit formula for $K\{\Xi(m,n)\}$ of CHART IV coincides with Binet's formula for F_{n+1} (CHART II) for $m=1$. In consistence with these features the above polynomials reproduce the Fibonacci numbers (from $F_1 = 1$) for $m=1$.

6.5 CATACONDENSED ALL-BENZENOIDS AND RELATED SYSTEMS

6.5.1 *Some General Properties and Some Examples*

Catacondensed all-benzenoids (see Paragraph 2.2.8 for definition and basic references) are in general highly kinked and branched systems. Assume that such a system is stripped for its terminal (mode L_1) hexagons (which always are full). The remaining system is referred to as the backbone. *Triphenylene* (Fig. 2.5) has the trivial backbone of one hexagon. Otherwise the backbone can only consist of short segments, actually holding two or three hexagons each. They are called 2-segments and 3-segments, respectively. A backbone may be branched or unbranched. In the former case a branching (mode A_3) hexagon may be either full or empty.

A backbone, or a branch of a backbone, may consist of 2-segments exclusively or 3-segments exclusively. Systems with mixed 2-segments and 3-segments are also possible, but not with these two types in an arbitrary succession.

Figure 11 shows some examples of catacondensed all-benzenoids. The two systems with unbranched backbone are isoarithmic; the backbones have the same *LA*-sequence, viz. $L^2A^5LA^2L$.

6.5.2 *Class with Only 2-Segments in the Backbone and Related Classes*

CHART V defines a class of all-benzenoids, Ч(n), where the backbone is a single zigzag chain. The parameter n designates the number of empty hexagons in the backbone or in the whole benzenoid. The two related classes Ч'(n) and Ч"(n), which no longer consist of all-benzenoids, are generated by removing one terminal hexagon from either one end or both ends, respectively; cf. CHART V. The systems were analyzed by the methods treated above in this chapter.

The derived explicit K formulas are found in CHART V.

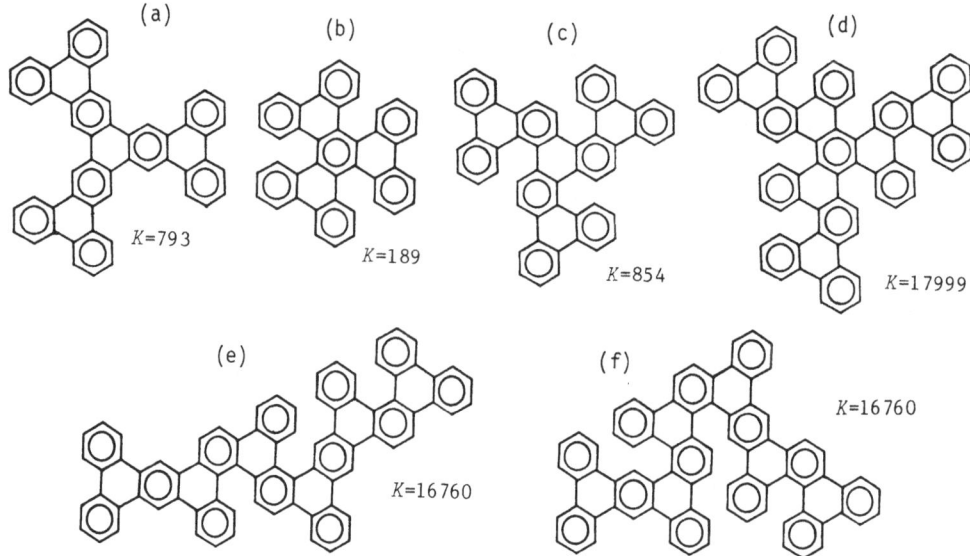

Fig. 6.11. Some catacondensed all-benzenoid systems. K numbers are given. Branched backbone (a)-(d): (a) and (c) branching hexagon empty; (b) and (d) branching hexagon full; (a) and (b) symmetry D_{3h}; (c) and (d) symmetry C_{3h}. Unbranched backbone (e), (f): Two isoarithmic systems with mixed occurrence of 2- and 3-segments. Their sequence is (in both cases) 3-2-2-2-2-3-2-2.

With the notation

$$K\{Ч(n)\} = Ч_n \qquad (6.36)$$

the recurrence relation reads

$$Ч_n = 5Ч_{n-1} - 2Ч_{n-2}; \qquad n \geq 2 \qquad (6.37)$$

The same form (37) is obeyed by the K numbers for the classes Ч' and Ч".

We use the abbreviations corresponding to (36) also for the classes Ч' and Ч". Then

$$Ч_n' = Ч_n - 2Ч_{n-1} \qquad (6.38)$$

and

$$Ч_n'' = Ч_n - 4Ч_{n-1} + 4Ч_{n-2} \qquad (6.39)$$

Table 3 gives the numerical values of the K numbers for Ч(n), Ч'(n) and Ч"(n) when $0 \leq n \leq 10$.

CHART 6-V. Classes of catacondensed all-benzenoids with only 2-segments in the backbone, with only 3-segments, and related (not all-benzenoid) classes

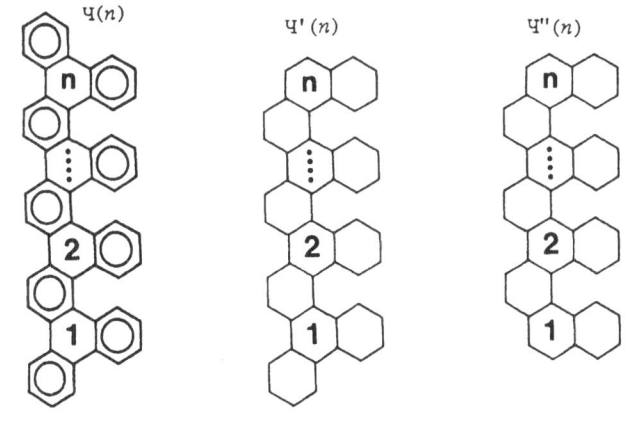

$$K\{Ч(n)\} = \frac{1}{\sqrt{17}}\left[(\sqrt{17}+4)\left(\frac{5+\sqrt{17}}{2}\right)^n + (\sqrt{17}-4)\left(\frac{5-\sqrt{17}}{2}\right)^n\right]$$

$$K\{Ч'(n)\} = \frac{1}{\sqrt{17}}\left[\left(\frac{5+\sqrt{17}}{2}\right)^{n+1} - \left(\frac{5-\sqrt{17}}{2}\right)^{n+1}\right]$$

$$K\{Ч''(n)\} = \frac{1}{2\sqrt{17}}\left[(\sqrt{17}+1)\left(\frac{5+\sqrt{17}}{2}\right)^n + (\sqrt{17}-1)\left(\frac{5-\sqrt{17}}{2}\right)^n\right]$$

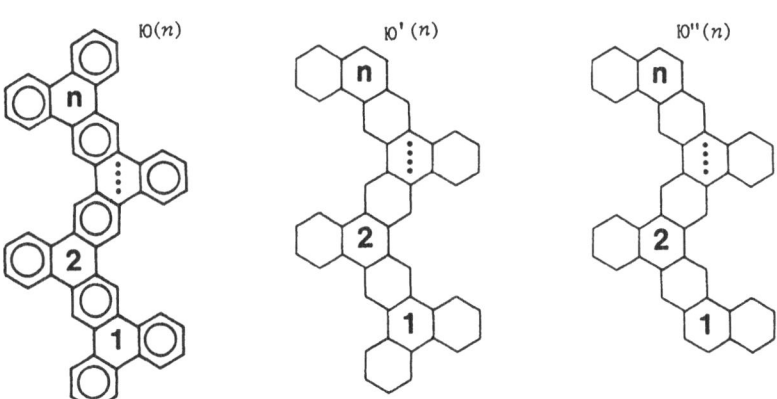

$$K\{Ю(n)\} = \frac{1}{4\sqrt{6}}\left[(2+\sqrt{6})^{n+2} - (2-\sqrt{6})^{n+2}\right]$$

$$K\{Ю'(n)\} = \frac{1}{4}\left[(2+\sqrt{6})^{n+1} + (2-\sqrt{6})^{n+1}\right]$$

$$K\{Ю''(n)\} = \frac{3}{2\sqrt{6}}\left[(2+\sqrt{6})^n - (2-\sqrt{6})^n\right]$$

Table 6.3. Numerical values of K numbers for the benzenoids Ч(n), Ч$'(n)$, Ч$''(n)$ and Ю(n), Ю$'(n)$, Ю$''(n)$.

n	Ч$_n$	Ч$_n'$	Ч$_n''$	Ю$_n$	Ю$_n'$	Ю$_n''$
0	2	1	1	2	1	0
1	9	5	3	9	5	3
2	41	23	13	40	22	12
3	187	105	59	178	98	54
4	853	479	269	792	436	240
5	3891	2185	1227	3524	1940	1068
6	17749	9967	5597	15680	8632	4752
7	80963	45465	25531	69768	38408	21144
8	369317	207391	116461	310432	170896	94080
9	1684659	946025	531243	1381264	760400	418608
10	7684661	4315343	2423293	6145920	3383392	1862592

6.5.3 *Class with Only 3-Segments in the Backbone and Related Classes*

CHART V includes the definition of Ю(n), another class of catacondensed all-benzenoids. Here the backbone is a single chain of 3-segments in a zigzag arrangement. Again the parameter n designates the number of empty hexagons in the backbone or the whole benzenoid. CHART V shows the derived explicit K formula for Ю(n) along with the corresponding K formulas for related (not all-benzenoid) classes Ю$'(n)$ and Ю$''(n)$. Introduce the notation

$$K\{Ю(n)\} = Ю_n \tag{6.40}$$

and the corresponding notations for the classes Ю$'$ and Ю$''$. The form of the recurrence relation, applicable to all of these three classes, is

$$Ю_n = 4Ю_{n-1} + 2Ю_{n-2}; \qquad n \geq 2 \tag{6.41}$$

The K numbers for Ю$'$ and Ю$''$ in terms of those for the class Ю appear to have the same forms as eqns. (38), (39);

$$Ю_n' = Ю_n - 2Ю_{n-1} \tag{6.42}$$

$$Ю_n'' = Ю_n - 4Ю_{n-1} + 4Ю_{n-2} \tag{6.43}$$

The numerical values of K numbers for Ю(n), Ю$'(n)$ and Ю$''(n)$ are included in Table 3.

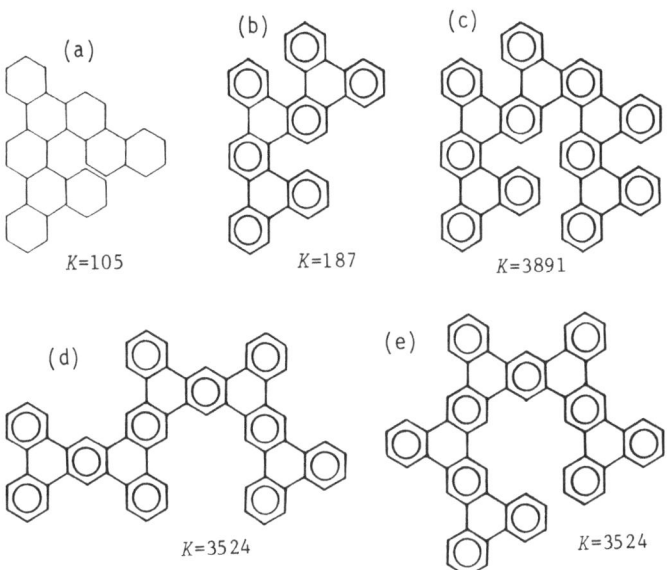

Fig. 6.12. (a) Helicenic system, which is isoarithmic with Ч'(3). (b)-(e) All-benze-noid systems isoarithmic with: (b) Ч(3); (c) Ч(5); (d) and (e) Ю(5).

6.5.4 Other All-Benzenoids and Related Systems

Unbranched Backbone. Many isoarithmic variants of the benzenoids Ч(n), Ч'(n), Ю(n), etc. can be constructed. Figure 12 shows some examples, which include a heli-cenic system.

The 2- and 3-segments of the backbone of a catacondensed all-benzenoid may alternate in many ways (but not arbitrarily; cf. Paragraph 6.5.1). One example of a class with mixed 2- and 3-segments seems to be sufficient here. The class "horse-shoes on a string", say U(n), has the sequence of segments like 2-2-3-2-2-3-2-2-.. ..-2-2. The parameter n is defined so that the backbone has $2n$ empty hexagons. Figure 13 shows the member U(3). With the notation $K\{U(n)\} = U_n$ the recurrence re-lation may be written

$$U_n = 20U_{n-1} + 4U_{n-2}; \qquad n \geq 2 \tag{6.44}$$

As initial conditions one has $U_1 = $ Ч$_2$ = 41, and $U_0 = K\{L(1)\} = 2$ for a degenerate case. The following explicit formula was derived.

$$U_n = \frac{1}{8\sqrt{26}}\left[(\sqrt{26} - 1)(10 + 2\sqrt{26})^{n+1} + (\sqrt{26} + 1)(10 - 2\sqrt{26})^{n+1}\right] \tag{6.45}$$

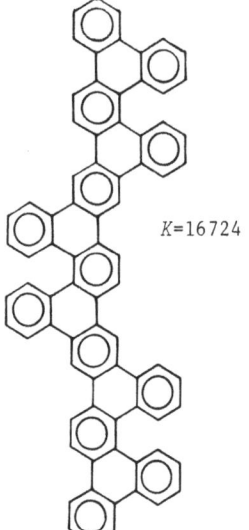

$K = 16724$

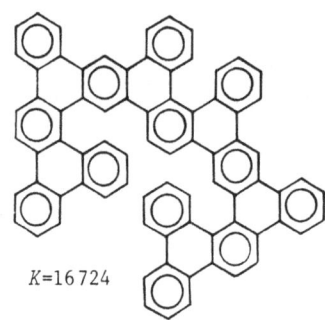

$K = 16724$

Fig. 6.13. A member of the class $U(n)$:
"horse-shoes on a string". The case
$n=3$ is depicted, along with one of its
isoarithmic partners.

Branched Backbone. Figure 11 (a)-(d) shows some examples of catacondensed all-benzenoids with branched backbone. In Fig. 14 a more advanced example is depicted. The figure indicates how the K number was determined by a fragmentation scheme according to the general methods (see especially Section 6.3).

Finally in this chapter we give the results of an analysis of the class of "catacondensed all-benzenoid ladders", say $3(n)$, which is defined in CHART VI. The

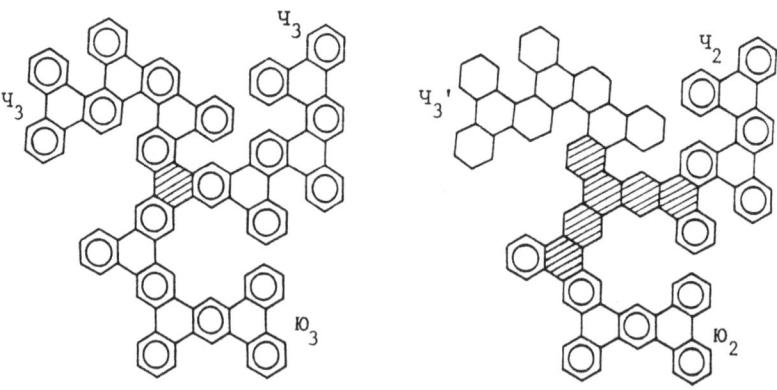

Fig. 6.14. The method of fragmentation applied to a catacondensed all-benzenoid with branched backbone. The total number of Kekulé structures:

$$K = Ч_3^2 \cdot Ю_3 + 2^2 \cdot Ч_3{}' \cdot Ч_2 \cdot Ю_2 = 6913282$$

benzenoid $3(n)$ has $2n$ empty hexagons. With the notation $K\{3(n)\} = 3_n$ one finds the recurrence relation

$$3_n = 18\ 3_{n-1} + 36\ 3_{n-2}; \qquad n \geq 2 \tag{6.46}$$

As initial conditions one has $3_1 = 10_2 = 40$ and $3_2 = 10_4 = 792$, but also $3_0 = 2$ for a degenerate case, and furthermore the nominal values $3_{-1} = \frac{1}{9}$ and $3_{-2} = 0$. The explicit formula is given in CHART VI.

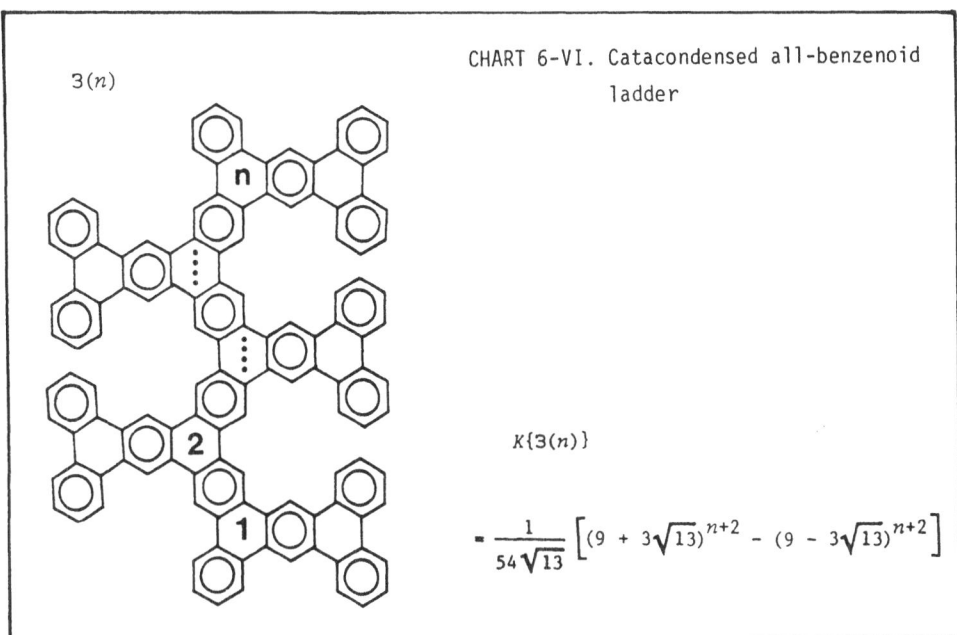

3(n)

CHART 6-VI. Catacondensed all-benzenoid
ladder

$K\{3(n)\}$

$$= \frac{1}{54\sqrt{13}}\left[(9 + 3\sqrt{13})^{n+2} - (9 - 3\sqrt{13})^{n+2}\right]$$

6.6 LIMIT VALUES INVOLVING K NUMBERS

For a single zigzag chain, $A(n)$; cf. CHART II, one finds a nonvanishing limit value of the quantity $(\ln K)/n$ when n tends to infinity:

$$\lim_{n \to \infty} \frac{\ln K\{A(n)\}}{n} = \lim_{n \to \infty} \frac{\ln F_{n+1}}{n} = \ln\left(\frac{1 + \sqrt{5}}{2}\right) \tag{6.47}$$

The result is consistent with known properties of the Fibonacci numbers.

The above case is contrasted by the behaviour of single linear chains (cf. CHART I), for which one has

$$\lim_{n \to \infty} \frac{\ln K\{L(n)\}}{n} = \lim_{n \to \infty} \frac{\ln(n+1)}{n} = 0 \qquad (6.48)$$

Nonvanishing limit values for quantities involving K as in (47) may be computed for several benzenoid classes already encountered in this chapter. We give only, as representative examples, the results for the classes of all-benzenoids and related classes defined in CHART V. It is found

$$\lim_{n \to \infty} \frac{\ln K\{Ч(n)\}}{n} = \lim_{n \to \infty} \frac{\ln K\{Ч'(n)\}}{n} = \lim_{n \to \infty} \frac{\ln K\{Ч''(n)\}}{n} = \ln\left(\frac{5 + \sqrt{17}}{2}\right) \qquad (6.49)$$

Notice that the limit value is the same for $Ч(n)$, $Ч'(n)$ and $Ч''(n)$, which all obey a recurrence relation of the same form (37); the limit value does not depend on the particular initial conditions. As the last example we give:

$$\lim_{n \to \infty} \frac{\ln K\{Ю(n)\}}{n} = \lim_{n \to \infty} \frac{\ln K\{Ю'(n)\}}{n} = \lim_{n \to \infty} \frac{\ln K\{Ю''(n)\}}{n} = \ln(2 + \sqrt{6}) \qquad (6.50)$$

7.1 DEFINITIONS

Two benzenoid units, say P_1 and P_2, are by definition said to be fused together (or P_1 fused to P_2) when they share exactly one edge (cf. also Paragraph
2.2.9). If at least one of the units is pericondensed the resulting benzenoid, which
may be symbolized by $P_1:P_2$, will be composite according to the definition of this
term in Paragraph 2.3.1.

Assume a system of fused units, B = P:C, where C is a catacondensed benzenoid.
Usually we assume that P is pericondensed, but the general formalism holds even
when P (and consequently B) is catacondensed. We shall refer to B as an annelated
benzenoid and say (by definition) that C is annelated to P, or P is annelated to C.
The simplest (and most important) cases occur when C is a single (unbranched) chain
of hexagons. The exposition of this chapter is limited to such cases. However, we
will allow for cases where several chains, C_1, C_2, ..., are simultaneously annelated to a benzenoid P.

The annelation of P to C may be called a one-sided annelation. If P_2 is annelated to $P_1:C$ so that it shares exactly one edge with C, the resulting benzenoid,
viz. B = $P_1:C:P_2$, is again said to be annelated. In this case we speak about a two-
sided annelation.

We wish to emphasize that the concept of annelation, which is a special case of
fusion, has been defined here in mathematical terms. (It should not be confused with
chemical annelations of benzenoid hydrocarbons, although there is a clear connection
here.)

7.2 PREVIOUS WORK

Gutman (1982b) gave the first general treatment for the numbers of Kekulé
structures under one-sided and two-sided annelations to a single linear chain. The
same author (Gutman 1985a) also treated the annelation of two linear chains to the
same benzenoid unit. The theory of one-sided and two-sided annelations to a single
zigzag chain was given by Cyvin and Gutman (1985). Cyvin (1987a) produced an algorithm for a benzenoid annelated to an arbitrary single chain.

A few more works are cited in the subsequent sections in connection with special benzenoid systems.

7.3 ANNELATION TO A LINEAR CHAIN

7.3.1 *Introduction*

Eilfeld and Schmidt (1981) reported formulas for the number of Kekulé structures (K) of some special classes of the forms P:L(n) or P:L(n):Q, where P and Q designate some definite benzenoids. All these formulas showed to be linear functions of n, viz. $K = An + B$, where A and B are constants. Gutman (1982) found this feature to be quite general, irrespective of the choice for P and Q. His results are reported in the following. Also the results with a zigzag chain are quoted, and the algorithm is explained (for references, see above).

7.3.2 *One-Sided Annelation*

Figure 1 illustrates the case P:L(n), a linear chain of n hexagons annelated to P. P' and P" are derived from P by deleting the edge of annelation (e_R) or the two vertices of annelation (v_R and w_R), respectively. Here the subscript R stands for right-hand side, which of course is arbitrarily chosen. The numbers of Kekulé structures $K\{P'\}$ and $K\{P"\}$ may be found by assigning a single or double bond, respectively, to the edge of annelation, as is indicated in the frames of Fig. 1. One

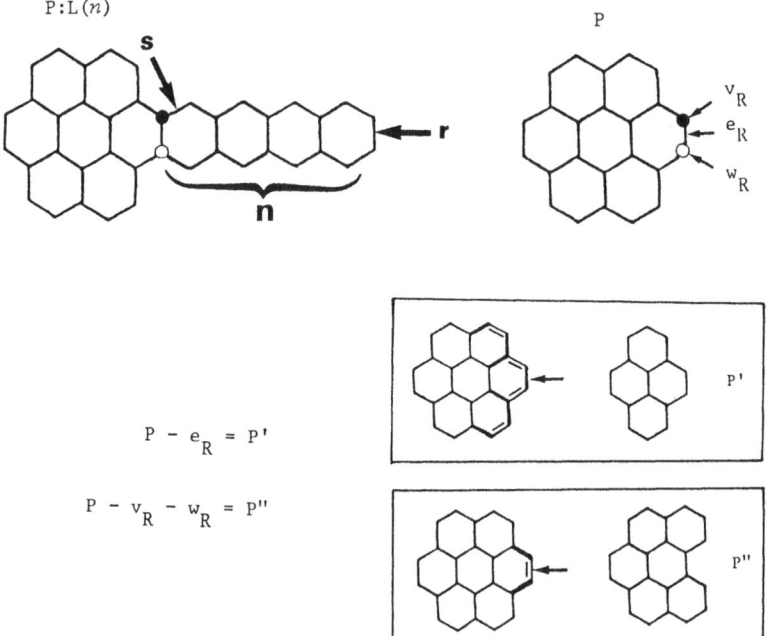

$$P - e_R = P'$$

$$P - v_R - w_R = P"$$

Fig. 7.1. The one-sided annelated system P:L(n), and some basic definitions. If P = *coronene* (as the depicted example), then $K\{P\} = 20$, $K\{P'\} = 6$, $K\{P"\} = 14$; cf. also Fig. 4.3. For $n=4$, $K=76$.

has, by virtue of eqn. (3.1) or its equivalent (4.1) in the frames of the method of fragmentation:

$$K\{P\} = K\{P'\} + K\{P''\} \tag{7.1}$$

It is instructive to derive the formula for $K\{P:L(n)\}$ in two ways. We may use the method of fragmentation and attack either the bond marked r in Fig. 1 or the bond s.

In the first case (bond r) one obtains

$$K\{P:L(n)\} = K\{P:L(n-1)\} + K\{P''\}; \qquad n \geq 1 \tag{7.2}$$

This is a recurrence relation. By means of the initial condition $K\{P:L(0)\} = K\{P\}$ one arrives at

$$K\{P:L(n)\} = K\{P\} + nK\{P''\} \tag{7.3}$$

In the second case (bond s) one obtains

$$K\{P:L(n)\} = (n+1)K\{P''\} + K\{P'\} \tag{7.4}$$

On eliminating $K\{P'\}$ by means of (1) the identity with eqn. (3) is established.

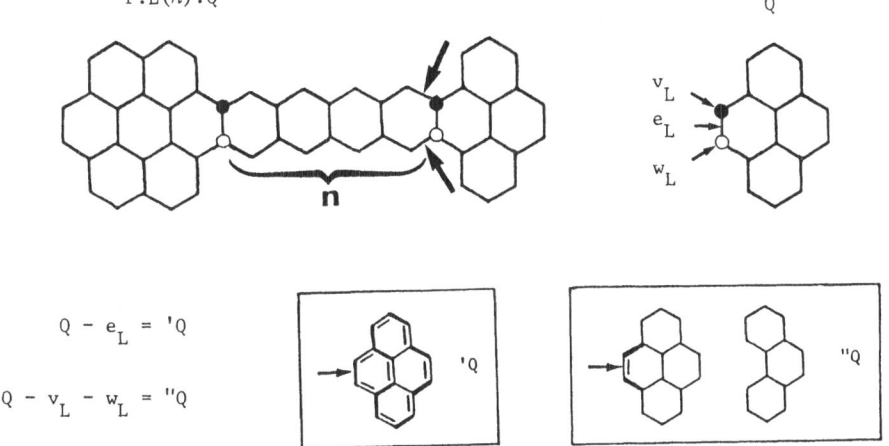

Fig. 7.2. The two-sided annelated system $P:L(n):Q$, and some supplementary basic definitions (see also Fig. 7.1). If $Q = pyrene$ (as the depicted example), then $K\{Q\} = 6$, $K\{'Q\} = 1$, $K\{''Q\} = 5$. For $n=4$, $K = 394$.

7.3.3 Two-Sided Annelation

Figure 2 illustrates a two-sided annelation to a linear chain of n hexagons, viz. $P:L(n):Q$. It also shows the definitions of $'Q$ and $''Q$, i.e. the systems obtained by deleting the edge e_L or the two vertices v_L and w_L, respectively. Here the subscript L stands for left-hand side. The analogous equation to (1) reads

$$K\{Q\} = K\{'Q\} + K\{''Q\} \tag{7.5}$$

In order to get the equation for $K\{P:L(n):Q\}$ we introduce some additional auxiliary systems derived from P and Q. Define P_v as the system obtained by deleting v_R from P, and correspondingly for P_w, $_vQ$ and $_wQ$; cf. Fig. 3. It is also natural to denote by P_{vw} the system obtained by deleting both v_R and w_R from P; then P_{vw} becomes synonymous with P''. Correspondingly $_{vw}Q = {}''Q$.

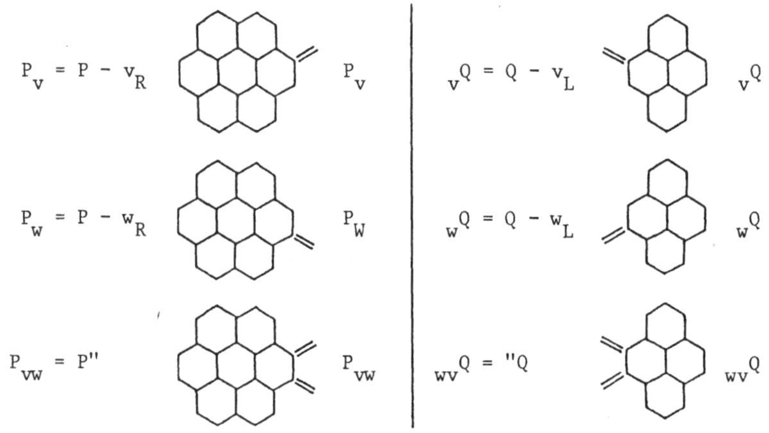

Fig. 7.3. Auxiliary systems derived from P and Q; see also Figs. 7.1 and 7.2.

If P is Kekuléan, then P_v and P_w are obvious non-Kekuléan. Also if Q is Kekuléan, then $_vQ$ and $_wQ$ are non-Kekuléan. However, we wish to arrive at a general result; hence these auxiliary systems are introduced although they cancel from the final result of the present paragraph. Furthermore, they will prove to be useful in subsequent expositions.

We employ the method of fragmentation to both of the edges marked by thick arrows in Fig. 2. The process may be executed stepwise by attacking one edge at a time, or, as indicated in Fig. 4, by considering the four possible bonding schemes for the two edges in question. The result is:

$$K\{P:L(n):Q\} = K\{P:L(n-1)\}K\{''Q\} + K\{P''\}K\{Q\} + K\{P_v\}K\{_wQ\} + K\{P_w\}K\{_vQ\} \tag{7.6}$$

It follows

$$K\{P:L(n):Q\} - K\{P:L(n-1):Q\} = [K\{P:L(n-1)\} - K\{P:L(n-2)\}]K\{''Q\}; \qquad n \geq 2 \tag{7.7}$$

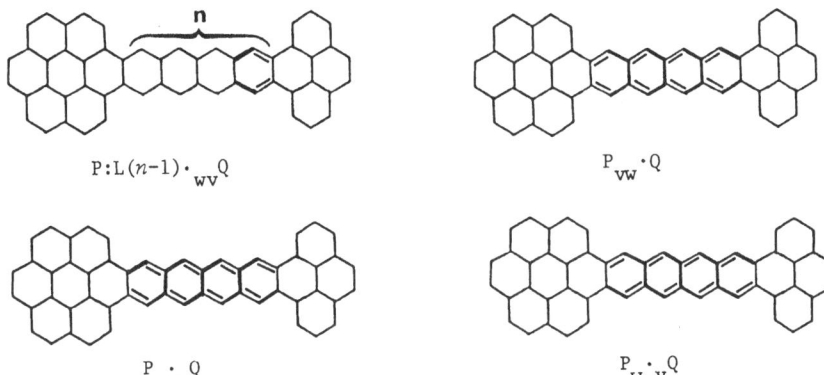

$$P:L(n-1)\cdot_{wv}Q \qquad\qquad P_{vw}\cdot Q$$

$$P_v\cdot{}_wQ \qquad\qquad P_w\cdot{}_vQ$$

Fig. 7.4. The method of fragmentation applied to the two edges of P:L(n):Q marked by arrows in Fig. 7.2. See also Fig. 7.3 for definitions.

where all the n-independent terms in (6) have vanished.

When the two terms in brackets are expanded according to eqn. (3), one arrives at

$$K\{P:L(n):Q\} = K\{P:L(n-1):Q\} + K\{P''\}K\{''Q\}; \qquad n \geq 1 \qquad (7.8)$$

This is a recurrence relation, from which it follows

$$K\{P:L(n):Q\} = K\{P:Q\} + nK\{P''\}K\{''Q\} \qquad (7.9)$$

Both equations (3) and (9) represent the linear form $K = An + B$.

7.4 ANNELATION TO A ZIGZAG CHAIN

For a benzenoid P annelated to a single zigzag chain of n hexagons (cf. Fig. 5) it was found, by the same methods as above, and using the same notation,

P:A(n)

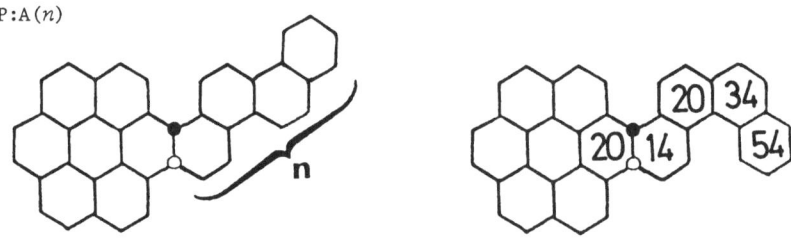

Fig. 7.5. The one-sided annelated system P:A(n) and one of its isoarithmic partners. When P = *coronene* and $n=4$, then $K = 142$. The right-hand side drawing is supplied with algorithm numerals.

$$K\{P:A(n)\} = K\{P:A(n-1)\} + K\{P:A(n-2)\}; \qquad n \geq 2 \qquad (7.10)$$

and

$$K\{P:A(n)\} = F_n K\{P\} + F_{n-1} K\{P''\}; \qquad n \geq 1 \qquad (7.11)$$

The single chain needs not necessarily be a zigzag chain; any all-kinked chains will give the same result (11). The feature is illustrated in Fig. 5, which shows two iso-arithmic systems (cf., e.g. Fig. 6.1). Eqn. (10) is a Fibonacci-type recurrence relation. In (11) F_n designates a Fibonacci number according to the definition of Section 4.4 or CHART 6-II.

For some few of the smallest n values (cf. also Fig. 4.8) eqns. (3) and (10) yield:

$$K\{P:L(0)\} = K\{P:A(0)\} = K\{P\}$$

$$K\{P:L(1)\} = K\{P:A(1)\} = K\{P\} + K\{P''\}$$

$$K\{P:L(2)\} = K\{P\} + 2K\{P''\}, \qquad K\{P:A(2)\} = 2K\{P\} + K\{P''\}$$

Now we also take the two-sided annelation to a single zigzag chain (Fig. 6) into account. For the smallest values of n the following formulas hold.

$$K\{P:L(0):Q\} = K\{P:A(0):Q\} = K\{P:Q\}$$

$$K\{P:L(1):Q\} = K\{P:A(1):Q\} = K\{P:Q\} + K\{P''\} \{''Q\}$$

$$K\{P:L(2):Q\} = K\{P:Q\} + 2K\{P''\}K\{''Q\}, \qquad K\{P:A(2):Q\} = K\{P:Q\} + K\{P\}K\{Q\} + K\{P''\}K\{''Q\}$$

For higher values of n it was found:

$$K\{P:A(n):Q\} = K\{P:Q\} + F_{n-1}K\{P\}K\{Q\} + (F_{n-2} - 1)[K\{P\}K\{''Q\} + K\{P''\}K\{Q\}]$$

$$+ (F_{n-3} + 1)K\{P''\}K\{''Q\}; \qquad n \geq 3 \qquad (7.12)$$

7.5 FURTHER DEVELOPMENTS

7.5.1 *Some Auxiliary Results*

Consider the system P:Q of two arbitrary benzenoids fused together. By a fragmentation scheme similar to the one of Fig. 4 it was obtained

$$K\{P:Q\} = K\{P'\}K\{''Q\} + K\{P''\}K\{Q\} + K\{P_v\}K\{_wQ\} + K\{P_w\}K\{_vQ\} \qquad (7.13)$$

After substituting $K\{P'\}$ according to (1) the result (13) may be written

$$K\{P:Q\} = X - K\{P''\}K\{''Q\} + Y \qquad (7.14)$$

P:A(n):Q

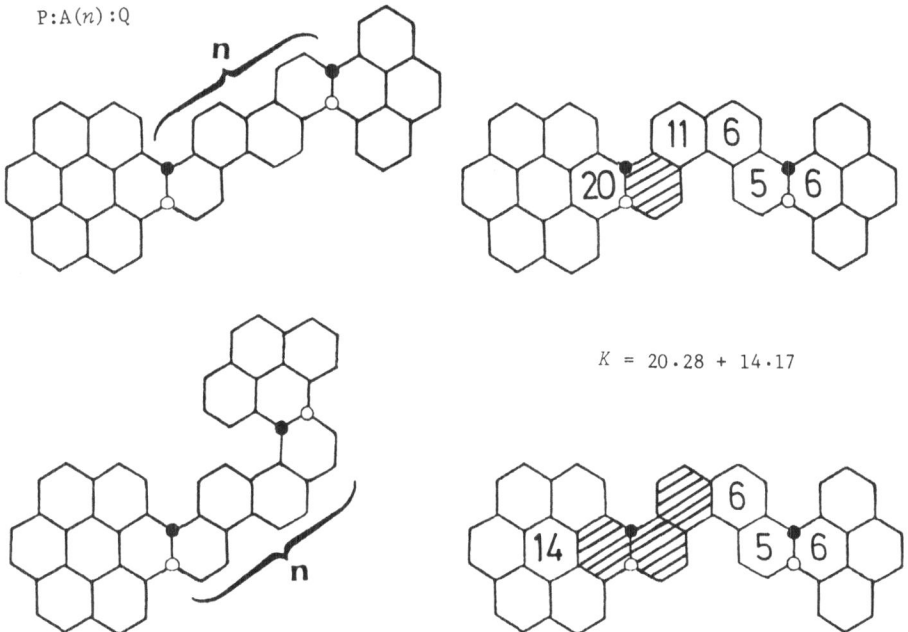

$K = 20 \cdot 28 + 14 \cdot 17$

Fig. 7.6. The two-sided annelated system P:A(n):Q and some of its isoarithmic part-
ners. When P = *coronene*, Q = *pyrene* (as in the depicted examples) and $n=4$, then $K = 798$. The right-hand side drawings are supplied with algorithm numerals.

where

$$X = K\{P\}K\{''Q\} + K\{P''\}K\{Q\} \tag{7.15}$$

and

$$Y = K\{P_v\}K\{_wQ\} + K\{P_w\}K\{_vQ\} \tag{7.16}$$

Another useful form of eqn. (13) is obtained when $K\{''Q\}$ and $K\{P''\}$ are substituted according to (5) and (1), respectively;

$$K\{P:Q\} = K\{P\}K\{Q\} - K\{P'\}K\{'Q\} + Y \tag{7.17}$$

7.5.2 *Further Developments of the Formulas for Two-Sided Annelations*

On inserting $K\{P:Q\}$ from eqn. (14) into (9) and (12) one obtains

$$K\{P:L(n):Q\} = X + (n-1)K\{P''\}K\{''Q\} + Y \tag{7.18}$$

and

$$K\{P:A(n):Q\} = F_{n-1}K\{P\}K\{Q\} + F_{n-2}X + F_{n-3}K\{P''\}K\{''Q\} + Y; \qquad n \geq 3 \tag{7.19}$$

respectively. For $n=1$ and $n=2$ one has:

$$K\{P:L(1):Q\} = K\{P:A(1):Q\} = X + Y$$

$$K\{P:L(2):Q\} = X + K\{P''\}K\{''Q\} + Y, \qquad K\{P:A(2):Q\} = K\{P\}K\{Q\} + X + Y$$

7.6 DISCUSSION OF THE FORMULAS

7.6.1 *Even and Odd Systems*

It is useful to distinguish between cases when the benzenoids P and Q have even or odd numbers of vertices. For the sake of brevity we shall speak about "even" and "odd" systems, respectively.

When P is an odd system, then $K\{P\} = K\{P'\} = K\{P''\} = 0$. Consequently, also $X=0$; cf. eqn. (15).

When P is an even system, then $K\{P_v\} = K\{P_w\} = 0$. Consequently, also $Y=0$; cf. eqn. (16).

Corresponding rules are valid for Q. These rules are understood by realizing that a system with an odd number of vertices always is obvious non-Kekuléan (cf. Chapter 5).

A Kekuléan system is necessarily even. But this is also the case for a concealed non-Kekuléan (cf. Chapter 5). Furthermore, an obvious non-Kekuléan may also be an even system as is the case, for instance, with *triangulene* (Fig. 2.3).

7.6.2 *One-Sided Annelations*

If P is an odd system, then $K\{P:L(n)\} = 0$ and $K\{P:A(n)\} = 0$ in consistence with eqns. (3) and (11), respectively.

Also the following more general statements hold. If P is Kekuléan, then also P:L(n) and P:A(n) are Kekuléan. If P is non-Kekuléan, then also P:L(n) and P:A(n) are non-Kekuléan. The statements may be generalized to P:C, where C is an arbitrary catacondensed system, or even an arbitrary Kekuléan system.

7.6.3 *Two-Sided Annelations*

More interesting features are revealed by a corresponding discussion of two-sided annelations.

P *and* Q *both even.* Assume first that both P and Q are even systems. Then Y vanishes, and the formulas of the preceding section yield:

$$K\{P:Q\} = K\{P\}K\{''Q\} + K\{P''\}K\{Q\} - K\{P''\}K\{''Q\} = K\{P\}K\{Q\} - K\{P'\}K\{'Q\} \qquad (7.20)$$

$$K\{P:L(n):Q\} = K\{P\}K\{''Q\} + K\{P''\}K\{Q\} + (n-1)K\{P''\}K\{''Q\} \qquad (7.21)$$

$$K\{P:A(n):Q\} = F_{n-1}K\{P\}K\{Q\} + F_{n-2}(K\{P\}K\{''Q\} + K\{P''\}K\{Q\}) + F_{n-3}K\{P''\}K\{''Q\};$$

$$n \geq 3 \quad (7.22)$$

For $n=2$ the last term in (22) should be deleted. From eqn. (22) it is deduced that a Fibonacci-type recurrence relation holds for the K numbers in question when P and Q are even systems:

$$K\{P:A(n):Q\} = K\{P:A(n-1):Q\} + K\{P:A(n-2):Q\}; \qquad n \geq 2 \qquad (7.23)$$

In all the examples of Figs. 1-7 P and Q were chosen as Kekuléan benzenoids; hence eqns. (20)-(23) hold in these cases.

Provided that both P and Q are even, all the expressions for $K\{P:Q\}$, $K\{P:L(n):Q\}$ and $K\{P:A(n):Q\}$ are symmetrical in such a way that they do not change if one of the benzenoids, say Q, is flipped around the edge of fusion. This is a manifestation of isoarithmicity (Section 3.7) and an illustration of *Theorem 17* (3.19). In Fig. 7 we give an example for P:Q (in addition to Fig. 3.2).

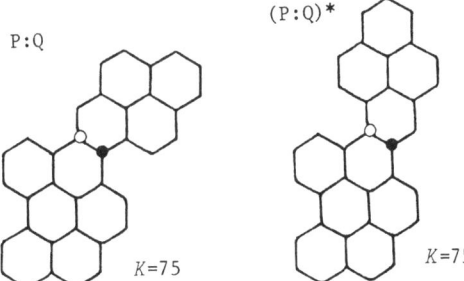

Fig. 7.7. Two isoarithmic systems.

P *odd and* Q *even, or vice versa.* If either P or Q, but not both are odd, then $K\{P:Q\} = 0$, $K\{P:L(n):Q\} = 0$ and $K\{P:A(n):Q\} = 0$. Then all these systems are obvious non-Kekuléan. In a more general case the single chain may be arbitrary.

P *and* Q *both odd.* If both P and Q are odd, the considered systems may be non-Kekuléan, but it is not necessarily so. The analysis of our equations yields

$$K\{P:Q\} = K\{P:L(n):Q\} = K\{P:A(n):Q\} = K\{P_v\}K\{_wQ\} + K\{P_w\}K\{_vQ\} \qquad (7.24)$$

where the right-hand side is identical with Y; cf. eqn. (16). The K numbers of all the considered systems are independent of n.

A simple example where eqn. (23) is applicable and the K numbers do not vanish, is furnished by the homologs of *zethrene* (Figs. 2.4 and 2.8), as was first pointed out by Gutman (1982b). In the present context these systems should be interpreted as two *phenalenes* (cf. Fig. 4.4) annelated to a linear chain. Cyvin and Gutman (1985) pointed out the same behaviour by two *phenalenes* annelated to a zigzag chain.

P:M(*LLLL*):Q

P:M(*LLAL*):Q

P:M(*LAAL*):Q

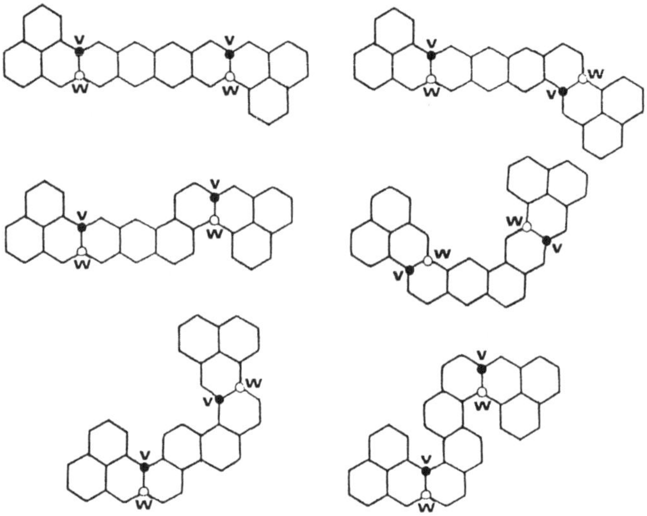

Fig. 7.8. Two *phenalene* units annelated to different four-membered single chains. In all cases $K = 9$.

Furthermore the authors found that the *LA*-sequence of the chain is quite arbitrary. It is only required that the *phenalene* units are annelated in such a way that their internal vertices acquire different colors (black and white). In that case only one of the two products in Y vanishes. If the vertices v and w are chosen as shown in Fig. 8, then $K\{P_v\} = K\{_wQ\} = 3$, while $K\{P_w\} = K\{_vQ\} = 0$. This is understood from the treatment of *phenalene* in Paragraph 4.2.3. Figure 8 shows different versions of the systems in question, viz. P:M:Q, where M is a single chain. Single chains with four hexagons and different *LA*-sequences are depicted; the notation conforms the one in Paragraph 6.2.4. However, the notation is still seen to be ambiguous for a characterization of the whole annelated benzenoid.

The systems of Fig. 8 are essentially disconnected (cf. Chapter 5).

7.7 ALGORITHM

Here an algorithm is explained for determining the K numbers of the following systems. (a) One-sided annelations P:M(n) and (b) two-sided annelations P:M(n):Q, where P and Q are Kekuléan, while M(n) is an arbitrary single chain of n hexagons. The first examples of the algorithm are included in (a) Fig. 5 and (b) Fig. 6. It is useful to relate the below algorithmic rules to those for a single chain (Paragraph 6.2.2; see especially Fig. 6.3).

Subgraph Method. One of the rules for determining algorithm numerals of a single chain (illustrated by the bottom-right part of Fig. 6.3) conforms to the subgraph method formulated below in more general terms. It is applicable to a single chain

(M) alone, or to cases when M is annelated to a benzenoid (P).

Algorithm numerals are (as usual) filled into hexagons according to a building-up process. In case of P:M the process is directed away from P. An algorithm numeral, X, associated with a hexagon X is determined in the following way.

1. Erase X and all hexagons preceding to it until the first kink is encountered.

2. Take X as the K number of the remaining benzenoid. It may be obtained by adding numerals already determined during the building-up process.

(a) *One-Sided Annelation*, P:M(n).

1. Start with $K\{P\}$ and $K\{P''\}$ on the two sides of the edge of annelation.

2. Continue with $K\{P''\}$ as long as a linear annelation is experienced, i.e. until the first kink.

3. After the first kink use the subgraph method (see above).

4. From now on either the subgraph method or the (simpler) rules for a single chain are applicable.

Figure 9 shows an example (in addition to Fig. 5).

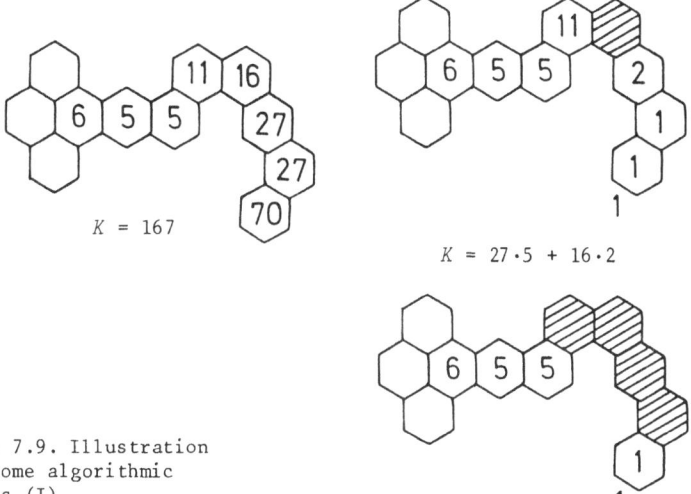

$K = 167$

$K = 27 \cdot 5 + 16 \cdot 2$

Fig. 7.9. Illustration of some algorithmic rules (I).

(b) *Two-Sided Annelation*, P:M(n):Q. The present rules may be related to those for catacondensed systems with one branching hexagon (Paragraph 6.3.1).

Assume that M has at least one angularly annelated (A_2-mode) hexagon, and choose one of them. The total K number is obtained as the sum of the K numbers for the two disconnected benzenoids generated in the following way. (1) Delete the chosen hexagon only. (2) Delete the chosen hexagon, and continue deleting hexagons in both

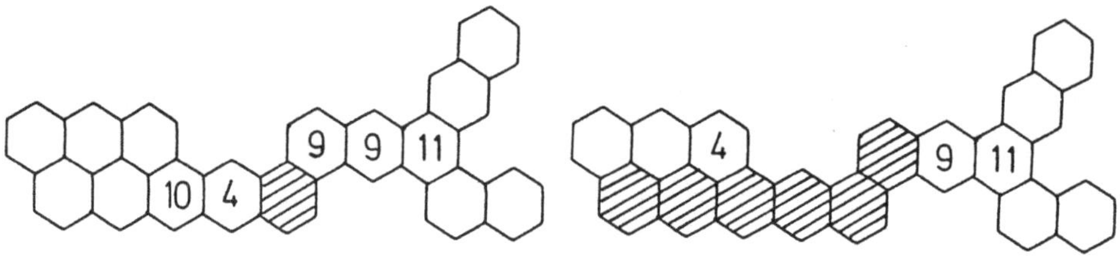

$$K = 14 \cdot 29 + 4 \cdot 20 = 486$$

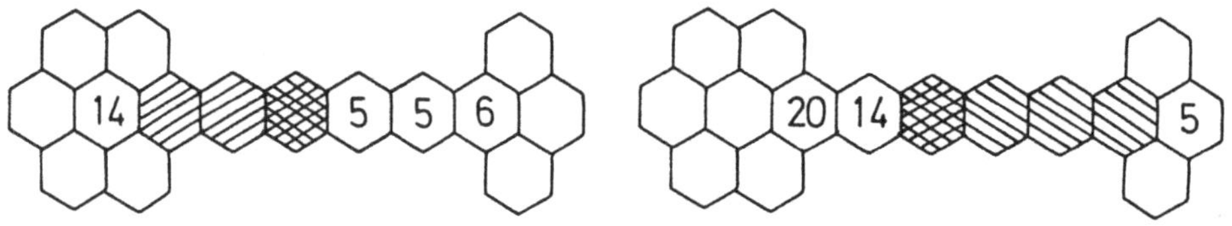

$$K = 14 \cdot 16 + 34 \cdot 5 = 394$$

Fig. 7.10. Illustration of some algorithmic rules (II).

directions until the first kink, or eventually beyond the edge of annelation in P or Q until only P" or "Q is left, respectively.

This procedure is also applicable to one-sided annelation P:M, as is shown in a part of Fig. 9. For two-sided annelations, see Figs. 6 and 10. In one of the examples (Fig. 10) one of the benzenoids annelated to the single chain is a single chain itself. It shows that it may be expedient to interpret branched catacondensed systems as annelated. The system of Fig. 6.8, for instance, is a two-sided annelation to a linear chain.

The procedure for P:M:Q described above must be modified for cases where M is a linear chain (no A_2 modes). We do not go into details here, but refer to the last example in Fig. 10 (see also Fig. 2).

7.8 DICTIONARY OF K NUMBERS WITH RELEVANCE TO ANNELATION

We have seen that $K\{P''\}$ plays an important role in the theory of annelations to P. For definitions and details, the preceding sections of this chapter should be consulted.

To take an example, P = *anthanthrene* ($K\{P\} = 10$) has three non-equivalent edges which may be involved in annelations. If one hexagon (*benzene*) is annelated, the following three systems emerge.

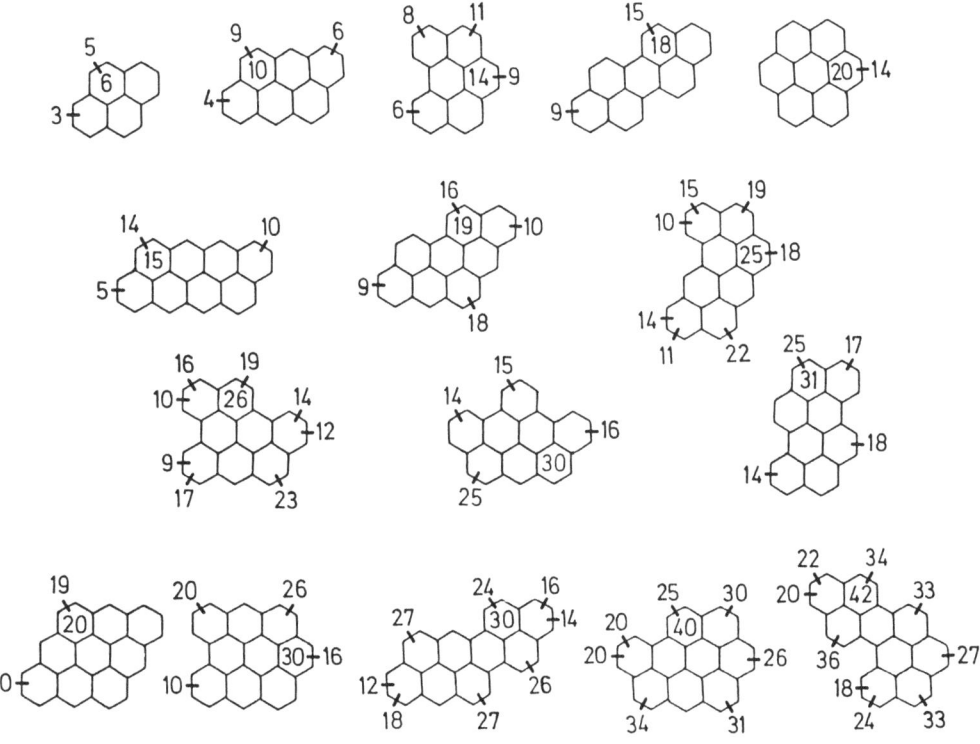

Benzo[b]anthanthrene   Benzo[a]anthanthrene   Benzo[e]anthanthrene

$K = 14$ $K = 16$ $K = 19$

The three numbers of $K\{P''\}$ for P = *anthanthrene* are entered into the respective (annelated) hexagons in the above drawings. As a specific information from the drawings one has $K = K\{P\} + K\{P''\}$ for the whole system, viz. *benzoanthanthrene*. The information of the above three drawings is compressed into one drawing of Fig. 11.

Figure 11 contains altogether the values of $K\{P''\}$ for all positions of annelation in normal basic benzenoids P with $h \leq 10$. Out of these systems there exist (Brunvoll, Cyvin SJ and Cyvin 1987) 1, 2, 2, 6, 5 and 20 with h = 4, 6, 7, 8, 9 and 10, respectively.

Fig. 7.11. Tabulation of $K\{P\}$ and $K\{P''\}$ for all normal basic benzenoids P with h = 4, 6, 7, 8, 9 and (next page) 10.

Fig. 7.11 (continued)

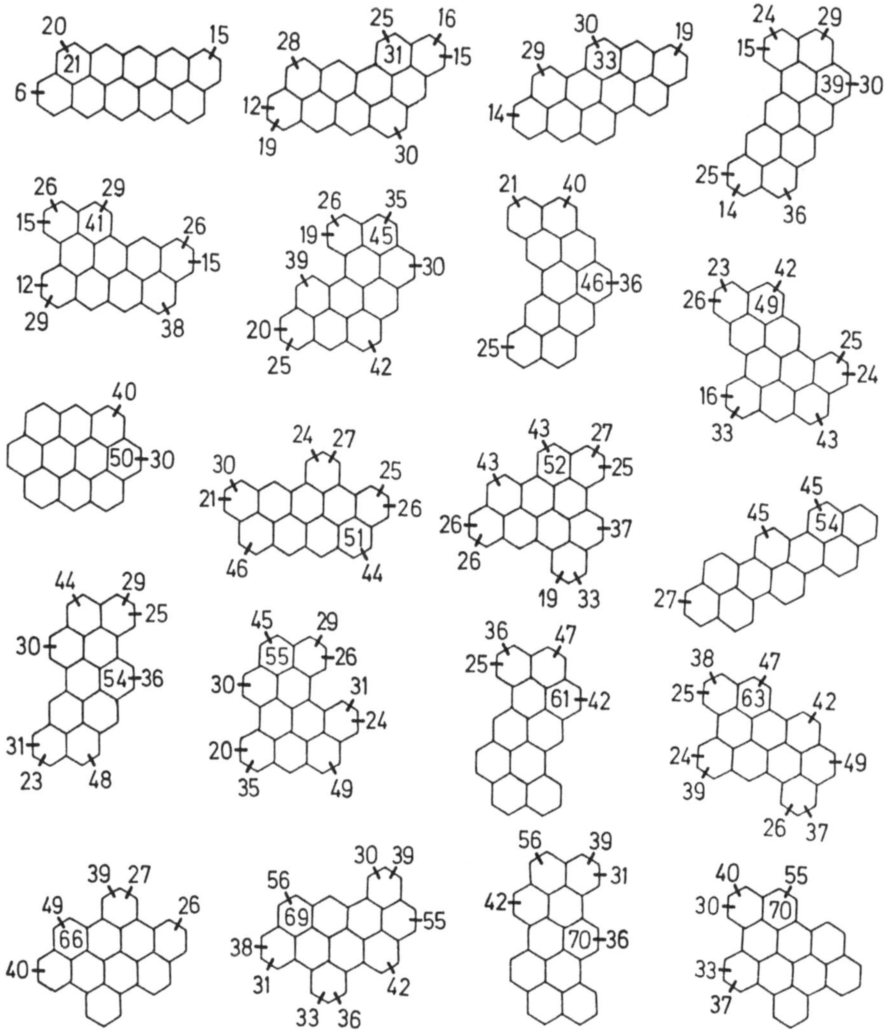

7.9 ANNELATION OF TWO SINGLE CHAINS

7.9.1 *Introduction*

Assume that a benzenoid system has at least two non-adjacent edges which may be involved in annelations. In this section we shall only consider systems of the kind M(*m*):P:M(*n*), where M(*m*) and M(*n*) are single chains with *m* and *n* hexagons, respectively. If P is Kekuléan, then the whole (annelated) system is Kekuléan; if P is non-Kekuléan, then the whole system is non-Kekuléan.

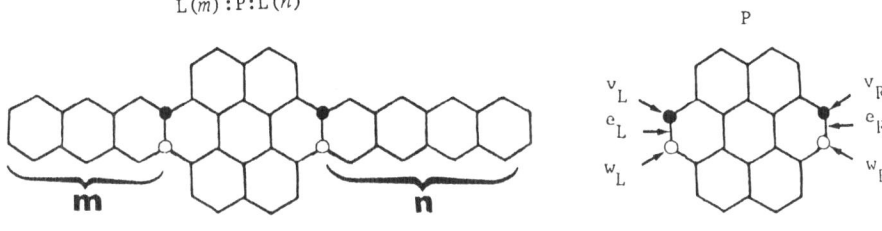

$$P - e_L - e_R = \text{'P'}$$

$$P - v_L - w_L$$
$$\quad - v_R - w_R = \text{"P"}$$

Fig. 7.12. The annelated system $L(m):P:L(n)$, and some supplementary basic definitions (see also Figs 7.1 and 7.2). If $P = coronene$ (as the depicted example), then
$K\{\text{'P'}\} = 1$, $K\{\text{"P"}\} = 9$. For $m=3$ and $n=4$, $K = 226$.

7.9.2 *Annelation of Two Linear Chains*

Consider the case of $L(m):P:L(n)$ as illustrated in Fig. 12. As an extension of
eqn. (3) it is found:

$$K\{L(m):P:L(n)\} = K\{P\} + mK\{\text{"P}\} + nK\{P\text{"}\} + mnK\{\text{"P"}\} \qquad (7.25)$$

CHART I shows some examples of applications of this formula. Additional formulas of this type are found in Cyvin SJ, Cyvin and Gutman (1986).

7.9.3 *Annelation of Two Chains of Which at Least One is a Zigzag Chain*

Consider the systems $L(m):P:A(n)$ and $A(m):P:A(n)$. As extensions of eqns. (3)
and (11) it is found:

$$K\{L(m):P:A(n)\} = F_nK\{P\} + mF_nK\{\text{"P}\} + F_{n-1}K\{P\text{"}\} + mF_{n-1}K\{\text{"P"}\}; \qquad n \geq 1 \quad (7.26)$$

and

$$K\{A(m):P:A(n)\} = F_mF_nK\{P\} + F_{m-1}F_nK\{\text{"P}\} + F_mF_{n-1}K\{P\text{"}\} + F_{m-1}F_{n-1}K\{\text{"P"}\};$$
$$m \geq 1, \; n \geq 1 \quad (7.27)$$

Again the zigzag chain may be replaced by other all-kinked chains without affecting
the K numbers.

Eqns. (26) and (27) could be exemplified by a great variety of annelated systems. Here we give only two cases; see CHART II.

7.9.4 *Extended Application of the Algorithm*

The example of Fig. 13, viz. L(3):P:A(5), where P = *benzo*[ghi]*perylene*, is covered by eqn. (26). The figure illustrates how the K enumeration problem also is solved by means of the algorithm (Section 7.7), if it only is employed in more steps.

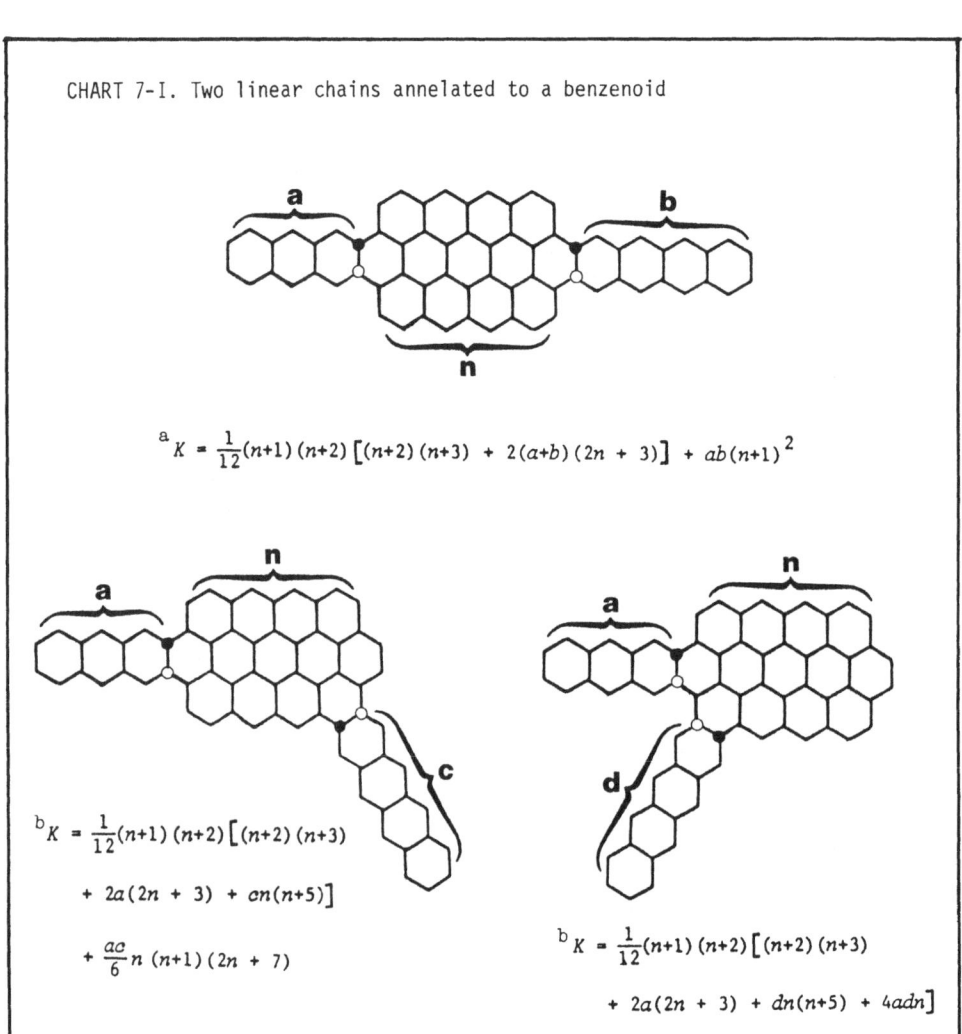

CHART 7-I. Two linear chains annelated to a benzenoid

$$^a K = \frac{1}{12}(n+1)(n+2)\left[(n+2)(n+3) + 2(a+b)(2n+3)\right] + ab(n+1)^2$$

$$^b K = \frac{1}{12}(n+1)(n+2)\left[(n+2)(n+3)\right.$$
$$\left. + 2a(2n+3) + cn(n+5)\right]$$
$$+ \frac{ac}{6}n(n+1)(2n+7)$$

$$^b K = \frac{1}{12}(n+1)(n+2)\left[(n+2)(n+3)\right.$$
$$\left. + 2a(2n+3) + dn(n+5) + 4adn\right]$$

[a]Gutman I (1985). Coll Sci Papers Fac Sci Kragujevac 6: 35; correction[b]
[b]Cyvin SJ, Cyvin BN, Gutman I (1986). Coll Sci Papers Fac Sci Kragujevac 7:5

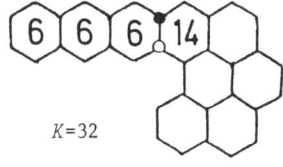

$K=32$

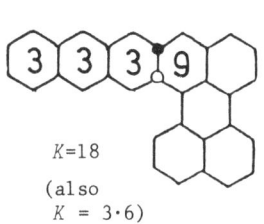

$K=18$
(also
$K = 3 \cdot 6$)

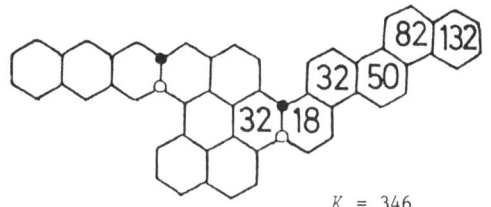

$K = 346$

Fig. 7.13. Application of the algorithm (in three steps) to deduce the K number for L(3):P:A(5), where P = *benzo*[ghi]*perylene*.

CHART 7-II. Systems with (a) one linear and one all-kinked chain, and (b) two all-kinked chains annelated to a benzenoid

(a)

a

b

n

$$K = \frac{1}{12}(n+1)(n+2)[F_b(n+2)(n+3) + 2(aF_b + F_{b-1})(2n + 3)] + aF_{b-1}(n+1)^2$$

(b)

a

b

n

$$K = \frac{1}{12}(n+1)(n+2)[F_a F_b(n+2)(n+3) + 2(F_{a-1}F_b + F_a F_{b-1})(2n + 3)] + F_{a-1}F_{b-1}(n+1)^2$$

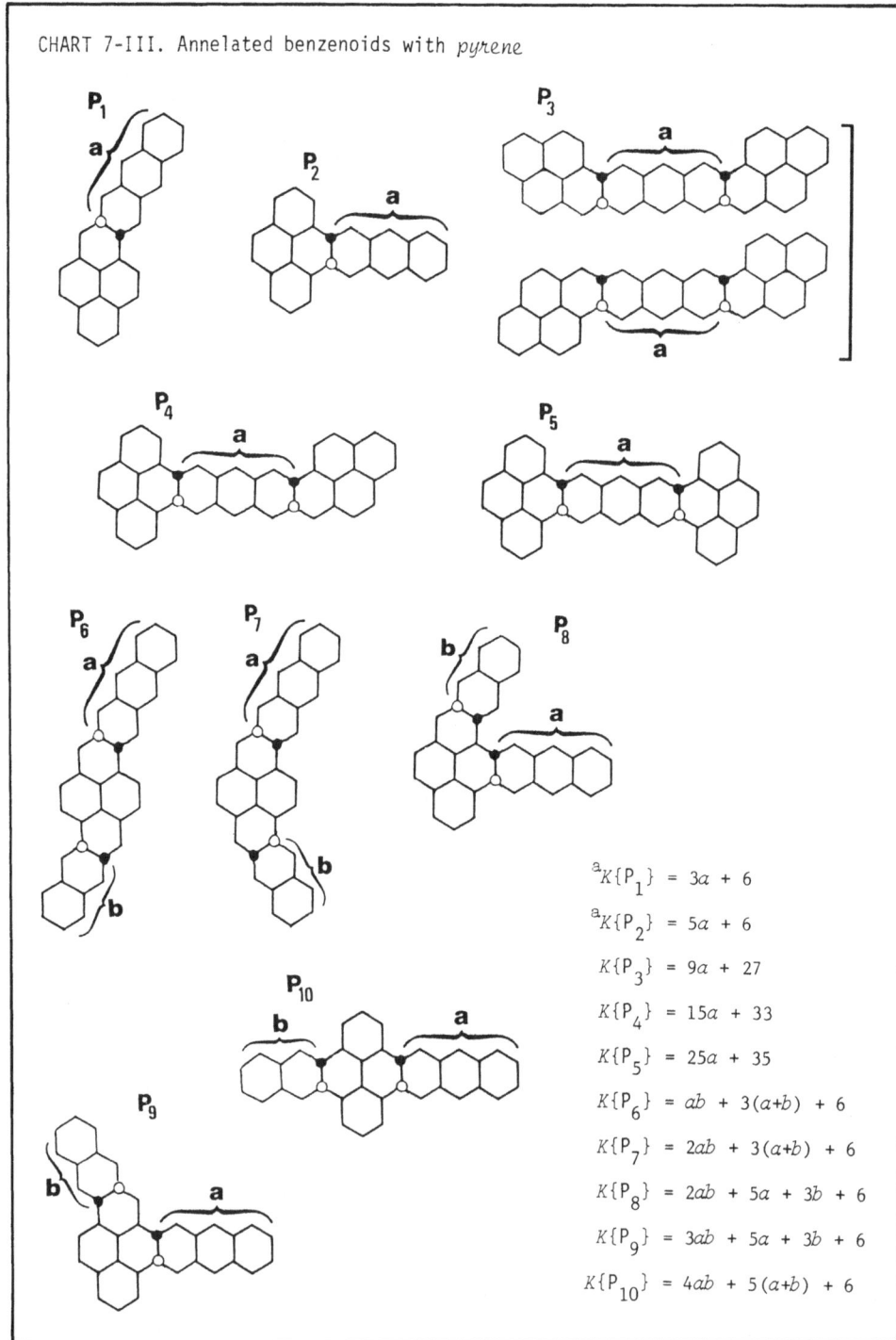

CHART 7-III. Annelated benzenoids with *pyrene*

$$^a K\{P_1\} = 3a + 6$$

$$^a K\{P_2\} = 5a + 6$$

$$K\{P_3\} = 9a + 27$$

$$K\{P_4\} = 15a + 33$$

$$K\{P_5\} = 25a + 35$$

$$K\{P_6\} = ab + 3(a+b) + 6$$

$$K\{P_7\} = 2ab + 3(a+b) + 6$$

$$K\{P_8\} = 2ab + 5a + 3b + 6$$

$$K\{P_9\} = 3ab + 5a + 3b + 6$$

$$K\{P_{10}\} = 4ab + 5(a+b) + 6$$

[a]Kiang YS (1980). Internat J Quant Chem 18: 331

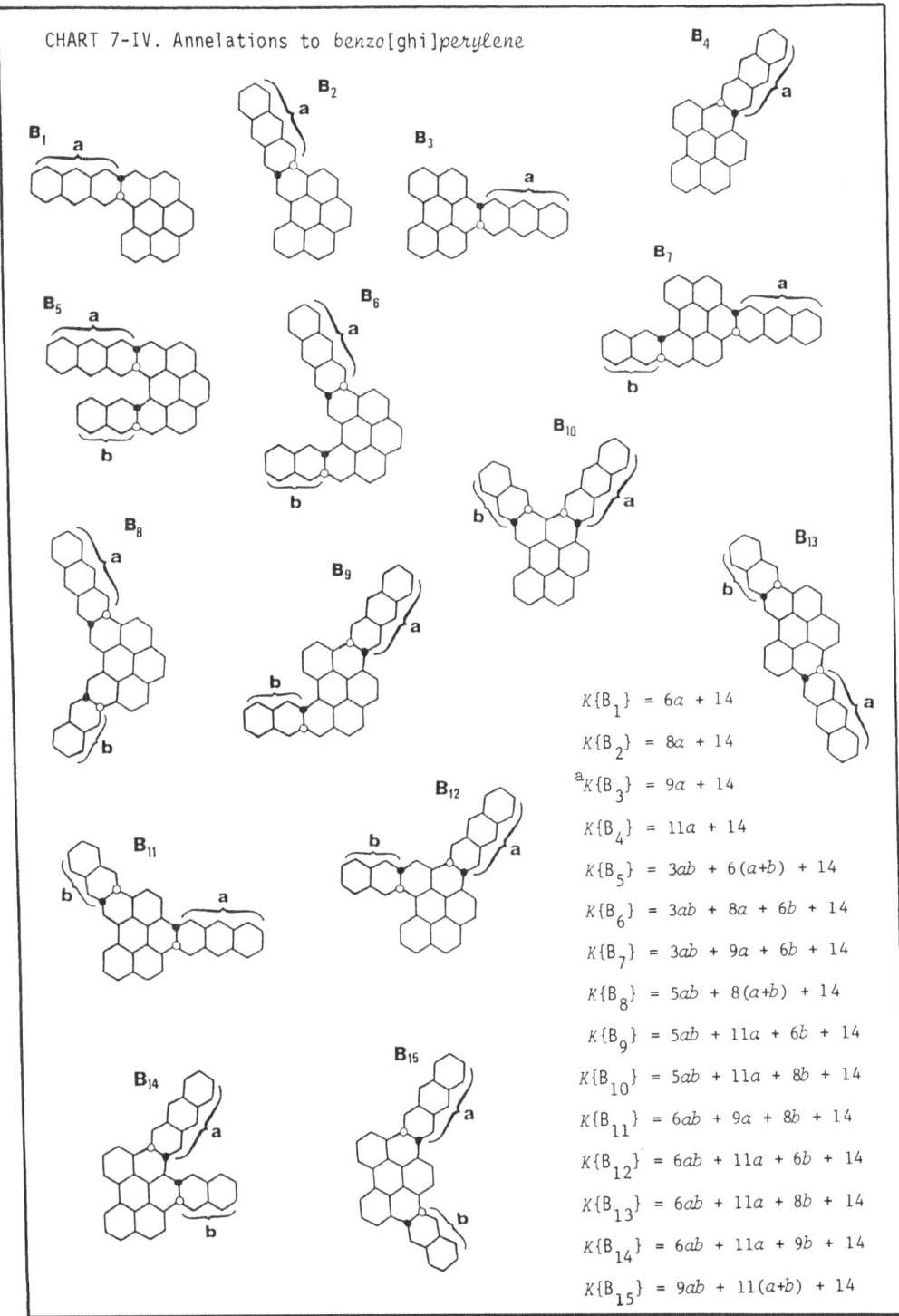

CHART 7-IV. Annelations to *benzo[ghi]perylene*

$K\{B_1\} = 6a + 14$

$K\{B_2\} = 8a + 14$

$^a K\{B_3\} = 9a + 14$

$K\{B_4\} = 11a + 14$

$K\{B_5\} = 3ab + 6(a+b) + 14$

$K\{B_6\} = 3ab + 8a + 6b + 14$

$K\{B_7\} = 3ab + 9a + 6b + 14$

$K\{B_8\} = 5ab + 8(a+b) + 14$

$K\{B_9\} = 5ab + 11a + 6b + 14$

$K\{B_{10}\} = 5ab + 11a + 8b + 14$

$K\{B_{11}\} = 6ab + 9a + 8b + 14$

$K\{B_{12}\} = 6ab + 11a + 6b + 14$

$K\{B_{13}\} = 6ab + 11a + 8b + 14$

$K\{B_{14}\} = 6ab + 11a + 9b + 14$

$K\{B_{15}\} = 9ab + 11(a+b) + 14$

[a]Eilfeld P, Schmidt W (1981). J Electr Spectr Related Phenom 24: 101

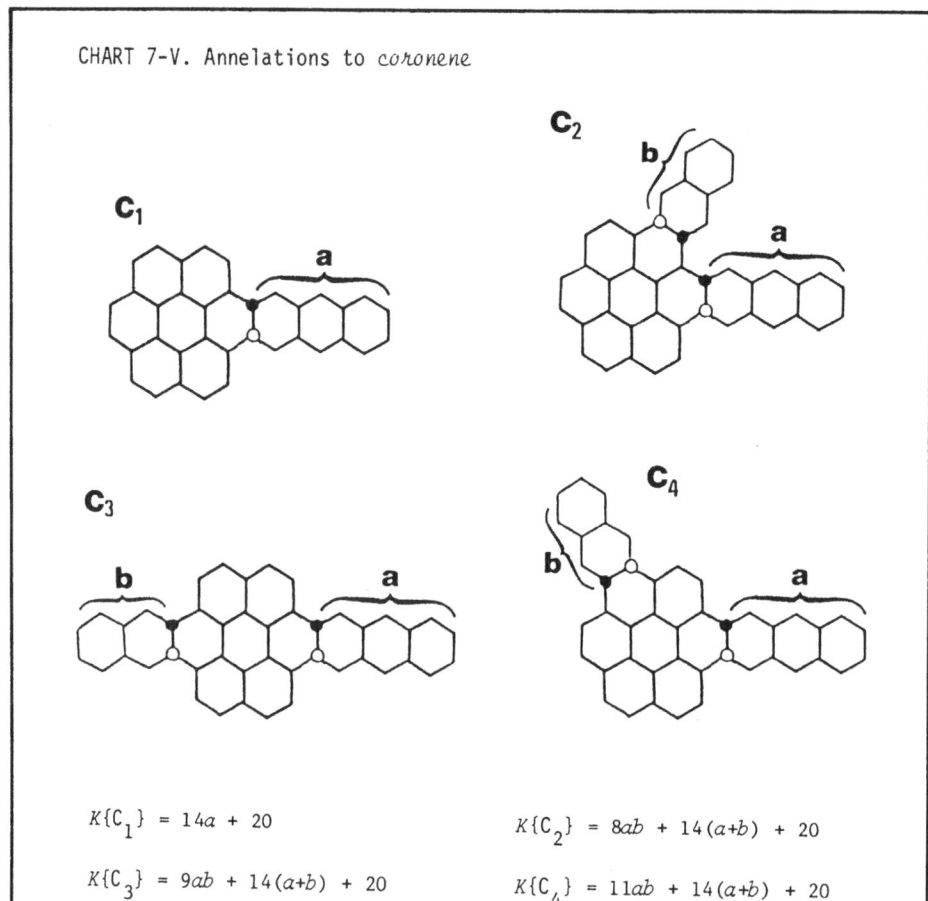

CHART 7-V. Annelations to *coronene*

$$K\{C_1\} = 14a + 20$$

$$K\{C_3\} = 9ab + 14(a+b) + 20$$

$$K\{C_2\} = 8ab + 14(a+b) + 20$$

$$K\{C_4\} = 11ab + 14(a+b) + 20$$

7.10 ANNELATIONS OF SPECIAL BENZENOIDS

7.10.1 *Introduction*

Here we report some formulas for annelated systems with the special benzenoids *pyrene*, *benzo*[ghi]*perylene* and *coronene*. Only annelations of or to single linear chains are considered. All the formulas are special cases of the more general results of the preceding section. They are supposed to be useful in studies of a variety of phenomena in organic chemistry (cf. Section 1.3) and have been selected by this criterion.

Many experimental works in organic chemistry on annelations of *pyrene* are available; we give references only to some of them, to which the numbers of Kekulé structures are especially relevant: Zander (1979); Bräuchle et al. (1980); Biermann and Schmidt (1980a); Eilfeld and Schmidt (1981); Bräuchle et al. (1982). Annelations to the other two benzenoids mentioned above have not been studied so extensively as the *pyrene* annelations, but still a number of works are available, both for *benzo-[ghi]perylene* (Zander 1978a; Eilfeld and Schmidt 1981) and for *coronene* (Braeuchle et al. 1978; Eilfeld and Schmidt 1981; Bräuchle et al. 1982).

7.10.2 *Tabulation of Formulas*

CHART III gives the K formulas for (a) the two classes of *pyrene* annelated (one-sided) to a linear chain, (b) the four classes of two *pyrenes* annelated (two-sided) to a linear chain, where two isoarithmic versions of a class (P_3) are counted twice, and (c) the five existing cases of two linear chains annelated to *pyrene*.

CHART IV gives the K formulas for *benzo[ghi]perylene* which correspond to the categories (a) and (b) above, and the same is given for *coronene* in CHART V.

CHAPTER 8
CLASSES OF BASIC BENZENOIDS (I)

8.1 INTRODUCTION

In the present chapter we are giving the first part of an account of some classes of benzenoids. In general they are pericondensed and basic, but for small parameters they degenerate to catacondensed systems. Some benzenoids related to those of
the main classes are also treated. They are also basic in general, but may degenerate
to composite or catacondensed systems, and even disconnected. Trivial cases of no
hexagons with $K=1$ are also encountered.

The main classes considered in this first part are: hexagons, chevrons, ribbons
and parallelograms. They have in common that the combinatorial formulas of K for
their members are known. A hexagon, chevron or ribbon is defined by three parameters,
a parallelogram by two.

8.2 HEXAGON

8.2.1 *Definition*

A hexagon-shaped benzenoid or simply hexagon, $O(k,m,n)$, defined by the parameters k, m and n, is shown in CHART I. The three parameters are permutable:

$$O(k,m,n) = O(m,n,k) = O(n,k,m) = O(m,k,n) = O(k,n,m) = O(n,m,k)$$

For one of the parameters equal to unity, say $k=1$, the hexagon degenerates to
a parallelogram: $O(1,m,n) = L(m,n)$. If one of the parameters vanishes one obtains
the trivial case of no hexagons and $K=1$.

A hexagon with all the (nonvanishing) parameters different is centrosymmetrical
(C_{2h}). If $k = m \neq n$ it is dihedral (D_{2h}). Finally it is regular hexagonal (D_{6h}) when
the parameters are all equal $(k=m=n)$. The notation is simplified according to

$$O(m,n) = O(m,m,n), \qquad O(n) = O(n,n,n)$$

8.2.2 *Previous Work and General Formulas*

Gordon and Davison (1952) have credited R.M. Everett for a formula of K (the
number of Kekulé structures) of a regular hexagonal hexagon, $O(n)$, and M. Woodger
for the dihedral hexagon, $O(m,n)$. The first study of centrosymmetrical hexagons is
due to Ohkami and Hosoya (1983). The general three-parameter formula of $K\{O(k,m,n)\}$
was given by Cyvin (1986d); see CHART I. A number of relations in the following are

CHART 8-I. Hexagon and related benzenoids

$O(k,m,n)$

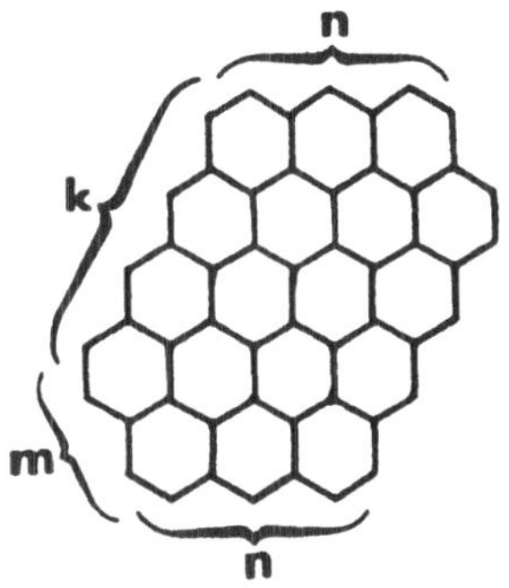

$Oa(k,m,n)$

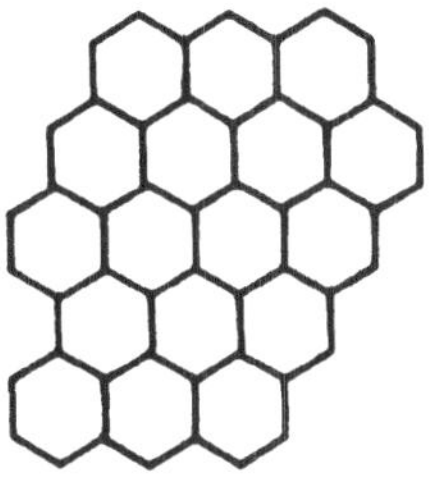

$Ob(k,m,n)$

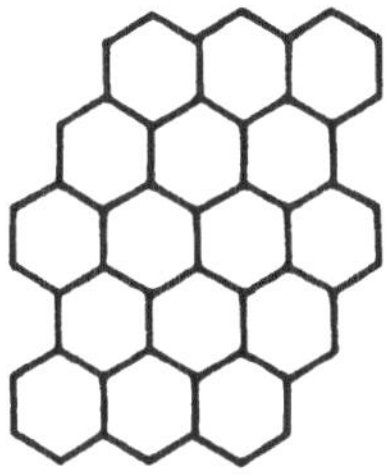

$$^{a}K\{O(k,m,n)\} = \prod_{i=0}^{k-1} \frac{\binom{m+n+i}{n}}{\binom{n+i}{n}} \qquad \text{(i)}$$

$$K\{Oa(k,m,n)\} = A \prod_{i=1}^{k-1} \frac{\binom{m+n+i-1}{m}}{\binom{m+i}{m}} = A \prod_{i=1}^{n-1} \frac{\binom{k+m+i-1}{m}}{\binom{m+i}{m}} ,$$

$$\text{where} \qquad A = \binom{k+m+n-1}{m} - \binom{m+n-1}{m} \qquad \text{(ii)}$$

$$K\{Ob(k,m,n)\} = B \frac{\displaystyle\prod_{i=1}^{k-2} \binom{m+n+i-1}{m}}{\displaystyle\prod_{i=1}^{k-1} \binom{m+i}{m}} = B \frac{\displaystyle\prod_{i=1}^{n-2} \binom{k+m+i-1}{m}}{\displaystyle\prod_{i=1}^{n-1} \binom{m+i}{m}} , \qquad \text{where}$$

$$B = \binom{k+m+n-1}{m}\binom{k+m+n-2}{m} - 2\binom{k+m+n-2}{m}\binom{m+n-1}{m} + \binom{m+n-1}{m}\binom{m+n-2}{m} \qquad \text{(iii)}$$

[a]Cyvin SJ (1986). Monatsh Chem 117: 33

also due to Cyvin (1986d). The formulas of CHART I were applied by Cyvin SJ, Cyvin and Gutman (1986) in their studies of linear chains annelated to a hexagon.

A hexagon $O(k,m,n)$ has

$$h = (k+m)(n-1) + km - n + 1 \tag{8.1}$$

Normally we would refer to this number as the number of hexagons. In order to avoid confusion we will refer to it as the number of hexagonal units when we speak about the class of hexagons (hexagon-shaped benzenoids).

A non-degenerate hexagon (i.e. a hexagon with all parameters greater than unity) has six corners. They are defined as the L_3-mode hexagonal units. The benzenoid $Oa(k,m,n)$ is referred to as a hexagon without corner; cf. CHART I. The removed corner is conventionally taken as the meeting of the m-chain (linear chain of m hexagonal units) and the k-chain, where it is referred to the first two parameters. In this system only the two first parameters (k,m) are permutable, while n is unique. One has

$$K\{Oa(k,m,n)\} = K\{O(k,m,n)\} - K\{O(k, m, n-1)\}; \quad n \geq 1 \tag{8.2}$$

On combining with eqn. (i) it was attained at the formula (ii) entered in CHART I. Notice that the last index in the two products depend either on k or n, which are non-equivalent.

When two opposite corners are removed from $O(k,m,n)$ as shown in CHART I, one obtains $Ob(k,m,n)$. For this benzenoid one has

$$K\{Ob(k,m,n)\} = K\{Oa(k,m,n)\} - K\{Oa(k, m, n-1)\}$$
$$= K\{O(k,m,n)\} - 2K\{O(k, m, n-1)\} + K\{O(k, m, n-2)\}; \quad n \geq 2 \tag{8.3}$$

The explicit formulas (iii) were achieved by means of (i).

In eqns. (ii) and (iii) of CHART I, if the lower index i is smaller than the upper one (cases for k or n equal to 1 and occasionally 2), the appropriate product should be put equal to 1. The numerical values pertaining to $k=4$, $m=2$ and $n=3$ (as chosen in CHART I) are: $K\{O(4,2,3)\} = 490$, $K\{O(4,2,2)\} = 105$, $K\{O(4,2,1)\} = 15$, $K\{Oa(4,2,3)\} = 385$, $K\{Oa(4,2,2)\} = 90$, $K\{Ob(4,2,3)\} = 295$.

For the sake of conformity we introduce the following notations in the symmetrical cases of $k=m$.

$$Oa(m,n) = Oa(m,m,n), \qquad Ob(m,n) = Ob(m,m,n)$$

Eqn. (2) is a recurrence relation for $O(k,m,n)$. It may be put into the form

$$K\{0(k+1,\ m,\ n)\} - K\{0(k,m,n)\} = \left[\binom{k+m+n}{n} - \binom{k+n}{n}\right]\prod_{i=0}^{k-1}\frac{\binom{m+n+i}{n}}{\binom{n+i+1}{n}}$$

$$= \left[\binom{k+m+n}{n} - \binom{k+n}{n}\right]\prod_{i=1}^{m-1}\frac{\binom{k+n+i}{n}}{\binom{n+i}{n}} \tag{8.4}$$

A recurrence formula for the quotient is simpler:

$$\frac{K\{0(k+1,\ m,\ n)\}}{K\{0(k,m,n)\}} = \frac{\binom{k+m+n}{n}}{\binom{k+n}{n}} = \prod_{i=1}^{n}\frac{k+n+i}{k+i} \tag{8.5}$$

8.2.3 *Dihedral and Regular Hexagonal Hexagons*

Dihedral. On inserting $k=m$ into eqn. (i) of CHART I one obtains a formula for the special case of dihedral hexagons. Different forms are produced when taking advantage of the possible permutations of the parameters. Firstly, one has

$$K\{0(m,n)\} = \prod_{i=0}^{m-1}\frac{\binom{m+n+i}{n}}{\binom{n+i}{n}} = \prod_{i=0}^{m-1}\frac{\binom{m+n+i}{m}}{\binom{m+i}{m}} \tag{8.6}$$

Secondly, one obtains

$$K\{0(m,n)\} = \prod_{i=0}^{n-1}\frac{\binom{2m+i}{m}}{\binom{m+i}{m}} \tag{8.7}$$

where the last index of the product depends on n rather than m. These two parameters are not equivalent and therefore can not be permuted. Both forms (6) and (7) may have their advantages depending on the actual situation.

A recurrence relation for dihedral hexagons in terms of a quotient reads

$$\frac{K\{0(m+1,\ n)\}}{K\{0(m,n)\}} = \frac{(2m + n + 1)\left(\dfrac{2m+n}{n}\right)^2}{(2m + 1)\left(\dfrac{m+n}{n}\right)^2} \tag{8.8}$$

A formula for $K\{0(m,n)\}$ in terms of linear factors reads

$$K\{0(m,n)\} = \left(\frac{n+m}{m}\right)^m \prod_{i=1}^{m-1}\left[\frac{(n+i)(n + 2m - i)}{i(2m - i)}\right]^i ; \qquad m > 1 \tag{8.9}$$

It is illustrated below for $m = 2$, 3 and 4 (Ohkami and Hosoya 1983).

$$K\{0(2,n)\} = \frac{(n+1)(n+2)^2(n+3)}{1\cdot 2^2\cdot 3}$$

$$K\{0(3,n)\} = \frac{(n+1)(n+2)^2(n+3)^3(n+4)^2(n+5)}{1\cdot 2^2\cdot 3^3\cdot 4^2\cdot 5}$$

$$K\{0(4,n)\} = \frac{(n+1)(n+2)^2(n+3)^3(n+4)^4(n+5)^3(n+6)^2(n+7)}{1\cdot 2^2\cdot 3^3\cdot 4^4\cdot 5^3\cdot 6^2\cdot 7}$$

Numerical values of K for dihedral hexagons are given in Table 1.

Regular Hexagonal. A further specialization is represented by the regular hexagonal hexagons, $0(n)$. In this case one has

$$K\{0(n)\} = \prod_{i=0}^{n-1}\frac{\left(\dfrac{2n+i}{n}\right)}{\left(\dfrac{n+i}{n}\right)} \tag{8.10}$$

Table 8.1. Numerical values of $K\{0(m,n)\}$.

n / m	1	2	3	4	5	6	7	8	9	10
1	2	3	4	5	6	7	8	9	10	11
2	6	20	50	105	196	336	540	825	1210	1716
3	20	175	980	4116	14112	41580	108900	259545	572572	1184183
4	70	1764	24696	232848	1646568	9343620	44537922	184225041	677352676	
5	252	19404	731808	16818516	267227532					
6	924	226512	24293412							
7	3432	2760615	877262100							
8	12870	34763300								
9	48620	449141836								
10	184756									

and the recurrence relation:

$$\frac{K\{0(n+1)\}}{K\{0(n)\}} = \frac{(3n + 1)^2 (3n + 2)\binom{3n}{n}^3}{(2n + 1)^3 \binom{2n}{n}^3} \qquad (8.11)$$

Table 1 includes the numerical values of $K\{0(n)\}$ up to $n=5$.

8.2.4 Limit Values Involving K Numbers

Gutman (1981b) proved the relation

$$\lim_{n \to \infty} \frac{\ln K\{0(n)\}}{n^2} = \frac{3}{2} \ln \frac{27}{16} \qquad (8.12)$$

Similar limiting properties were derived for the less symmetrical hexagons, viz.

$$\lim_{m \to \infty} \frac{\ln K\{0(m,n)\}}{m} = 2n \qquad (8.13)$$

and

$$\lim_{k \to \infty} \frac{\ln K\{0(k,m,n)\}}{\ln k} = mn \qquad (8.14)$$

8.3 CHEVRON

8.3.1 *Definition*

A chevron-shaped benzenoid (chevron), Ch(k,m,n), contains three parameters (k,m,n); see CHART II. For $n=1$ the system degenerates to a single chain with two segments, M(k,m); cf. Paragraph 6.2.4. In the general case the chevron consists of n condensed single chains M(k,m) properly aligned so that the system also can be interpreted as $k+m-1$ condensed rows of n hexagons each. Hence

$$h = (k+m-1) n \qquad (8.15)$$

A chevron is an example of a multiple (n-tuple) chain. We write it in terms of the *LA*-sequence as

$$Ch(k,m,n) = M_n(L^{k-1}AL^{m-1})$$

It is assumed $k \geq 1$ and $m \geq 1$. The parameters k and m are permutable in the sense that

$$Ch(k,m,n) = Ch(m,k,n)$$

For $k=1$ (or $m=1$) the chevron degenerates into the $m\times n$ (or $k\times n$) parallelogram: Ch$(1,m,n)$ = L(m,n). The parameter n may be zero; then the chevron degenerates into the trivial case of no hexagons with $K=1$.

A non-degenerate chevron is mirror-symmetrical (C_{2v}) when $k=m$ and unsymmetrical (C_s) when $k \neq m$. In the former case the notation is simplified according to

$$Ch(m,n) = Ch(m,m,n)$$

8.3.2 *Previous Work and General Formulas*

Gordon and Davison (1952) have given the formula for the number of Kekulé structures (K) of Ch(k,m,n). The slightly modified version of this formula (Cyvin 1985), which implies a summation to n, is reproduced in CHART II (i). The results of Jiang (1980), who considered the algebraic form of the determinant of the adjacency matrix, imply the special case of this formula for $k=m$. The alternative formula (ii), which contains a summation to k, is due to Cyvin (1985). The summation therein may always start from $i=1$, but starting from $i = k-n$ when $k > n+1$ avoids some terms equal to zero. Several of the other results of this section are also taken from Cyvin (1985). The formula apparatus (CHART II) was used by Cyvin SJ, Cyvin and Gutman (1986) in their studies of a linear chain annelated to a chevron.

A non-degenerate chevron has three corners (L_3-mode hexagons). The corner where the k- and m-chains meet is referred to as the apex. The anti-apex is the (unique) L_5-mode hexagon. CHART II includes additional K formulas, which pertain to: (iii) chevron without apex, $Ca(k,m,n)$; (iv) chevron without anti-apex or with a fjord, $Cf(k,m,n)$; (v) chevron without the corner where the k- and n-chains meet, $Cc(k,m,n)$. For the sake of conformity we may also use the following abbreviated notation in the symmetrical cases $(k=m)$.

CHART 8-II. Chevron and related benzenoids

$Ch(k,m,n)$ $Ca(k,m,n)$ $Cf(k,m,n)$ $Cc(k,m,n)$

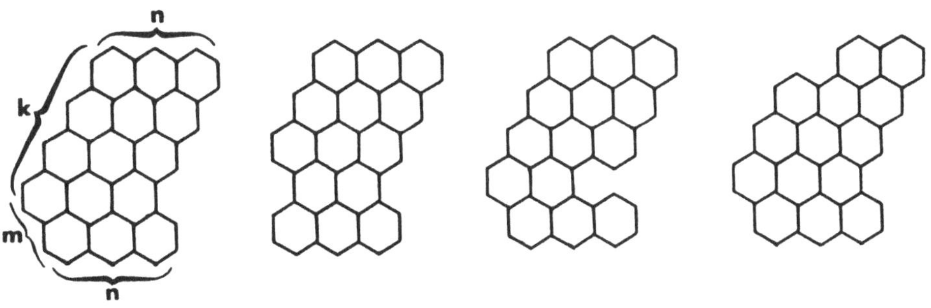

$$^{a,b}K\{Ch(k,m,n)\} = \sum_{i=0}^{n}\binom{k+i-1}{i}\binom{m+i-1}{i} = \sum_{i=0}^{n}\binom{k+i-1}{k-1}\binom{m+i-1}{m-1} \tag{i}$$

$$^{b}K\{Ch(k,m,n)\} = \sum_{i=\max(1,k-n)}^{k}(-1)^{k-i}\binom{n+i-1}{i-1}\binom{m+n}{k+m-i}$$

$$= \sum_{i=\max(1,k-n)}^{k}(-1)^{k-i}\binom{n+i-1}{n}\binom{m+n}{n-k+i} \tag{ii}$$

$$K\{Ca(k,m,n)\} = \binom{k+n-1}{n}\binom{m+n-1}{n} \tag{iii}$$

$$K\{Cf(k,m,n)\} = \sum_{i=1}^{n}\binom{k+i-1}{i}\binom{m+i-1}{i} \tag{iv}$$

$$K\{Cc(k,m,n)\} = \binom{m+n-1}{n}\left[\binom{k+n-1}{n} - 1\right] + \sum_{i=0}^{n-1}\binom{k+i-1}{i}\binom{m+i-1}{i} \tag{v}$$

---

[a] Gordon M, Davison WHT (1952). J Chem Phys 20: 428

[b] Cyvin SJ (1985). J Mol Struct (Theochem) 133: 211

$$Ca(m,n) = Ca(m,m,n), \qquad Cf(m,n) = Cf(m,m,n)$$

Below is a simple recurrence relation for chevrons.

$$K\{Ch(k,m,\ n+1)\} - K\{Ch(k,m,n)\} = K\{Ca(k,m,\ n+1)\} = \binom{k+n}{n+1}\binom{m+n}{n+1} \qquad (8.16)$$

When one of the other parameters (k or m) is increased instead, the recurrence properties are not so simple. The below equation still contains a summation in contrast to (16).

$$K\{Ch(k+1,m,\ n)\} - K\{Ch(k,m,n)\} = \sum_{i=1}^{n} \binom{k+i-1}{i-1}\binom{m+i-1}{i} \qquad (8.17)$$

Alternative with summation to k :

$$K\{Ch(k+1,m,\ n)\} - K\{Ch(k,m,n)\} = \sum_{i=0}^{k} (-1)^{k-i}\binom{n+i-1}{i}\binom{m+n}{k+m-i} \qquad (8.18)$$

8.3.3 *Algorithm*

An algorithm for the derivation of the K number of a chevron is available. Figure 1 shows the example of $Ch(4,2,3)$; see also CHART II. Thin arrows show the presumably easiest way to produce the numerals within each hexagon. A building-up process starts with the bottom row and proceeds upwards. Up to the kink (occurring after the apex) the system is simply a parallelogram, for which the algorithm was explained in Paragraph 4.8.2. So far the rows were built up from left to right when

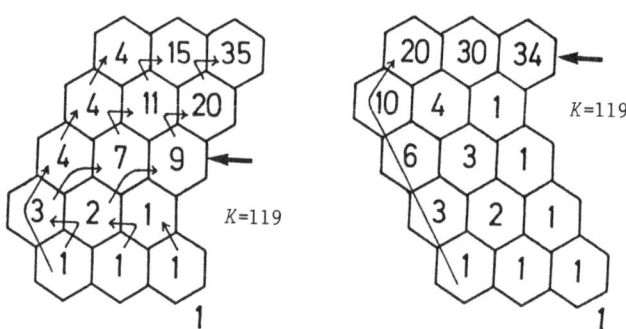

Fig. 8.1. The chevron $Ch(4,2,3) = Ch(2,4,3)$ with algorithm numerals filled in. The total K number is the sum of all numerals. The same number is obtained from eqns. (i) and (ii) of CHART 8-II:

$$K\{Ch(4,2,3)\} = \binom{3}{0}\binom{1}{0} + \binom{4}{1}\binom{2}{1} + \binom{5}{2}\binom{3}{2} + \binom{6}{3}\binom{4}{3} = 119$$

$$K\{Ch(4,2,3)\} = K\{Ch(2,4,3)\} = -\binom{3}{0}\binom{7}{5} + \binom{4}{1}\binom{7}{4} = 119$$

the apex (as usual) is chosen to be the extreme left hexagon. All the subsequent rows are built up from right to left; this change in direction is indicated by the external arrow. The numerals for each row are filled in the opposite direction of the building-up process. The first row after the kink is special. Here the first numeral to be filled in is found in a way which conforms the rules for single chains (Paragraph 6.2.2). An instructive illustration of this particular rule is furnished by the right-hand drawing of Fig. 1, where the same system is represented in a different orientation.

This algorithm, as well as the algorithms encountered in the pertinent parts of Chapters 4 and 6, are all special cases of a more general algorithm, which will be treated in more details in a subsequent chapter.

The legend of Fig. 1 includes an example of the computation of a K number from two equations.

8.3.4 *Derivation of Formulas*

Application of the Method of Fragmentation. The fundamental equation (i) of CHART II is most easily derived by the extended method of fragmentation with two special features: (a) one of the fragments is a benzenoid of the same kind as the original one (a chevron), only with one of the parameters (n) lowered by unity; (b) the other fragment is an essentially disconnected benzenoid. Figure 2 illustrates the process, where the fragmentation is supposed to be executed n times. The total number of Kekulé structures for $Ch(k,m,n)$ is the sum of the K numbers of the $n+1$ systems (including the trivial one with no hexagons) in the right column of Fig. 2. The parts of these essentially disconnected benzenoids are parallelograms (occasionally degenerated to linear single chains as exemplified in the figure). Hence

$$K\{Ch(k,m,n)\} = \sum_{i=0}^{n} K\{L(k-1,\ i)\} \cdot K\{L(m-1,\ i)\} \tag{8.19}$$

With the aid of the K formula of parallelograms (4.26) one readily attains at eqn. (i).

It is not necessary to conduct this derivation in such an elaborate way by repeated application of the fragmentation method. The first step is sufficient to give the recurrence formula

$$K\{Ch(k,m,n)\} = K\{Ch(k,m,\ n-1)\} + \binom{k+n-1}{n}\binom{m+n-1}{n}; \qquad n \geq 1 \tag{8.20}$$

which is virtually equivalent to eqn. (16). Here we notice that the chevron without apex is essentially disconnected. The rest of the derivation may be conducted analytically when the initial condition $K\{Ch(k,m,0)\} = 1$ is supposed to be known. Then a repeated application of (20) leads to the desired formula (i).

Also eqns. (iv) and (v) of CHART II are easily derived by the method of frag-

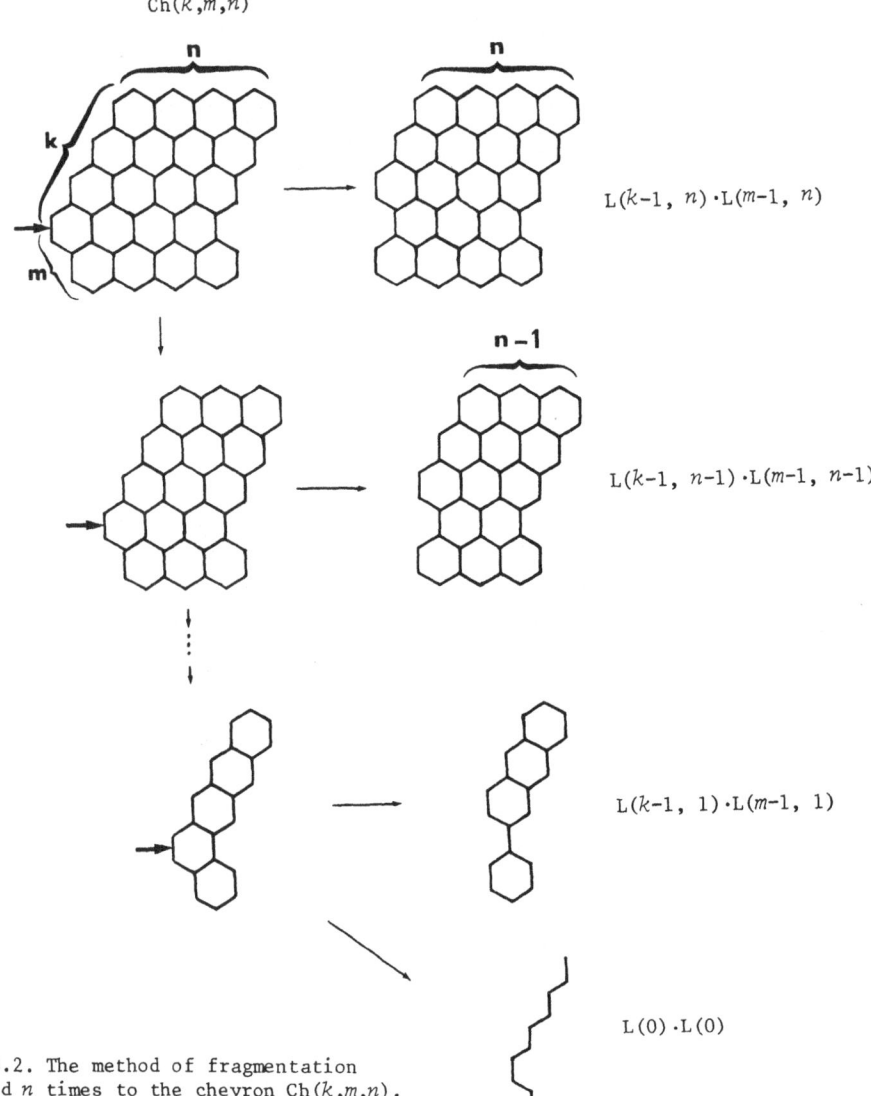

Fig. 8.2. The method of fragmentation
applied n times to the chevron Ch(k,m,n).

mentation. The formula for $K\{Cf(k,m,n)\}$ is obtained on attacking the free edge of
the anti-apex of the corresponding chevron; cf. Fig. 3. The corresponding bond is
found to be double in exactly one Kekulé structure. The property leads to

$$K\{Ch(k,m,n)\} = K\{Cf(k,m,n)\} + 1 \qquad (8.21)$$

Consequently

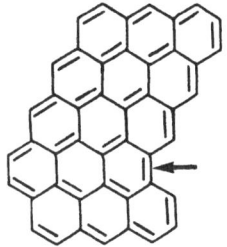

Fig. 8.3. A unique Kekulé
structure encountered during
the derivation of the K formula
for a chevron without anti-apex.

$$K\{Cf(k,m,n)\} = \sum_{i=0}^{n} \binom{k+i-1}{i}\binom{m+i-1}{i} - 1 \qquad (8.22)$$

or the formula (iv) entered into CHART II.

The chevron without corner, Cc, is defined in CHART II. The method of fragmentation leads to

$$K\{Ch(k,m,n)\} = K\{Cc(k,m,n)\} + K\{L(m-1, n)\} \qquad (8.23)$$

Hence

$$K\{Cc(k,m,n)\} = \sum_{i=0}^{n} \binom{k+i-1}{i}\binom{m+i-1}{i} - \binom{m+n-1}{n} \qquad (8.24)$$

which leads to eqn. (v) of CHART II.

Addition of Algorithm Numerals. This method was used to derive eqn. (ii) of CHART II in the original work (Cyvin 1985). Here we show some details of a derivation of both (i) and (ii) according to this method.

Figure 4 shows a chevron where the algorithm numerals are expressed in terms of n integers (a, b, c and d in the depicted example). The idea is to add the numerals in each row and finally take the gross sum in order to obtain the K number. The symbol $\tau_0 = \binom{m+n}{n}$ is used to denote the sum of algorithm numerals of the first m rows, i.e. the K number of the $m \times n$ parallelogram. In the general case the integers a, b, c, ... are: $a = m = \binom{m}{m-1}$, $b = \binom{m+1}{m-1}$, $c = \binom{m+2}{m-1}$, etc.; altogether n terms up to $\binom{m+n-1}{m-1}$. Notice that the last integer (d in the example of Fig. 4) is equal to $K\{L(m-1, n)\}$. Hence (cf. Fig. 4)

$$\tau_0 = 1 + \binom{m}{m-1} + \binom{m+1}{m-1} + \ldots + \binom{m+n-1}{m-1} = \sum_{i=0}^{n} \binom{m+i-1}{m-1} \qquad (8.25)$$

Similarly for the last row:

$$\tau_{k-1} = \sum_{i=1}^{n} \binom{k+i-2}{k-1}\binom{m+i-1}{m-1} \tag{8.26}$$

On adding all rows together the K number of the whole chevron is obtained in consistence with eqn. (i):

$$K = \sum_{j=0}^{k-1} \tau_j = 1 + \sum_{i=1}^{n} \binom{k+i-1}{k-1}\binom{m+i-1}{m-1} \tag{8.27}$$

The next equation, (ii), is derived in a similar way by adding algorithm numerals in the rows, but differences between binomial coefficients should be employed during some of the summations in the process. Figure 5(a) shows the algorithm numerals in Ch(4,3,4); they are identical to those of Fig. 4. The arrows indicate how the numerals of the three top rows may be deduced. For the two top rows the procedure deviates from the one of Fig. 1; it is an alternative leading to the same re-

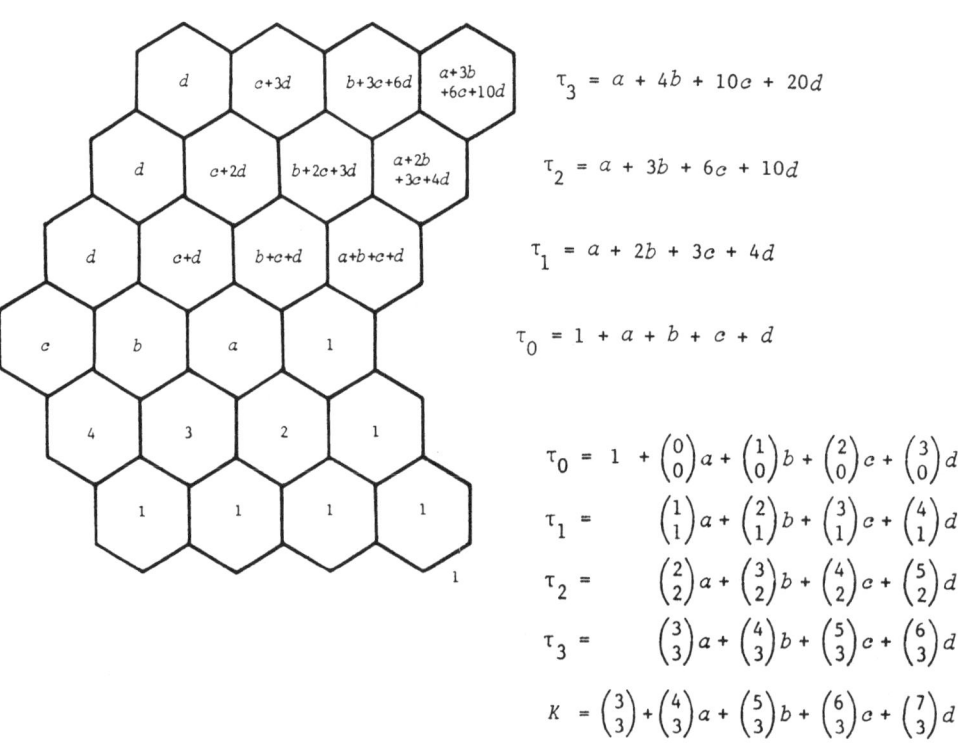

$$\tau_3 = a + 4b + 10c + 20d$$

$$\tau_2 = a + 3b + 6c + 10d$$

$$\tau_1 = a + 2b + 3c + 4d$$

$$\tau_0 = 1 + a + b + c + d$$

$$\tau_0 = 1 + \binom{0}{0}a + \binom{1}{0}b + \binom{2}{0}c + \binom{3}{0}d$$

$$\tau_1 = \binom{1}{1}a + \binom{2}{1}b + \binom{3}{1}c + \binom{4}{1}d$$

$$\tau_2 = \binom{2}{2}a + \binom{3}{2}b + \binom{4}{2}c + \binom{5}{2}d$$

$$\tau_3 = \binom{3}{3}a + \binom{4}{3}b + \binom{5}{3}c + \binom{6}{3}d$$

$$K = \binom{3}{3} + \binom{4}{3}a + \binom{5}{3}b + \binom{6}{3}c + \binom{7}{3}d$$

Fig. 8.4. The chevron Ch(4,3,4) with algorithm numerals; in this example $a=3$, $b=6$, $c=10$ and $d=15$. τ_0 is the sum of numerals for the bottom three rows, i.e. $K\{L(3,4)\} = 35$. τ_1 τ_2 and τ_3 are the numeral sums for the respective rows at the top. K is the total number of Kekulé structures for Ch(4,3,4).

(a)

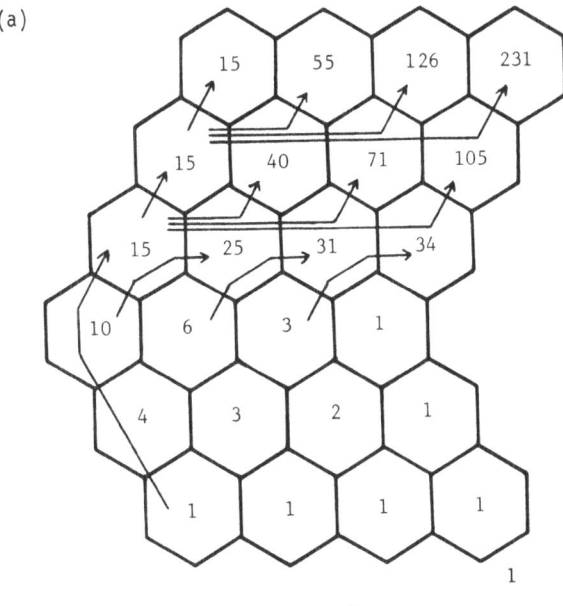

$K = 798$

(b)

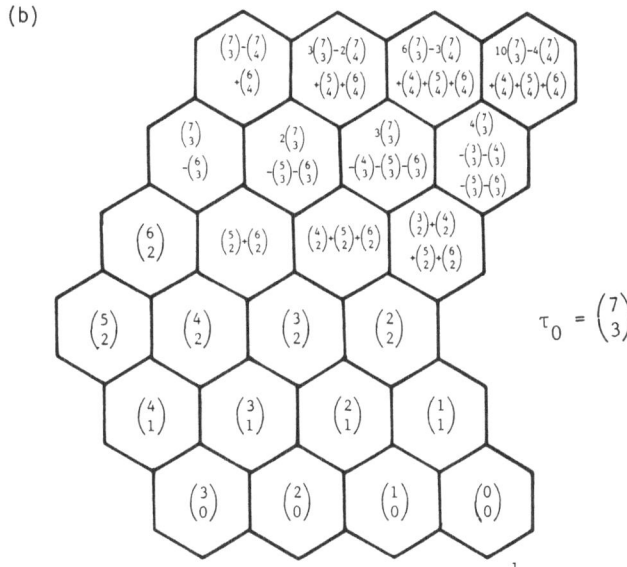

Fig. 8.5 (a) and (b); for (c), see next page. The chevron Ch(4,3,4) with algorithm numerals: (a) integers; (b) and (c) different versions in terms of binomial coefficients; see the text for details. The same example is found in Fig. 8.4.

sult (cf. Cyvin and Gutman 1986b and subsequent parts of the present book). In Fig. 5(b) the numerals of the bottom parallelogram are given in terms of binomial coefficients. Also the sums in the 4-th row (from the bottom) are easily filled in. These numerals were transferred to differences between binomial coefficients as shown in Fig. 5(c), using well-known mathematical identities. Next the 5-th row of Fig. 5(b) was completed by means of the expressions of the 4-th row of Fig. 5(c), transferred to the 5-th row of Fig. 5(c), and so on. The systematic procedure implies binomial coefficients $\binom{r}{s} = 0$ where $r < s$; they are omitted in the figures. The τ values

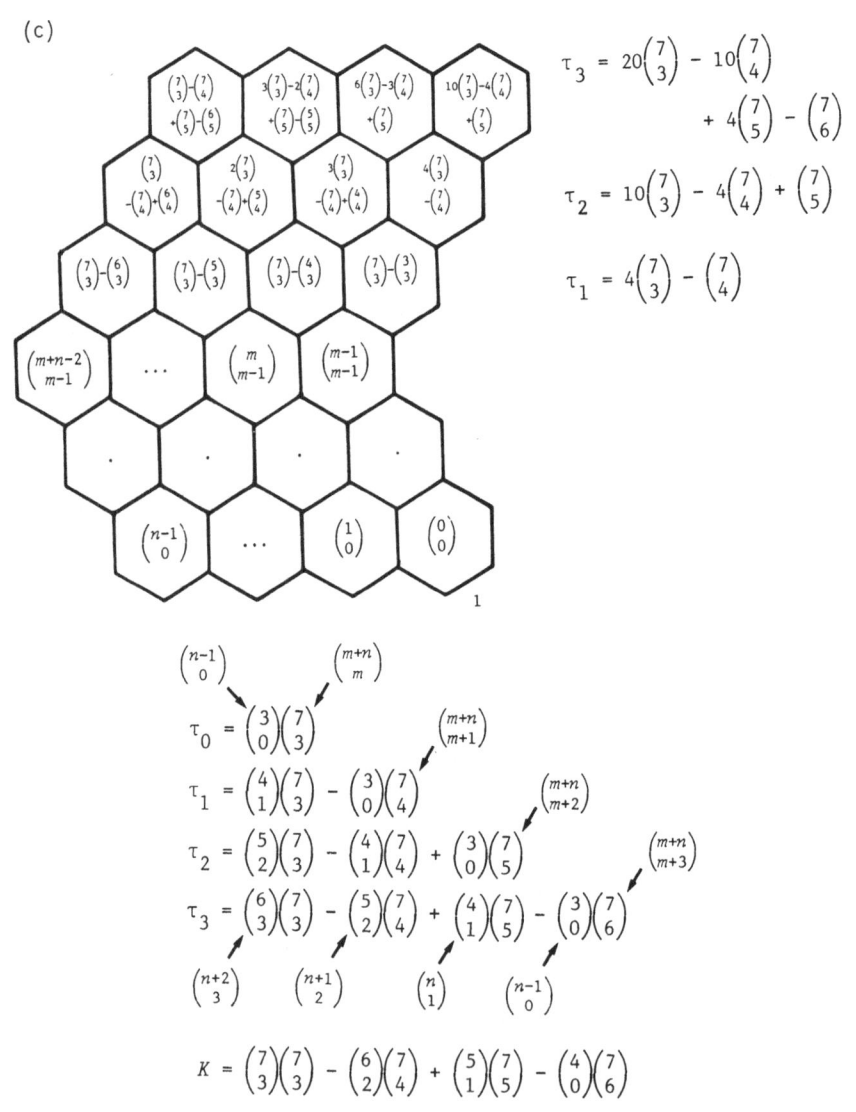

$$\tau_3 = 20\binom{7}{3} - 10\binom{7}{4}$$
$$+ 4\binom{7}{5} - \binom{7}{6}$$

$$\tau_2 = 10\binom{7}{3} - 4\binom{7}{4} + \binom{7}{5}$$

$$\tau_1 = 4\binom{7}{3} - \binom{7}{4}$$

$$\tau_0 = \binom{3}{0}\binom{7}{3}$$

$$\tau_1 = \binom{4}{1}\binom{7}{3} - \binom{3}{0}\binom{7}{4}$$

$$\tau_2 = \binom{5}{2}\binom{7}{3} - \binom{4}{1}\binom{7}{4} + \binom{3}{0}\binom{7}{5}$$

$$\tau_3 = \binom{6}{3}\binom{7}{3} - \binom{5}{2}\binom{7}{4} + \binom{4}{1}\binom{7}{5} - \binom{3}{0}\binom{7}{6}$$

$$K = \binom{7}{3}\binom{7}{3} - \binom{6}{2}\binom{7}{4} + \binom{5}{1}\binom{7}{5} - \binom{4}{0}\binom{7}{6}$$

introduced above are indicated. Figure 5(c) explains the general form of the numerals of the $m \times n$ parallelogram when it is adherred to $k=4$. Also in the list of τ
values the general forms of the binomial coefficients are indicated. When added together, the K number of a chevron with $k=4$, viz. $Ch(4,m,n)$, is obtained as

$$K = \sum_{j=0}^{3} \tau_j = -\binom{n}{0}\binom{m+n}{m+3} + \binom{n+1}{1}\binom{m+n}{m+2} - \binom{n+2}{2}\binom{m+n}{m+1} + \binom{n+3}{3}\binom{m+n}{m} \qquad (8.28)$$

This expression is found to be the special case of eqn. (ii) of CHART II for $k=4$.

8.3.5 *Mirror-Symmetrical Chevrons*

The formula apparatus for chevrons (CHART II and the above text) is adapted to
$Ch(m,n)$ on inserting $k=m$. Eqn. (i) of CHART II gives in particular

$$K\{Ch(m,n)\} = \sum_{i=0}^{n} \binom{m+i-1}{i}^2 \qquad (8.29)$$

It shows that the K formulas for fixed values of n are polynomials in m. Those of n
up to 5 are given below.

$$K\{Ch(m,1)\} = m^2 + 1$$

$$K\{Ch(m,2)\} = \tfrac{1}{4}(m^4 + 2m^3 + 5m^2 + 4)$$

$$K\{Ch(m,3)\} = \tfrac{1}{36}(m^6 + 6m^5 + 22m^4 + 30m^3 + 49m^2 + 36)$$

$$K\{Ch(m,4)\} = \tfrac{1}{576}(m^8 + 12m^7 + 74m^6 + 240m^5 + 545m^4 + 612m^3 + 820m^2 + 576)$$

$$K\{Ch(m,5)\} = \tfrac{1}{14400}(m^{10} + 20m^9 + 195m^8 + 1100m^7 + 4123m^6 + 9980m^5 + 17805m^4 + 17700m^3 + 21076m^2 + 14400)$$

The expressions of $K\{Ch(m,n)\}$ for fixed values of m are polynomials in n. They
are most easily found by means of eqn. (ii) of CHART II (with $k=m$).

Numerical values of K for mirror-symmetrical chevrons are given in Table 2.

8.3.6 *Generalized Chevron*

The formula (i) of CHART II may be generalized to a four-parameter system, viz.
$Ch(k, m, n_1, n_2)$, where $n_1 \leq n_2$. It is a generalized chevron, where different thicknesses are allowed for the two branches; cf. Fig. 6. This system consists of the two
parallelograms $k \times n_1$ and $(m-1) \times n_2$. The method of fragmentation leads to the formula

$$K\{Ch(k, m, n_1, n_2)\} = \sum_{i=0}^{n_1} \binom{k+i-1}{i}\binom{m+n_2-n_1+i-1}{n_2-n_1+i} \qquad (8.30)$$

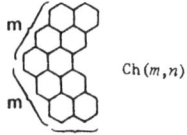

Table 8.2. Numerical values of $K\{Ch(m,n)\}$.

m \ n	1	2	3	4	5	6	7	8	9	10
1	2	3	4	5	6	7	8	9	10	11
2	5	14	30	55	91	140	204	285	385	506
3	10	46	146	371	812	1596	2892	4917	7942	12298
4	17	117	517	1742	4878	11934	26334	53559	101959	183755
5	26	251	1476	6376	22252	66352	175252	420277	931502	1933503
6	37	478	3614	19490	82994	296438	923702	2580071	6588075	15606084
7	50	834	7890	51990	265434	1119210	4063866	13081875	38131900	102259964
8	65	1361	15761	124661	751925	3696581	15475205	56884430	187758030	565982734
9	82	2107	29332	274357	1930726	10948735	52357960	217994860	808970960	
10	101	3126	51526	562751	4570755	29620780	160494380	751470480		

The following recurrence formula was derived (cf. the legend of Fig. 6).

$$K\{Ch(k+1, m, n_1+1, n_2)\} = K\{Ch(k, m, n_1+1, n_2)\} + K\{Ch(k+1, m, n_1, n_2)\} \quad (8.31)$$

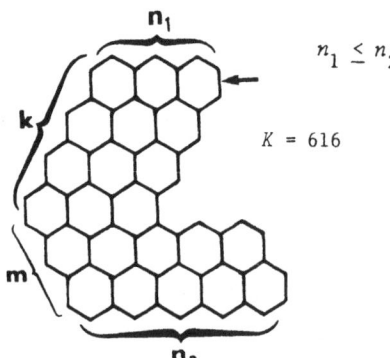

$n_1 \le n_2$

$K = 616$

Fig. 8.6. The generalized chevron, $Ch(k,m,n_1,n_2)$. The bond marked with an arrow may be attacked in the method of fragmentation in order to derive eqn. (8.31).

8.4 RIBBON

8.4.1 *Definition*

A ribbon (V-shaped benzenoid) is denoted $V(k,m,n)$ in terms of its three parameters (k,m,n); see CHART III. It is said to have m rows and n columns. It is assumed $m \geq 1$ and $n \geq 1$. These two parameters are permutable in the sense that

$$V(k,m,n) = V(k,n,m)$$

The first parameter (k) is restricted to $k \leq \min(m,n)$. For the terminal value,
viz. $k = \min(m,n)$, the ribbon degenerates to the $L(m,n)$ parallelogram. For $k=1$ the
ribbon reduces to a single chain identical with a chevron: $V(1,m,n) = Ch(m,n,1)$.
Finally we may have $k=0$, which results in the trivial case of no hexagons $(K=1)$.

8.4.2 *Previous Work and General Formulas*

The algebra of ribbons is analogous in many respects to the one of chevrons,
but curiously "inverted" in a way which will become apparent through the following
treatment when compared to the treatment of chevrons (Section 8.3). Firstly, one
should notice that the "thickness" of the ribbon, defined by the first parameter
(k), in many ways corresponds to the last parameter (n) in the chevron. This is of
course only a formal difference due to the present conventions, which it nevertheless seemed natural to adopt. More fundamental differences between chevrons and
ribbons were pointed out by Cyvin (1985) with special emphasis on those with the
thickness of 2 hexagons: $Ch(k,m,2)$ and $V(2,m,n)$.

The general formula (i) of CHART III for the number of Kekulé structures (K)
of $V(k,m,n)$ was first derived by Cyvin and Gutman (1986b) by means of the addition
of algorithm numerals. In the next paragraph we show a simpler derivation of (i).
The restriction $\min(m,n) \geq k$ is unimportant in eqn. (i). The formula gives still
correct results by virtue of the fact that $\binom{m}{i} = 0$ for $m < i$ and $\binom{n}{i} = 0$ for $n < i$.
An example will elucidate this feature. For $k=2$ eqn. (i) gives

$$K\{V(2,m,n)\} = 1 + \binom{m}{1}\binom{n}{1} + \binom{m}{2}\binom{n}{2} = 1 + mn + \frac{mn}{4}(m-1)(n-1) \qquad (8.32)$$

For $m=1$ or $n=1$ eqn. (32) gives $K = 1 + n$ and $K = 1 + m$, respectively. These numbers
are consistent with the formula for $k=1$ (cf. also CHART 6-I):

$$K\{V(1,m,n)\} = 1 + mn \qquad (8.33)$$

CHART III includes the combinatorial K formulas for some benzenoids related
to the ribbon: (iii) ribbon without anti-apex or with a fjord, $Vf(k,m,n)$; (iv)

CHART 8-III. Ribbon and related benzenoids

V(k,m,n)　　　　Vf(k,m,n)　　　　Va(k,m,n)　　　　Vc(k,m,n)

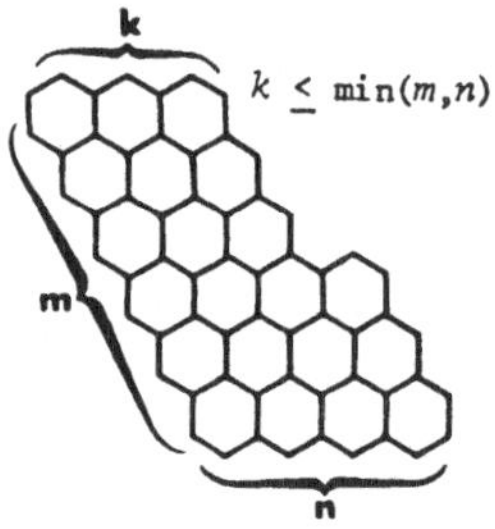

$k \leq \min(m,n)$

$$^{a}K\{V(k,m,n)\} = \sum_{i=0}^{k} \binom{m}{i}\binom{n}{i} \qquad (i)$$

$$K\{V(k,m,n)\} = \binom{m+n}{m} - \sum_{i=1}^{\min(m,n)-k} \binom{m}{k+i}\binom{n}{k+i} \qquad (ii)$$

$$K\{Vf(k,m,n)\} = \binom{m}{k}\binom{n}{k} \qquad (iii)$$

$$K\{Va(k,m,n)\} = \sum_{i=1}^{k} \binom{m}{i}\binom{n}{i} \qquad (iv)$$

$$K\{Vc(k,m,n)\} = \binom{n}{k}\left[\binom{m}{k} - 1\right] + \sum_{i=0}^{k-1} \binom{m}{i}\binom{n}{i} \qquad (v)$$

For equations (vi) and (vii), see after eqn. (8.43)

[a]Cyvin SJ, Gutman I (1986). Match 19: 229

ribbon without apex, Va(k,m,n); (v) ribbon without the corner where the m- and k-chains meet.

8.4.3 Derivation of Formulas

Formulas of CHART III. The fundamental equation (i) of CHART III is most easily derived by the method of fragmentation in the same way as eqn. (i) of CHART II was derived. Figure 7 shows the scheme, which leads to

$$K\{V(k,m,n)\} = K\{V(k-1, m, n)\} + K\{Vf(k,m,n)\}; \qquad k \geq 1 \qquad (8.34)$$

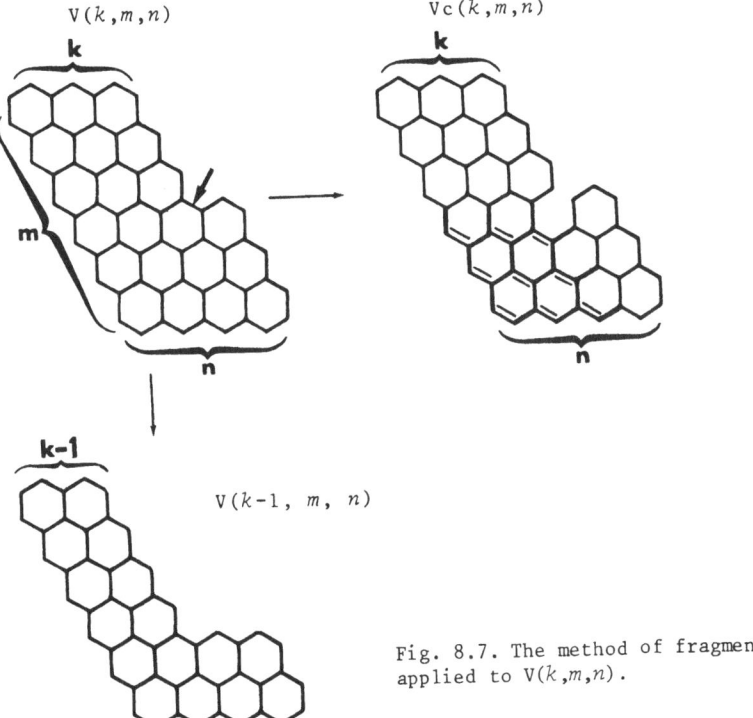

$V(k,m,n)$

k

$Vc(k,m,n)$

k

m

n

n

k-1

$V(k-1, m, n)$

Fig. 8.7. The method of fragmentation
applied to $V(k,m,n)$.

Here the benzenoid Vf is essentially disconnected; cf. Fig. 7. Notice that the two
units are joined by another ribbon, viz. $V(k-1, m-k, n-k)$. The joined units are
parallelograms (one of them being degenerated to a single linear chain in the
example of Fig. 7). It is obtained

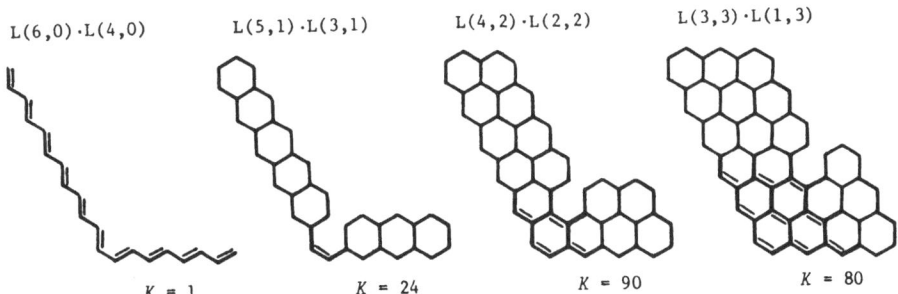

$L(6,0)\cdot L(4,0)$

$L(5,1)\cdot L(3,1)$

$L(4,2)\cdot L(2,2)$

$L(3,3)\cdot L(1,3)$

$K = 1$

$K = 24$

$K = 90$

$K = 80$

Fig. 8.8. Illustration of a summation property for ribbons. The K number for
$V(3,6,4)$ of Fig. 8.7 is the sum of the four K values in the present figure,
viz. $K\{V(3,6,4)\} = 195$.

$$K\{V(k,m,n)\} - K\{V(k-1, m, n)\} = K\{L(m-k, k)\} \cdot K\{L(n-k, k)\} = \binom{m}{k}\binom{n}{k} ; \quad k \geq 1 \quad (8.35)$$

Here the well-known K formula for a parallelogram was used. Eqn. (35) is a recurrence relation for $K\{V(k,m,n)\}$. Along with the initial condition $K\{V(0,m,n)\} = 1$ (see above) one obtains readily eqn. (i) of CHART III. Figure 8 gives a further illustration of this formula.

Eqn. (i) may be partitioned as

$$K\{V(k,m,n)\} = \sum_{i=0}^{m} \binom{m}{i}\binom{n}{i} - \sum_{i=k+1}^{m} \binom{m}{i}\binom{n}{i} = \binom{m+n}{m} - \sum_{i=k+1}^{m} \binom{m}{i}\binom{n}{i} \quad (8.36)$$

On substituting the summation index by $j = i-k$ it is arrived at

$$K\{V(k,m,n)\} = \binom{m+n}{m} - \sum_{j=1}^{m-k} \binom{m}{k+j}\binom{n}{k+j} \quad (8.37)$$

It is sufficient to execute the above summation until $j = \min(m,n) - k$. Yet it is not necessary to impose the restriction $m \leq n$ upon eqn. (37). In the opposite case $(m > n)$ the extra terms are vanishing since they contain factors $\binom{n}{s}$ with $s > n$. Eqn. (37) is consistent with (ii) of CHART III.

Fig. 8.9. A unique Kekulé structure encountered during the derivation of the K formula for a ribbon without apex.

Eqn. (iv) for Va, the ribbon without apex, is also readily obtained by the method of fragmentation. It is to be applied to an edge of the apex of the corresponding ribbon, as indicated by an arrow in Fig. 9. The corresponding bond appears to be single in exactly one Kekulé structure. The outcome of this feature is:

$$K\{V(k,m,n)\} = K\{Va(k,m,n)\} + 1 \quad (8.38)$$

Consequently

$$K\{Va(k,m,n)\} = \sum_{i=0}^{k} \binom{m}{i}\binom{n}{i} - 1 \quad (8.39)$$

or the formula (iv) entered into CHART III.

Finally we find the following relations for Vc, again by means of the method of fragmentation.

$$K\{V(k,m,n)\} = K\{Vc(k,m,n)\} + K\{L(n-k,\ k)\} \tag{8.40}$$

Hence

$$K\{Vc(k,m,n)\} = \sum_{i=0}^{k} \binom{m}{i}\binom{n}{i} - \binom{n}{k} \tag{8.41}$$

which leads to eqn. (v) of CHART III.

Recurrence Relations. Eqn. (35) is already a recurrence relation for the ribbon. It may be reformulated into

$$K\{V(k+1,\ m,\ n)\} - K\{V(k,m,n)\} = \binom{m}{k+1}\binom{n}{k+1} \tag{8.42}$$

The situation is not so simple when one of the other parameters (m or n) is increased instead of k. From eqn. (i) of CHART III it is obtained

$$K\{V(k,\ m+1,\ n)\} - K\{V(k,m,n)\} = \sum_{i=1}^{k} \binom{m}{i-1}\binom{n}{i} \tag{8.43}$$

The formula (43) still contains a summation to k. This index may be converted to $m-k$ as in the summation of eqn. (37); it was attained at

$$K\{V(k,\ m+1,\ n)\} - K\{V(k,m,n)\} = \binom{m+n}{m+1} - \sum_{j=0}^{m-k} \binom{m}{k+j}\binom{n}{k+j+1} \tag{8.44}$$

Here again it is sufficient to execute the summation until $j = \min(m,n) - k$.

Alternative Formulas. The techniques of adding numerals were treated in detail in the preceding section. When applied to the case of ribbons we arrived at interesting alternative formulas different from (i) and (ii). Their form is unsymmetrical in the parameters m and n, although these parameters of course still are permutable. The formulas are:

$$K\{V(k,m,n)\} = \sum_{i=0}^{k} \binom{m+k-i}{k}\binom{n-k+i-1}{i} \tag{vi}$$

$$K\{V(k,m,n)\} = \binom{m+n}{m} - \sum_{j=1}^{m-k} \binom{m-j}{k}\binom{n+j-1}{n-k-1} \tag{vii}$$

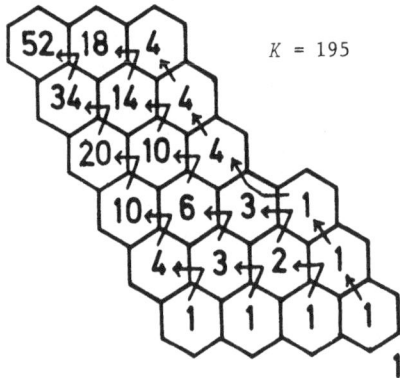

$$K = 195$$

Fig. 8.10. The ribbon V(3,6,4) with algorithm numerals. The K number is the sum of all numerals. Application of different formulas:

$$K\{V(3,6,4)\} = 195 = \binom{6}{0}\binom{4}{0} + \binom{6}{1}\binom{4}{1} + \binom{6}{2}\binom{4}{2} + \binom{6}{3}\binom{4}{3} \tag{i}$$

$$= \binom{10}{6} - \binom{6}{4}\binom{4}{4} \tag{ii}$$

$$= \binom{9}{3}\binom{0}{0} + \binom{8}{3}\binom{1}{1} + \binom{7}{3}\binom{2}{2} + \binom{6}{3}\binom{3}{3} \tag{vi}$$

$$= \binom{10}{6} - \binom{5}{3}\binom{4}{0} - \binom{4}{3}\binom{5}{0} - \binom{3}{3}\binom{4}{0} \tag{vii}$$

Notice that the summation index has different signs in the two factors of the summands in (vi) and (vii). The application of these equations (along with the main formulas) is exemplified in the legend of Fig. 10. The equations (vi) and (vii) may also be used to produce recurrence relations as alternatives to eqns. (43) and (44).

8.4.4 *Algorithm*

The algorithm for parallelograms (see Chapter 4) is applicable to ribbons with a slight addition (Cyvin and Gutman 1986b). In Fig. 10 the algorithm numerals are filled into the hexagons of V(3,6,4), and the arrows indicate how they are determined. The rules will be treated in more details in connection with a more general algorithm.

8.4.5 *Generalized Ribbon*

Figure 11 shows a generalized ribbon, where it has been allowed for different thicknesses of the two branches: $V(k_1, k_2, m, n)$; $k_1 \leq m$ and $k_2 \leq n$. The method of fragmentation leads to the following generalization of eqn. (i) of CHART III.

$$K\{V(k_1, k_2, m, n)\} = \sum_{i=0}^{k_1} \binom{m-k_1+k_2}{k_2-i}\binom{n+k_1-k_2}{k_1-i} \tag{8.45}$$

Here it is sufficient to execute the summation until $i = \min(k_1, k_2)$. Cyvin and Gutman (1986b) derived an alternative form by means of addition of algorithm numerals:

$$K\{V(k_1, k_2, m, n)\} = \sum_{i=0}^{k_2} \binom{m-k_1+i-1}{i}\binom{n+k_1-i}{k_1} \qquad (8.46)$$

On inserting $k_1 = k_2 = k$ and interchanging the parameters m and n this equation becomes identical with (vi).

The following recurrence formula was derived (cf. the legend of Fig. 11)

$$K\{V(k_1, k_2+1, m+1, n+1)\} = K\{V(k_1, k_2+1, m, n+1)\} + K\{V(k_1, k_2, m+1, n)\} \qquad (8.47)$$

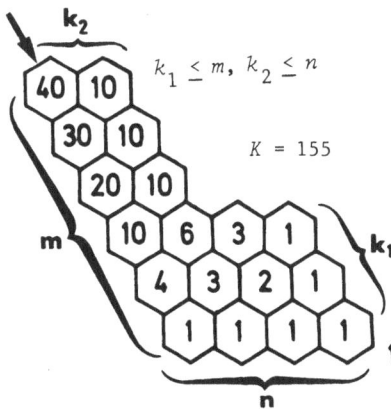

$k_1 \leq m, \; k_2 \leq n$

$K = 155$

Fig. 8.11. The generalized ribbon, $V(k_1, k_2, m, n)$. Algorithm numerals pertaining to V(3,2,6,4) are included for the sake of additional information. Their sum is the number of Kekulé structures (K). The edge marked with an arrow may be attacked in the method of fragmentation in order to derive eqn. (8.47).

8.5 PARALLELOGRAM

8.5.1 *Definition*

The parallelogram is a two-parameter (m,n) system, $L(m,n)$; cf. CHART IV and the treatment in Chapter 4. It consists of m rows of aligned linear chains of n hexagons each (n columns). Alternatively it may be designated $L(m{\times}n)$. The parallelogram is a multiple chain without kinks:

$$L(m,n) = M_n(L^m)$$

The two parameters, designating the numbers of rows and columns, are permutable:

$$L(m,n) = L(n,m)$$

For $m=1$ or $n=1$ the system degenerates to a linear single chain:

$$L(1,n) = L(n), \qquad L(m,1) = L(m)$$

Vanishing parameters ($m=0$ or $n=0$) are also possible; then the system degenerates to the trivial case of no hexagons:

$$L(0,n) = L(m,0) = L(0)$$

A parallelogram with $m > 1$ and $n > 1$ and $m \neq n$ is centrosymmetrical (C_{2h}). For $m = n > 1$ it is dihedral (D_{2h}) and referred to as a rhomb.

8.5.2 *Previous Work and General Formulas*

The parallelogram was treated early in the classical paper of Gordon and Davison (1952). These investigators reported two different approaches of combinatorial reasoning leading to the famous formula (4.26) for the number of Kekulé structures (K); cf. also (i) of CHART IV. Other derivations of this formula are due to Jiang (1980) and Ohkami and Hosoya (1983); see also Section 4.8.

The parallelogram is a special case of a hexagon, chevron and ribbon: $L(m,n)$ $= O(1,m,n) = Ch(1,m,n) = V(m,m,n)$; in the last case (ribbon) $m \le n$. The different formulas applied to this special case give the correct result in consistence with well-known mathematical properties;

$$K\{L(m,n)\} = \sum_{i=0}^{n} \binom{m+i-1}{i} = \sum_{i=0}^{\min(m,n)} \binom{m}{i}\binom{n}{i} = \binom{m+n}{n} \tag{8.48}$$

Let $La(m,n)$ denote the $m{\times}n$ parallelogram with one corner (L_3 mode hexagon) removed as shown in CHART IV. The effect on the number of Kekulé structures is a lowe-

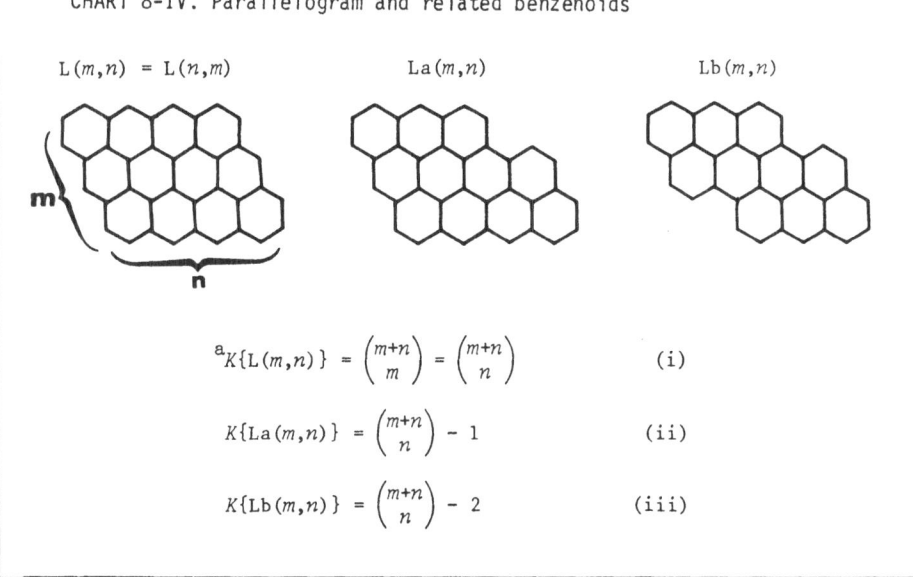

CHART 8-IV. Parallelogram and related benzenoids

$L(m,n) = L(n,m)$ $La(m,n)$ $Lb(m,n)$

$$^{a}K\{L(m,n)\} = \binom{m+n}{m} = \binom{m+n}{n} \qquad \text{(i)}$$

$$K\{La(m,n)\} = \binom{m+n}{n} - 1 \qquad \text{(ii)}$$

$$K\{Lb(m,n)\} = \binom{m+n}{n} - 2 \qquad \text{(iii)}$$

[a]Gordon M, Davison WHT (1952). J Chem Phys 20: 428

ring by unity;

$$K\{La(m,n)\} = K\{L(m,n)\} - 1 \qquad (8.49)$$

For both corners removed (cf. CHART IV) one has

$$K\{Lb(m,n)\} = K\{La(m,n)\} - 1 = K\{L(m,n)\} - 2 \qquad (8.50)$$

A recurrence relation for parallelograms is given by (4.33). The following recurrence property for the quotient is valid.

$$\frac{K\{L(m+1,\ n)\}}{K\{L(m,n)\}} = \frac{m+n+1}{m+n} \qquad (8.51)$$

8.5.3 *Rhomb*

A rhomb is a special parallelogram defined by $L(m,m)$. One has

$$K\{L(m,m)\} = \binom{2m}{m} = 2\binom{2m-1}{m} = 2K\{L(m-1,\ m)\}; \qquad m \geq 1 \qquad (8.52)$$

8.5.4 *Auxiliary Benzenoid Class*

The parallelogram with one corner removed, viz. $La(m,n)$, is a member of the auxiliary benzenoid class treated in Section 4.8.3 and designated $L(n,m,l)$. The class consists in general of $m \times n$ parallelograms augmented by a row of l hexagons. For the sake of systematization we have entered the K formula (4.28) into CHART V (i). For the special case of $m=1$ Cyvin (1987b) has given the K formula in an alternative form not obtained immediately from (i) of CHART V. One has

$$K\{L(n,1,l)\} = (n+1)(l+1) - \binom{l+1}{2} = \binom{n+2}{2} - \binom{n-l+1}{2} \qquad (8.53)$$

This alternative form is a special case of the formula (ii) of CHART V. The two forms corresponding to (ii) and (i) are referred to as the first and second representation, respectively.

Gutman and Cyvin (1987b), who considered the benzenoid class $L(n,m,l)$, also pointed out the existence of a similar class, viz. $L_o(n,m,l)$, consisting of non-Kekuléans; cf. Fig. 12. These systems are obvious non-Kekuléan benzenoids.

$$L_o(5,3,3)$$

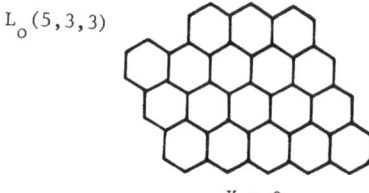

$$K = 0$$

Fig. 8.12. Example of a non-Kekuléan belonging to $L_o(n,m,l)$.

CHART 8-V. Parallelogram augmented with a row

$$L(n,m,l)$$

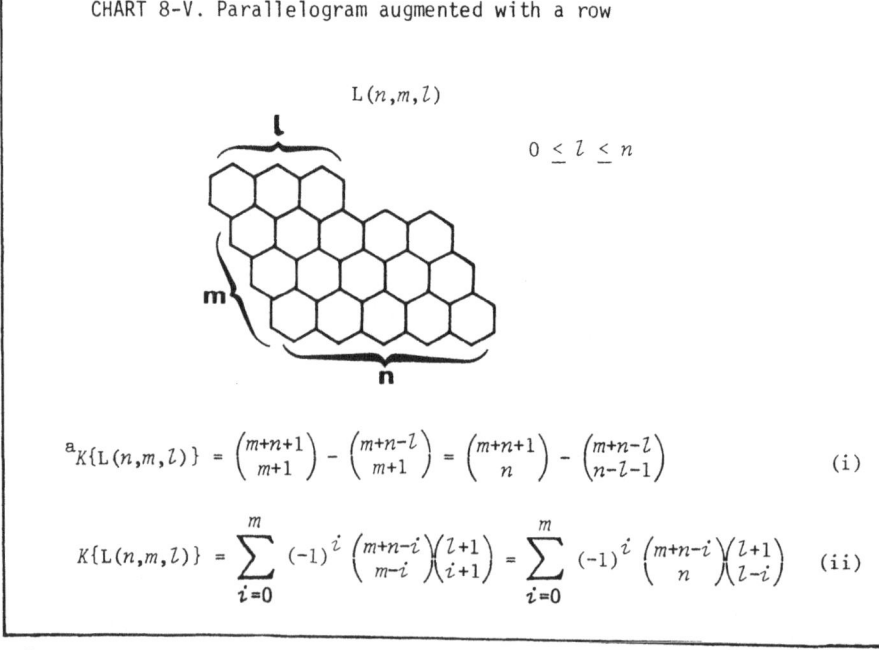

$$0 \leq l \leq n$$

$$^a K\{L(n,m,l)\} = \binom{m+n+1}{m+1} - \binom{m+n-l}{m+1} = \binom{m+n+1}{n} - \binom{m+n-l}{n-l-1} \qquad (i)$$

$$K\{L(n,m,l)\} = \sum_{i=0}^{m} (-1)^i \binom{m+n-i}{m-i}\binom{l+1}{i+1} = \sum_{i=0}^{m} (-1)^i \binom{m+n-i}{n}\binom{l+1}{l-i} \qquad (ii)$$

[a]Gutman I, Cyvin SJ (1987). Monatsh Chem 118: 541

8.5.5 *Algorithm*

Parallelogram L(n,m), *and the Class* L(n,m,l). The algorithm for a parallelogram-shaped benzenoid was presented in Section 4.8. The numerals, being binomial
coefficients, constitute a part of Pascal's triangle. The properties of these numerals were exploited in order to find the presumably easiest way to produce them. The
method was already employed for chevrons (Fig. 1) and ribbons (Fig. 10).

Consider the parallelogram L(4,3) = L(3,4) as shown in Fig. 4.10. The building-up process, which is associated with the algorithm, is here defined (as usual)
so that the rows are built up from the bottom, and each row is built up from the
left. Figure 12 shows one of the members of the auxiliary classes L(4,m,l). It is
encountered during the building-up of L(4,3). The figure indicates how the algorithm numerals most easily are found without invoking the complete parallelogram. A
special rule is necessary only for the last numeral of an incomplete row (the numeral 6 in Fig. 12). The K number for L(4,2,2) is found either by (a) adding all the
numerals of Fig. 12 or (b) adding the appropriate part of the numerals of Fig. 4.10,
which actually are the same.

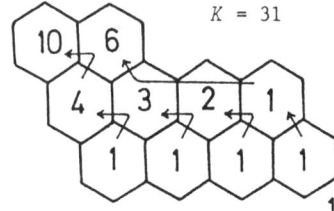

Fig. 8.12. A parallelogram augmented
with a row, L(4,2,2). Algorithm
numerals are displayed.

Truncated Parallelogram. The algorithm is easily extended to a truncated parallelogram as defined in the following (Cyvin and Gutman 1986b).

Consider a benzenoid with m condensed linear chains of the lengths r_1, r_2, ..
..., r_m in terms of their numbers of hexagons. These are the m rows. Assume the restrictions

$$r_{i+1} \leq r_i ; \qquad i = 1, 2, \ldots, m-1 \qquad (8.54)$$

The rows should be aligned so that the first hexagon (conventionally drawn to the
left) from all chains also form a linear chain, the first column.

Generalized ribbons (Paragraph 8.4.5), which include the parallelograms
augmented with a row (CHART V), are special cases of the truncated parallelogram
defined above.

Let the truncated parallelogram be denoted L(r_1; r_2; ....; r_m) with reference
to the above definition.

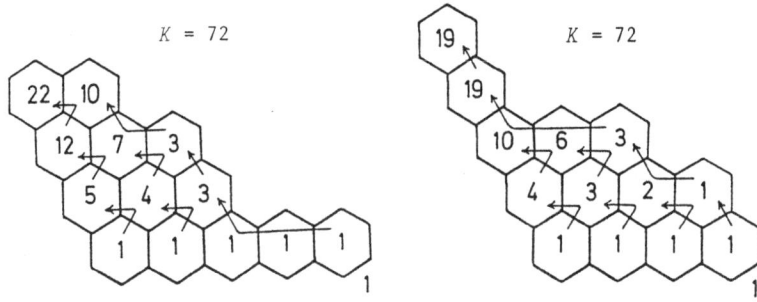

Fig. 8.13. The truncated parallelogram L(5; 2×3; 2) = L(2×4; 3; 2×1) with algorithm numerals filled in. The total K number is the sum of all numerals.

Then, for instance, L(4,2,2) = L(4;4;2); cf. Fig. 12. The notation may be abbreviated to L(2×4; 2). The generalized ribbon of Fig. 11 is equivalent to L(4;4;4;2;2;2) = L(3×4; 3×2). This notation is consistent with L(m×n) for the parallelogram L(m,n); cf. Paragraph 8.5.1.

Figure 13 exemplifies the algorithm for a truncated parallelogram, which is drawn in two orientation obtained by switching the rows and columns.

Prolate and Oblate Triangles. Triangle-shaped benzenoids or triangles are treated in more detail in a subsequent chapter. Here we only use these classes for further exemplifications of the algorithm.

A prolate triangle, $T^i(m)$, is a special truncated parallelogram. As such it is defined by

$$T^i(m) = L(m; m-1; m-2; ....; 1)$$

An oblate triangle, $T^j(m)$, is defined similarly as belonging to another class of truncated parallelograms:

$$T^j(m) = L(m; m; m-1; m-2; ....; 2)$$

Both systems $T^i(m)$ and $T^j(m)$ possess m rows and m columns.

Figure 14 shows the prolate and oblate triangles with $m=5$ as an example. The figure indicates how the K numbers for all members of these classes with smaller m values are found among the individual numerals.

Concluding Remarks. The algorithm of the present paragraph (8.5.5), together with most of the algorithms treated in the preceding sections, is a special case of a most general algorithm of this kind. A detailed treatment of it is found in Chapter 15.

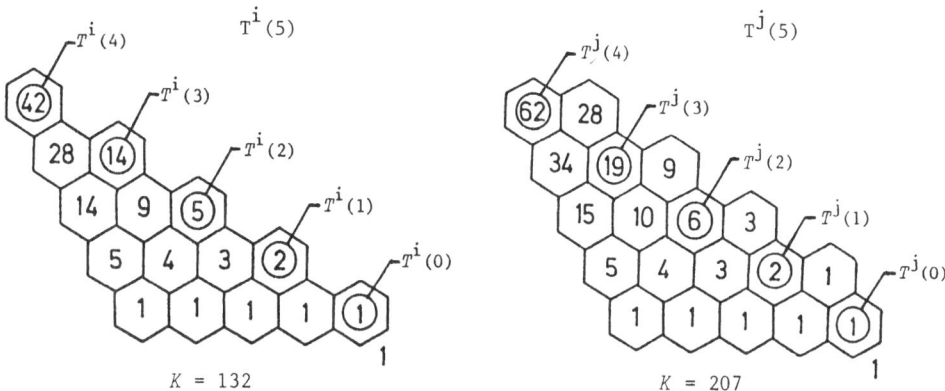

Fig. 8.14. Examples of a prolate triangle, $T^i(m)$, and an oblate triangle, $T^j(m)$, where $m=5$. Algorithm numerals are given. The encircled numerals indicate $T^i(k) = K\{T^i(k)\}$ and $T^j(k) = K\{T^j(k)\}$ for $0 \le k < m$ $(m=5)$.

CHAPTER 9

CLASSES OF BASIC BENZENOIDS (II): MULTIPLE ZIGZAG CHAIN

9.1 DEFINITION

A zigzag chain may be single or multiple. Such a system (a triple chain) is depicted in Fig. 1, where two different notations are explained, viz. A and Z. It is useful to have access to both of these definitions.

A zigzag chain, when written $Z(m,n)$, may be interpreted as m tier condensed linear chains (rows) of n hexagons each in a zigzag arrangement. The system is also identified by the symbol

$$Z(m,n) = M_n (LA^{m-2}L)$$

which is a generalization of the symbol used for single chains introduced in Paragraph 6.2.4. Here (in contrast to single chains) the kinks must go in the alternating manner as ...-left-right-left-... . For $m=2$ the system coincides with the parallelogram, $Z(2,n) = L(2,n)$. For $m=1$ it becomes a linear single chain,

$$Z(1,n) = L(n)$$

Finally the zigzag chain degenerates to no hexagons for $m=0$ or $n=0$ ($K=1$).

A zigzag chain, $Z(m,n)$, is centrosymmetrical (C_{2h}) for $m = 2, 4, 6, \ldots$ and mirror-symmetrical (C_{2v}) for $m = 3, 5, 7, \ldots$.

In Chapters 3 and 6 we have denoted a single zigzag chain of n hexagons by $A(n)$. A natural generalization leads to the definition of $A(m,n)$ as shown in the left part of Fig. 1. Alternatively it may be written $A(n \times m)$. For a single zigzag chain we have

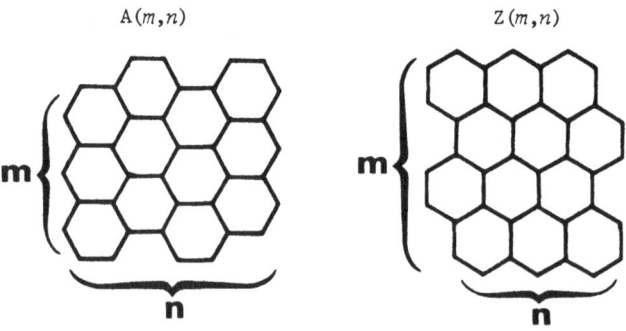

Fig. 9.1. The definition of $A(m,n)$ and $Z(m,n)$. The depicted example is $A(3,4) = Z(4,3)$; ($K=85$).

$$A(1,n) = A(n)$$

The connection between A and Z is given by

$$Z(m,n) = A(n,m)$$

Notice that $Z(m,n) = A(m \times n)$.

9.2 PREVIOUS WORK

The combinatorial K formula for $Z(3,n)$, which is identical to the chevron $Ch(3,n)$, has been obtained several times by different methods; cf. Chapter 8. A formula for $Z(m,n)$ with $m=4$ was first given by Cyvin (1986d), for $m=5$ by Cyvin SJ, Cyvin and Gutman (1985), and for $m=6$ by Cyvin and Gutman (1986a). The list of K formulas for multiple zigzag chains, $Z(m,n)$ with fixed values of m, was extended to $m=7$ and $m=8$ by Gutman and Cyvin (1987b).

For $A(m,n)$ with fixed values of m no explicit K formula is known unless $m=1$, i.e. for the single chain $A(1,n) = A(n)$. Randić (1980) studied the class of $A(2,n)$ benzenoids. The work of Ohkami and Hosoya (1983) implies a recurrence relation for this class and for $A(3,n)$. Corresponding recurrence relations up to $m=5$ have been reported (Cyvin and Gutman 1986a; Gutman and Cyvin 1987b).

Some formulas pertaining to linear chains annelated to $Z(5,n)$ were given by Cyvin SJ, Cyvin and Gutman (1986).

9.3 AUXILIARY BENZENOID CLASSES

Definitions. In the studies of multiple zigzag chains it is advantageous to introduce auxiliary benzenoid classes at an early stage. Let $A(n,m,l)$ denote the benzenoid $A(n,m) = Z(m,n)$ augmented by a row of l hexagons, where $0 \le l \le n$; cf. Fig. 2. The case may be compared with the definition of $L(n,m,l)$ under parallelograms (Chapter 8).

Gutman and Cyvin (1987b) also pointed out the existence of non-Kekuléan benzenoids, which are denoted by $A_o(n,m,l)$; they may be compared to $L_o(n,m,l)$ mentioned under parallelograms (Chapter 8). Figure 3 shows two examples, where the values of the parameters (n,m,l) are the same as in Fig. 2. Also $A_o(n,m,l)$ like $L_o(n,m,l)$ is obvious non-Kekuléan.

Special Cases. For the extremal values of l in $A(n,m,l)$ one has by virtue of definition

$$A(n,m,0) = A(n,m), \qquad A(n,m,n) = A(n, m+1)$$

It is useful to observe other trivial or degenerate cases of $A(n,m,l)$. For $m=0$ the benzenoid degenerates to $L(l)$; hence

$$K\{A(n,0,l)\} = l+1 \tag{9.1}$$

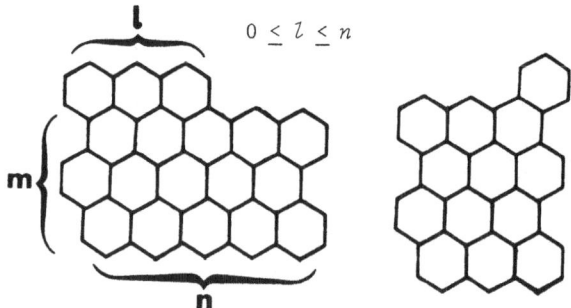

Fig. 9.2. The definition of $A(n,m,l)$. The depicted examples are $A(5,3,3)$ $(K=309)$ to the left and $A(3,4,1)$ $(K=160)$ to the right.

$A_o(5,3,3)$ $A_o(3,4,1)$

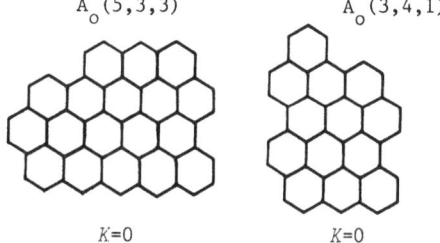

Fig. 9.3. Examples of non-Kekuléan benzenoids belonging to $A_o(n,m,l)$.

$K=0$ $K=0$

In this equation we may allow for $l=0$, which leads to $A(n,0,0) = A(n,0)$ as the case of no hexagons. The same is realized for $n=0$: $A(0,m,0) = A(0,m)$. In the latter case also l must necessarily vanish. In conclusion we write for the trivial case of no hexagons:

$$K\{A(0,m)\} = K\{A(n,0)\} = 1 \qquad (9.2)$$

with arbitrary m and n.

Another useful special case is

$$K\{A(n,1,0)\} = n+1 \qquad (9.3)$$

where the appropriate system is $A(n,1,0) = A(n,1) = L(n)$.

Recurrence Relations. The main recurrence formula for the considered benzenoids reads

$$K\{A(n,m,l)\} = K\{A(n, m, l-1)\} + K\{A(n, m-1, n-l)\}; \qquad m \geq 1 \qquad (9.4)$$

The relation is obtained by the method of fragmentation as shown in Fig. 4 and similar to the scheme of Fig. 4.11. The special case of (4) for $l=0$ is also useful. By means of $A(n,m,0) = A(n,m) = A(n, m-1, n)$ it is obtained

$$K\{A(n,m)\} = K\{A(n, m-1, n-1)\} + K\{A(n, m-2)\}; \qquad m \geq 2, \quad n \geq 1 \qquad (9.5)$$

Summation Formula. On repeated application of eqn. (4), together with $A(n,m,0) = A(n, m-1, n)$, it was obtained

$$K\{A(n,m,l)\} = \sum_{i=0}^{l} K\{A(n, m-1, n-i)\}; \qquad m \geq 1 \qquad (9.6)$$

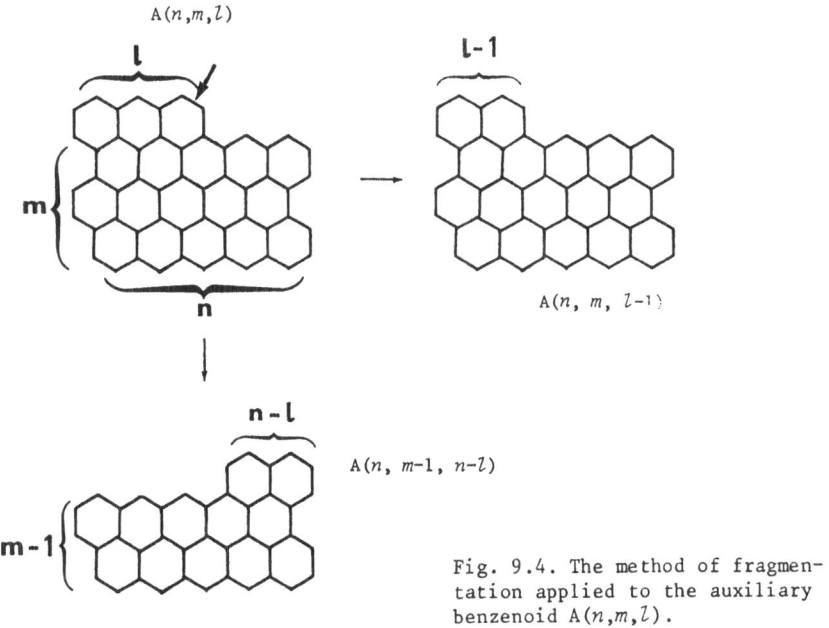

Fig. 9.4. The method of fragmentation applied to the auxiliary benzenoid $A(n,m,l)$.

9.4 RECURRENCE RELATIONS FOR $A(n,m)$ WITH FIXED VALUES OF n

On inserting $n=1$ into eqn. (5) one obtains

$$K\{A(1,m)\} = K\{A(1, m-1, 0)\} + K\{A(1, m-2)\}; \qquad m \geq 2 \qquad (9.7)$$

That gives the first equation of Table 1. This is the Fibonacci-type recurrence formula for single zigzag chains, equivalent to eqn. (4.4). By similar manipulations of eqn. (5) together with (4) the other recurrence relations collected in Table 1 were deduced. We will come back to this issue in Section 9.9.

9.5 COMBINATORIAL K FORMULAS FOR $A(n,m,l)$ WITH FIXED VALUES OF m

Introduction. In the below exposition we show how eqn. (6) may be used to derive $K\{A(n,m,l)\}$ successively for any value of m, while n and l are arbitrary ($0 \leq l \leq n$).

Table 9.1. Recurrence relations for $K\{A(n,m)\} = K\{Z(m,n)\} = Z_n(m)$.*

n	Recurrence relation		
1	$^a Z_1(m) = Z_1(m-1) + Z_1(m-2);$	$m \geq 2$	
2	$^{b,c} Z_2(m) = 2Z_2(m-1) + Z_2(m-2) - Z_2(m-3);$	$m \geq 3$	
3	$^{b,d} Z_3(m) = 2Z_3(m-1) + 3Z_3(m-2) - Z_3(m-3) - Z_3(m-4);$	$m \geq 4$	
4	$^d Z_4(m) = 3Z_4(m-1) + 3Z_4(m-2) - 4Z_4(m-3) - Z_4(m-4) + Z_4(m-5);$	$m \geq 5$	
5	$^d Z_5(m) = 3Z_5(m-1) + 6Z_5(m-2) - 4Z_5(m-3) - 5Z_5(m-4) + Z_5(m-5) + Z_5(m-6);$	$m \geq 6$	

*The same relations apply to K numbers of auxiliary benzenoids, $K\{A(n,m,l)\} = Z_n^{(l)}(m)$. The restrictions for m are released when allowance is made for nominal values. See Section 9.9.

$^a Z_1(m) = K\{A(m)\}.$
^{b}Ohkami N, Hosoya H (1983). Theor Chim Acta 64: 153.
^{c}Cyvin SJ, Gutman I (1986). Comp & Maths with Appls 12B: 859.
^{d}Gutman I, Cyvin SJ (1987). Monatsh Chem 118: 541

Formulas for m up to 3 have been derived previously (Cyvin and Gutman 1986a; Gutman and Cyvin 1987b; Cyvin 1987b).

First Set of Formulas. On inserting $m=1$ into eqn. (6) one obtains

$$K\{A(n,1,l)\} = \sum_{i=0}^{l} K\{A(n, 0, n-i)\} = \sum_{i=0}^{l} (n-i+1) \qquad (9.8)$$

where eqn. (1) has been applied. Here it is expedient to change the summation index to $j = n-i$ and take the terms in the reverse order; hence

$$K\{A(n,1,l)\} = \sum_{j=n-l}^{n} (j+1) = \sum_{j=0}^{n} (j+1) - \sum_{j=0}^{n-l-1} (j+1) = \binom{n+2}{2} - \binom{n-l+1}{2} \qquad (9.9)$$

The result is seen to be consistent with eqn. (4.29) and (i) of CHART 8-V when it is realized that $A(n,1,l)$ is identical with $L(n,1,l)$.

In the next step we insert $m=2$ into eqn. (6) and make use of (9). It follows

$$K\{A(n,2,l)\} = \sum_{i=0}^{l} K\{A(n, 1, n-i)\} = \sum_{i=0}^{l} \left[\binom{n+2}{2} - \binom{i+1}{2} \right] \qquad (9.10)$$

Under this summation n is constant; hence it is easily executed and gives

$$K\{A(n,2,l)\} = \binom{n+2}{2} \sum_{i=0}^{l} 1 - \sum_{i=1}^{l} \binom{i+1}{2} = \binom{n+2}{2}(l+1) - \binom{l+2}{3} \tag{9.11}$$

The next step, where $m=3$, runs in the following way.

$$K\{A(n,3,l)\} = \sum_{i=0}^{l} K\{A(n,\ 2,\ n-i)\} = \sum_{i=0}^{l} \left[\binom{n+2}{2}(n-i+1) - \binom{n-i+2}{3} \right]$$

$$= \binom{n+2}{2} \sum_{j=n-l}^{n} (j+1) - \sum_{j=n-l}^{n} \binom{j+2}{3} = \binom{n+2}{2}^2 - \binom{n+3}{4} - \binom{n+2}{2}\binom{n-l+1}{2} + \binom{n-l+2}{4} \tag{9.12}$$

This process may be continued without great difficulties, inasmuch as each summation never will contain more than one binomial coefficient.

Second Set of Formulas. An alternative set of formulas of $K\{A(n,m,l)\}$ is obtained if one starts from

$$K\{A(n,1,l)\} = (n+1)(l+1) - \binom{l+1}{2} \tag{9.13}$$

This equation is equivalent to (9), but gives the expression in an alternative form. Eqns. (13) and (9) correspond to the first and second representation of $K\{L(n,1,l)\}$, where $L(n,1,l) = A(n,1,l)$; cf. eqn. (8.53). Eqn. (6) is now applied according to the procedure above. It leads to

$$K\{A(n,2,l)\} = (n+1)\binom{n+2}{2} - \binom{n+2}{3} - (n+1)\binom{n-l+1}{2} + \binom{n-l+1}{3} \tag{9.14}$$

as an alternative to (11). We also give the result of the next step ($m=3$), where the below equation should be compared to (12).

$$K\{A(n,3,l)\} = \left[(n+1)\binom{n+2}{2} - \binom{n+2}{3} \right](l+1) - (n+1)\binom{l+2}{3} + \binom{l+2}{4} \tag{9.15}$$

Conclusion. In CHART I the combinatorial formulas of $K\{A(n,m,l)\}$ through $m=6$ are collected. They are taken partly from the first set of formulas ($m = 2, 4, 6$), and partly from the second ($m = 1, 3, 5$). We will come back to these auxiliary benzenoids in Section 9.9.

9.6 COMBINATORIAL K FORMULAS FOR $Z(m,n)$ WITH FIXED VALUES OF m

Basic Principle. Combinatorial formulas of K numbers for $Z(m,n) = A(n,m)$ are now obtained directly from the expressions of $K\{A(n,\ m',\ l)\}$; cf. CHART I and the above text - on inserting $l=0$ or $l=n$. Thus, for instance,

$$K\{A(n,4)\} = K\{A(n,3,n)\} = K\{A(n,4,0)\} = \binom{n+2}{2}^2 - \binom{n+3}{4} \tag{9.16}$$

140

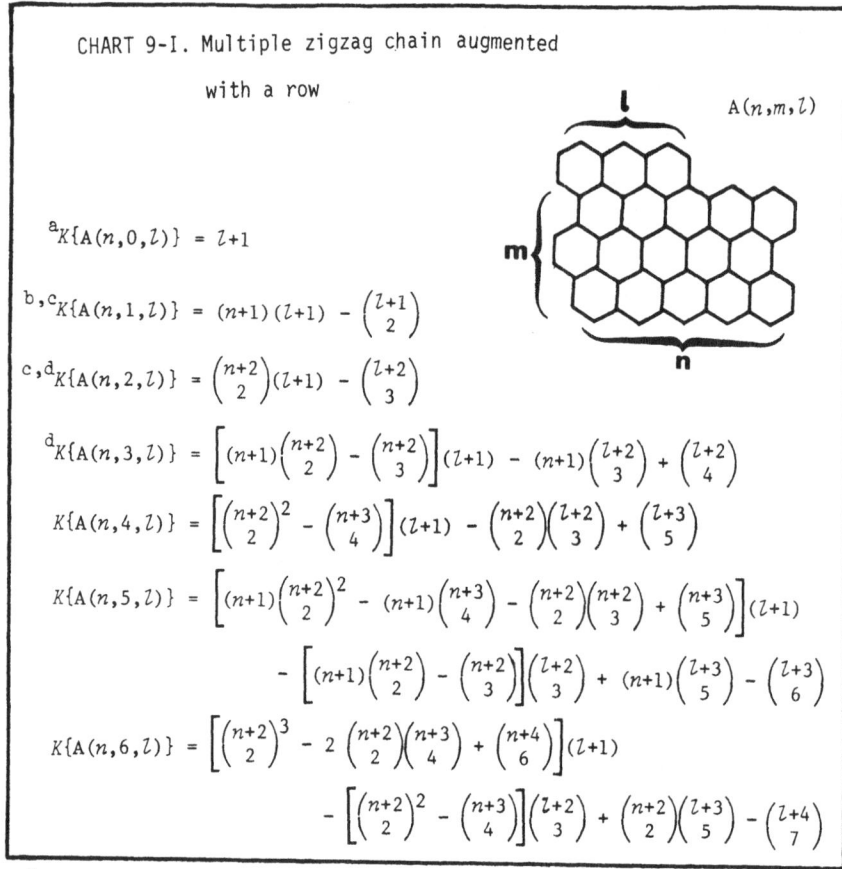

CHART 9-I. Multiple zigzag chain augmented
with a row

$A(n,m,l)$

$^{a}K\{A(n,0,l)\} = l+1$

$^{b,c}K\{A(n,1,l)\} = (n+1)(l+1) - \binom{l+1}{2}$

$^{c,d}K\{A(n,2,l)\} = \binom{n+2}{2}(l+1) - \binom{l+2}{3}$

$^{d}K\{A(n,3,l)\} = \left[(n+1)\binom{n+2}{2} - \binom{n+2}{3}\right](l+1) - (n+1)\binom{l+2}{3} + \binom{l+2}{4}$

$K\{A(n,4,l)\} = \left[\binom{n+2}{2}^2 - \binom{n+3}{4}\right](l+1) - \binom{n+2}{2}\binom{l+2}{3} + \binom{l+3}{5}$

$K\{A(n,5,l)\} = \left[(n+1)\binom{n+2}{2}^2 - (n+1)\binom{n+3}{4} - \binom{n+2}{2}\binom{n+2}{3} + \binom{n+3}{5}\right](l+1)$

$\quad - \left[(n+1)\binom{n+2}{2} - \binom{n+2}{3}\right]\binom{l+2}{3} + (n+1)\binom{l+3}{5} - \binom{l+3}{6}$

$K\{A(n,6,l)\} = \left[\binom{n+2}{2}^3 - 2\binom{n+2}{2}\binom{n+3}{4} + \binom{n+4}{6}\right](l+1)$

$\quad - \left[\binom{n+2}{2}^2 - \binom{n+3}{4}\right]\binom{l+2}{3} + \binom{n+2}{2}\binom{l+3}{5} - \binom{l+4}{7}$

$^{a}A(n,0,l) = L(l)$
$^{b}A(n,1,l) = L(n,1,l)$
[c] Cyvin SJ (1987). Monatsh Chem (in press)
[d] Cyvin SJ, Gutman I (1986). Comp & Maths with Appls 12B: 859

Here eqn. (12) was used for $K\{A(n,3,l)\}$, while $K\{A(n,4,l)\}$ was taken from CHART I. A different expression is obtained when (15) for $K\{A(n,3,l)\}$ is used:

$$K\{A(n,4)\} = (n+1)^2\binom{n+2}{2} - 2(n+1)\binom{n+2}{3} + \binom{n+2}{4} \tag{9.17}$$

The expressions (16) and (17) are equivalent.

The subsequent exposition is simplified and systematized by means of the abbreviation

$$A_m = K\{A(n,m)\} \tag{9.18}$$

Computations from the First Set of Formulas. By means of the first set of formulas of Section 9.5 it is found that A_2, A_4, A_6, ... may be expressed in terms of recurrence relations in the following way.

$$A_0 = 1, \qquad\qquad A_4 = A_2\binom{n+2}{2} - A_0\binom{n+3}{4},$$

$$A_2 = A_0\binom{n+2}{2}, \qquad A_6 = A_4\binom{n+2}{2} - A_2\binom{n+3}{4} + A_0\binom{n+4}{6}, \text{ etc.}$$

In general:

$$A_0 = 1, \qquad A_{2p} = \sum_{i=1}^{p} (-1)^{i-1} A_{2p-2i}\binom{n+i+1}{2i}; \qquad p = 1, 2, 3, \ldots \qquad (9.19)$$

A similar scheme is found for odd integers m:

$$A_1 = A_0(n+1), \qquad A_3 = A_2(n+1) - A_0\binom{n+2}{3}, \qquad A_5 = A_4(n+1) - A_2\binom{n+2}{3} + A_0\binom{n+3}{5}, \text{ etc.}$$

In general:

$$A_{2p+1} = \sum_{i=0}^{p} (-1)^{i} A_{2p-2i}\binom{n+i+1}{2i+1}; \qquad p = 0, 1, 2, \ldots \qquad (9.20)$$

All the A_m terms ($m = 0, 1, 2, \ldots$) are seen to be expressed in terms of A_0, A_2, A_4, ... Explicit formulas for some of these terms ($m = 2, 4, 6, 8, 10$) are given below.

$$A_2 = \binom{n+2}{2},$$

$$A_6 = \binom{n+2}{2}^3 - 2\binom{n+2}{2}\binom{n+3}{4} + \binom{n+4}{6},$$

$$A_4 = \binom{n+2}{2}^2 - \binom{n+3}{4},$$

$$A_8 = \binom{n+2}{2}^4 - 3\binom{n+2}{2}^2\binom{n+3}{4} + 2\binom{n+2}{2}\binom{n+4}{6} - \binom{n+5}{8} + \binom{n+3}{4}^2,$$

$$A_{10} = \binom{n+2}{2}^5 - 4\binom{n+2}{2}^3\binom{n+3}{4} + 3\binom{n+2}{2}^2\binom{n+4}{6} + \binom{n+2}{2}\left[3\binom{n+3}{4}^2 - 2\binom{n+5}{8}\right] + \binom{n+6}{10} - 2\binom{n+3}{4}\binom{n+4}{6}$$

Computations from the Second Set of Formulas. The application of the second set of formulas of Section 9.5 gives alternative schemes of recurrence relations. It was found:

$$A_1 = \qquad\qquad n+1,$$

$$A_3 = A_1\binom{n+2}{2} \qquad\qquad -\binom{n+2}{3},$$

$$A_5 = A_3\binom{n+2}{2} - A_1\binom{n+3}{4} \qquad +\binom{n+3}{5},$$

$$A_7 = A_5\binom{n+2}{2} - A_3\binom{n+3}{4} + A_1\binom{n+4}{6} - \binom{n+4}{7}, \text{ etc.}$$

In general:

$$A_1 = n+1, \quad A_{2p+1} = (-1)^p \binom{n+p+1}{2p+1} + \sum_{i=1}^{p} (-1)^{i-1} A_{2p-2i+1} \binom{n+i+1}{2i};$$

$$p = 1, 2, 3, \ldots \quad (9.21)$$

This equation resembles (19), but has an extra term outside the summation. The same feature is found when A_m for even integers m are expressed in terms of A_1, A_3, A_5, ...:

$$A_2 = A_1(n+1) \qquad\qquad - \binom{n+1}{2},$$

$$A_4 = A_3(n+1) - A_1\binom{n+2}{3} \qquad\qquad + \binom{n+2}{4},$$

$$A_6 = A_5(n+1) - A_3\binom{n+2}{3} + A_1\binom{n+3}{5} - \binom{n+3}{6}, \text{ etc.}$$

In general:

$$A_{2p+2} = (-1)^{p+1}\binom{n+p+1}{2p+2} + \sum_{i=0}^{p} (-1)^i A_{2p-2i+1}\binom{n+i+1}{2i+1}; \quad p = 0, 1, 2, \ldots \quad (9.22)$$

Here the A_m terms are expressed by A_1, A_3, A_5, ... (in contrast to A_0, A_2, A_4, ... above). For the sake of completeness we give the explicit formulas of A_m for $m = 1$, 3, 5, 7 and 9.

$$A_1 = n+1,$$

$$A_3 = (n+1)\binom{n+2}{2} - \binom{n+2}{3}, \qquad A_5 = (n+1)\left[\binom{n+2}{2}^2 - \binom{n+3}{4}\right] - \binom{n+2}{3}\binom{n+2}{2} + \binom{n+3}{5},$$

$$A_7 = (n+1)\left[\binom{n+2}{2}^3 - 2\binom{n+2}{2}\binom{n+3}{4} + \binom{n+4}{6}\right] - \binom{n+2}{3}\left[\binom{n+2}{2}^2 - \binom{n+3}{4}\right] + \binom{n+3}{5}\binom{n+2}{2} - \binom{n+4}{7},$$

$$A_9 = (n+1)\left[\binom{n+2}{2}^4 - 3\binom{n+2}{2}^2\binom{n+3}{4} + 2\binom{n+2}{3}\binom{n+4}{6} - \binom{n+5}{8} + \binom{n+3}{4}^2\right]$$
$$- \binom{n+2}{3}\left[\binom{n+2}{2}^3 - 2\binom{n+2}{2}\binom{n+3}{4} + \binom{n+4}{6}\right] + \binom{n+3}{5}\left[\binom{n+2}{2}^2 - \binom{n+3}{4}\right] - \binom{n+4}{7}\binom{n+2}{2} + \binom{n+5}{9}$$

Conclusion. The expressions of $A_m = K\{Z(m,n)\}$ through $m=10$ were transferred to the factored polynomial form and collected in CHART II. Table 2 gives a listing of numerical K numbers.

9.7 THE POLYNOMIAL $P_m(n) = K\{Z(m,n)\}$

From Section 9.6 it is easily perceived that $K\{Z(m,n)\}$ are polynomials of m-th degree in n, say $P_m(n)$. It is also easily seen that all the polynomials for $m \geq 2$ have the factor $(n+1)(n+2)$. We observe also (see CHART II) that $P_m(n)$ for $m = 3, 5$, 7 and 9 have the factor $(2n + 3)$. Below we demonstrate that this also is true for all higher odd integers m.

It is referred to eqn. (21) and the corresponding scheme of recurrence relations. We know that A_3 has the factor $(2n + 3)$. Hence A_5 will also possess this factor if $A_1\binom{n+3}{4} - \binom{n+3}{5}$ has it. In that way we can continue; the hypothesis will be proved if we can ascertain that the last term in the summation of eqn. (21) plus the extra term has the factor $(2n + 3)$ for any $p \geq 1$. This is true because

$$(n+1)\binom{n+p+1}{2p} - \binom{n+p+1}{2p+1} = \binom{n+p+1}{2p}\left[n + 1 - \frac{n-p+1}{2p+1}\right] = \frac{p}{2p+1}\binom{n+p+1}{2p}(2n+3) \qquad (9.23)$$

The above conditions are ideal for the application of the fully computerized

---

CHART 9-II. Multiple zigzag chain with
fixed values of m

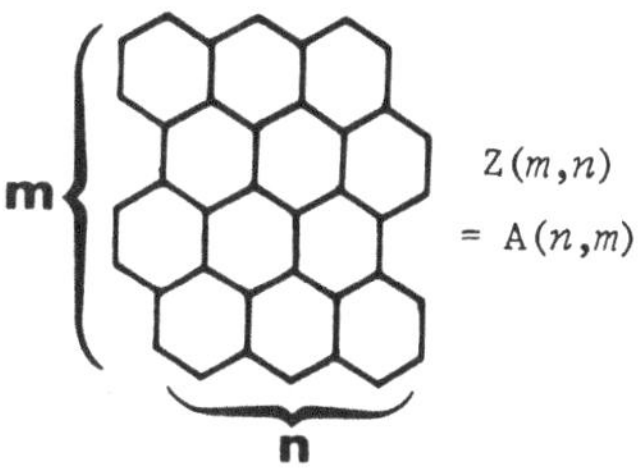

$Z(m,n)$
$= A(n,m)$

[a]$K\{Z(1,n)\} = n+1$

[b]$K\{Z(2,n)\} = \frac{1}{2}(n+1)(n+2)$

[c]$K\{Z(3,n)\} = \frac{1}{6}(n+1)(n+2)(2n + 3)$

[d,e]$K\{Z(4,n)\} = \frac{1}{24}(n+1)(n+2)(5n^2 + 15n + 12)$

[d,f]$K\{Z(5,n)\} = \frac{1}{30}(n+1)(n+2)(2n + 3)(2n^2 + 6n + 5)$

[d]$K\{Z(6,n)\} = \frac{1}{720}(n+1)(n+2)(61n^4 + 366n^3 + 845n^2 + 888n + 360)$

[g]$K\{Z(7,n)\} = \frac{1}{2520}(n+1)(n+2)(2n + 3)(68n^4 + 408n^3 + 949n^2 + 1011n + 420)$

[g]$K\{Z(8,n)\} = \frac{1}{40320}(n+1)(n+2)(1385n^6 + 12465n^5 + 47517n^4 + 98127n^3$
$+ 115810n^2 + 74136n + 20160)$

$K\{Z(9,n)\} = \frac{1}{11340}(n+1)(n+2)(2n + 3)(124n^6 + 1116n^5 + 4267n^4 + 8862n^3$
$+ 10549n^2 + 6834n + 1890)$

$K\{Z(10,n)\} = \frac{1}{3628800}(n+1)(n+2)(50521n^8 + 606252n^7 + 3221422n^6$
$+ 9895860n^5 + 19220389n^4 + 24175248n^3$
$+ 19240308n^2 + 8866800n + 1814400)$

---

[a]$Z(1,n) = L(n)$

[b]$Z(2,n) = L(2,n)$

[c]Gordon M, Davison WHT (1952). J Chem Phys 20: 428 ; also Ch(2,n)

[d]Cyvin SJ, Gutman I (1986). Comp & Maths with Appls 12B: 859

[e]Cyvin SJ (1986). Monatsh Chem 117: 33

[f]Cyvin SJ, Cyvin BN, Gutman I (1985). Z Naturforsch 40a: 1253

[g]Gutman I, Cyvin SJ (1987). Monatsh Chem 118: 541

Table 9.2. Numerical values of $K\{Z(m,n)\}$.

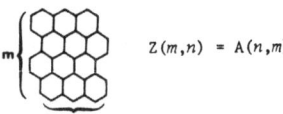

$Z(m,n) = A(n,m)$

n\m	1	2	3	4	5	6	7	8	9	10
1	2	3	4	5	6	7	8	9	10	11
2	3	6	10	15	21	28	36	45	55	66
3	5	14	30	55	91	140	204	285	385	.506
4	8	31	85	190	371	658	1086	1695	2530	3641
5	13	70	246	671	1547	3164	5916	10317	17017	26818
6	21	157	707	2353	6405	15106	31998	62349	113641	196119
7	34	353	2037	8272	26585	72302	173502	377739	760804	1437799
8	55	793	5864	29056	110254	345775	940005	2286648	5089282	10532302
9	89	1782	16886	102091	457379	1654092	5094220	13846117	34053437	77173602
10	144	4004	48620	358671	1897214	7911970	27604798	83833256	227837533	565424068

method (cf. Section 4.10) for the deduction of K formulas. We are not going to treat this method here, but only give some results for the two highest m values (viz. $m=9$, 10) of CHART II.

$$K\{Z(9,n)\} = (n+1)(n+2)(2n+3)\left[\frac{496}{63}\binom{n}{6} + \frac{1984}{63}\binom{n}{5} + \frac{15658}{315}\binom{n}{4}\right.$$
$$\left. + \frac{12254}{315}\binom{n}{3} + \frac{1622}{105}\binom{n}{2} + \frac{14}{5}n + \frac{1}{6}\right] \tag{9.24}$$

$$K\{Z(10,n)\} = (n+1)(n+2)\left[\frac{50521}{90}\binom{n}{8} + \frac{50521}{18}\binom{n}{7} + \frac{163285}{28}\binom{n}{6}\right.$$
$$\left. + \frac{136208}{21}\binom{n}{5} + \frac{20691}{5}\binom{n}{4} + \frac{3003}{2}\binom{n}{3} + \frac{1717}{6}\binom{n}{2} + \frac{47}{2}n + \frac{1}{2}\right] \tag{9.25}$$

When such large numbers are involved the forms of the type (24), (25) may be more convenient for practical applications than the plain polynomials (CHART II).

9.8 ALGORITHM

9.8.1 *Multiple Zigzag Chain*, A(n,m), *and the Class* A(n,m,l)

A simple algorithm exists for the K numbers of $A(n,m,l)$ benzenoids, which include the multiple zigzag chains $A(n,m)$. Here the building-up process changes its direction for every row, as is indicated by thick arrows in Fig. 5 (left part). The numerals are filled in as in the first row after the kink of a chevron; cf. Fig. 8.1. Thin arrows in Fig. 5 explain the procedure. The last numeral (during the

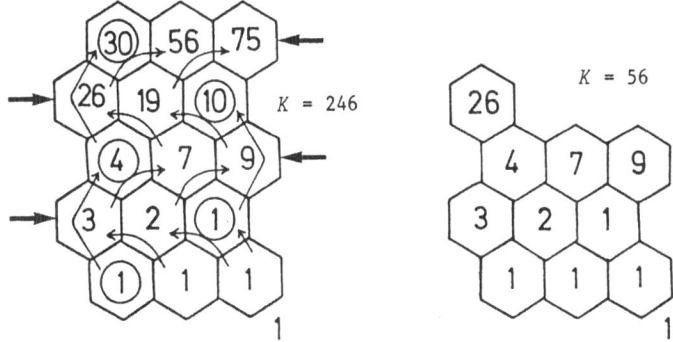

Fig. 9.5. The zigzag chain A(3,5) (left) and the benzenoids A(3,3,1) (right) with algorithm numerals filled in. The K number is the sum of algorithm numerals.

building-up process) of each row is encircled on the left-hand part of Fig. 5. That is the numeral to be put down first in the particular row.

Figure 5 illustrates how the K numbers are obtained as a sum of numerals, either for a multiple zigzag chain, $A(n,m)$, or an auxiliary benzenoid, $A(n,m,l)$, which appears during the building-up process. The K numbers themselves are sooner or later found among the individual numerals. Specifically is $K\{A(n,i)\}$ equal to the encircled numeral of the $(i+2)$-th row. With reference to Fig. 5 one has $K\{A(3,i)\}$ = 1, 4, 10 and 30 for i = 0, 1, 2 and 3, respectively. Furthermore, one finds for instance in the fourth row: $K\{A(3,2)\}$ = 10, $K\{A(3,2,1)\}$ = 19 and $K\{A(3,2,2)\}$ = 26. The number $K\{A(3,3,1)\}$ = 56 (see the right-hand part of Fig. 5) is found in the top row of the left-hand drawing.

9.8.2 *Truncated Multiple Zigzag Chain*

A slightly more complicated algorithm was deduced for a more general class of benzenoids, the truncated zigzag chain, where each row may be shortened in relation to the preceding one (during the building-up process). The $A(n,m,l)$ or $A(m{\times}n; l)$ benzenoids belong to this generalized class. Figure 6 shows how the notation is borrowed from the truncated parallelograms (Section 8.5). The extra problem here is to find the first (encircled) numeral of a truncated row: To obtain the (encircled) numeral Y in the i-th row, (a) start (as before) with the sum of the encircled numeral in the $(i-2)$-th row and the one right above at the end of that row, say X. (b) If the hexagon with X is not adjacent to the hexagon with Y, continue to add the numerals from X (which consequently is taken twice) along the row in the direction of the thick arrow until the last hexagon which is not adjacent to the one with the desired numeral (Y). Although this sounds somewhat complicated it is supposed to be quite easy when illustrated by the thin arrows of Fig. 6 (right-hand part). (Some reservations in the formulations of this general rule are necessary for the second row. That is not important, however, because the first two rows may be considered as a part of a truncated parallelogram.)

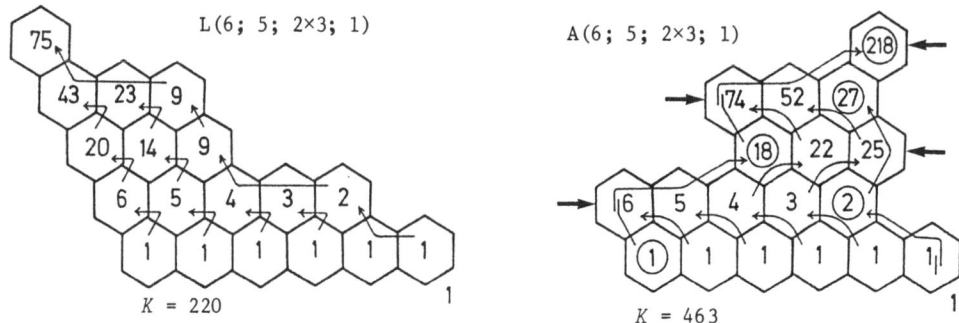

Fig. 9.6. A truncated parallelogram (left) and the corresponding truncated zigzag chain (right) with algorithm numerals.

9.9 SOME GENERAL FORMULATIONS

9.9.1 *Summation K Formulas in Terms of Auxiliary Benzenoids*

From eqn. (6) it is obtained (Cyvin and Gutman 1986a)

$$K\{A(n,m)\} = \sum_{i=0}^{n} K\{A(n, m-2, i)\}; \qquad m \geq 2 \qquad (9.26)$$

Here $l=n$ has been inserted, and the order of summation has been reversed.

Another, more general formula reads

$$K\{A(n,m)\} = \sum_{i=0}^{n} K\{A(n,p,i)\} \cdot K\{A(n,q,i)\}; \qquad p+q = m-3; \qquad m \geq 3 \qquad (9.27)$$

where p and q are positive integers or zero.

Here we shall exemplify eqns. (27) and (28) for $m=5$. Eqn. (27) gives

$$K\{A(n,5)\} = K\{Z(5,n)\} = \sum_{i=0}^{n} K\{A(n,3,i)\} \qquad (9.28)$$

The case with $n=3$ is illustrated in Fig. 7 (left column). Eqn. (27) gives rise to two possibilities: (1) $p=2$, $q=0$; since $A(n,0,l) = L(l)$ one obtains

$$K\{Z(5,n)\} = \sum_{i=0}^{n} K\{A(n,2,i)\} \cdot K\{L(i)\} \qquad (9.29)$$

(2) $p=q=1$; since $A(n,1,l) = L(n,1,l)$ one obtains

$$K\{Z(5,n)\} = \sum_{i=0}^{n} \left[K\{L(n,1,i)\} \right]^{2} \qquad (9.30)$$

Eqns. (30) and (31) for $n=3$ are illustrated in the middle and right-hand column of Fig. 7, respectively.

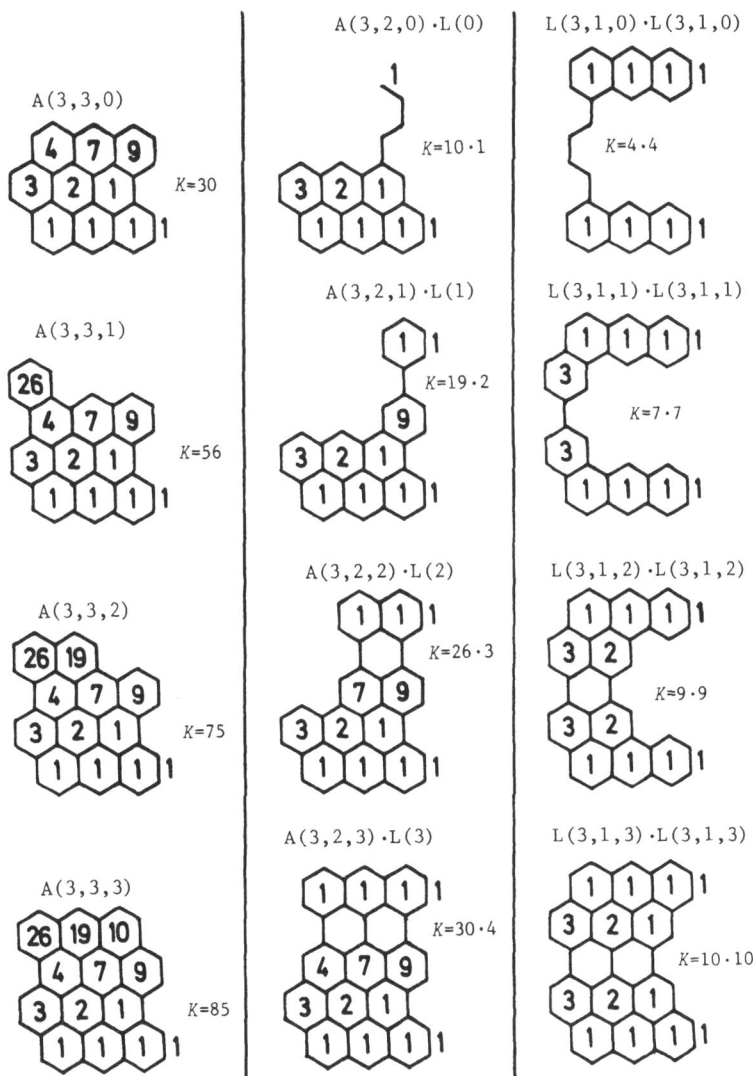

Fig. 9.7. Illustration of eqns. (28)-(30). The sum of K numbers for the four systems in each column is $K\{A(3,5)\} = K\{Z(5,3)\} = 246$. The algorithm numerals are included for the sake of additional information.

9.9.2 *Matrix Formulation*

Let us introduce the abbreviated notation (cf. Table 1)

$$K\{A(n,m,l)\} = Z_n^{(l)}(m) \tag{9.31}$$

Then, in consistence with the notation of Table 1 and the formulas of Section 9.3:

$$Z_n^{(0)}(m) = Z_n(m), \qquad Z_n^{(n)}(m) = Z_n(m+1) \tag{9.32}$$

Also:

$$Z_n^{(l)}(0) = l+1, \qquad Z_n^{(0)}(0) = 1 \tag{9.33}$$

In terms of this notation one has in consequence of eqn. (27):

$$Z_n(m+j) = \sum_{i=0}^{n} Z_n^{(i)}(j-1) \, Z_n^{(i)}(m-2); \quad m \geq 2; \quad j \geq 1 \tag{9.34}$$

When applied to $j = 1, 2, \ldots, n$ eqn. (34) yields in matrix notation:

$$
\begin{bmatrix}
Z_n(m+1) \\[6pt]
Z_n(m+2) \\[2pt]
\vdots \\[2pt]
Z_n(m+n)
\end{bmatrix}
= M
\begin{bmatrix}
Z_n^{(0)}(m-2) \\[6pt]
Z_n^{(1)}(m-2) \\[2pt]
\vdots \\[2pt]
Z_n^{(n)}(m-2)
\end{bmatrix}
\tag{9.35}
$$

where M is the square $(n+1)\times(n+1)$ matrix with the general element equal to

$$(M)_{rs} = Z_n^{(s-1)}(r-1) \tag{9.36}$$

Now it is advantageous to introduce, not only the degenerate and trivial cases pertaining to eqn. (33), but also nominal K values, for which the restrictions on m and j in (34) are released. Specifically we will take advantage of the cases for $m = -1, -2$ and -3. In order that the values shall fit into the formula apparatus, which is manifested in eqn. (34), one must have

$$Z_n^{(l)}(-1) = 1 \tag{9.37}$$

for all n and l. Furthermore:

$$Z_n^{(n)}(-2) = 1, \qquad Z_n^{(l)}(-2) = 0; \quad l < n \tag{9.38}$$

and

$$Z_n^{(0)}(-3) = 1, \qquad Z_n^{(l)}(-3) = 0; \qquad l > 0 \qquad (9.39)$$

9.9.3 The Case of n=3

Here we will exemplify the matrix formulation by the case of $n=3$. Furthermore we show a systematic derivation of the third recurrence relation (pertaining to $n=3$) of Table 1, in addition to other relations.

Table 3 shows the numerical values of $Z_3^{(l)}(m)$ for $m = -3, -2, -1, 0, 1, 2, \ldots \ldots, 10$. The values contain those of the third column of Table 2 (for $n=3$). It is noted that all the (not nominal) values of Table 3 are obtainable as algorithm numerals (cf. Section 9.8) during the building-up of a triple zigzag chain. It is referred to the figure in Table 3, where the chain is built up (contrary to usual practice) from the top in order to conform better the arrangement of the table. Cyvin SJ, Cyvin, Brunvoll and Gutman (1987) have given the corresponding tables for $n = 1, 2, 3, 4, 5, 6$.

The matrix formulation applied to $j = -2, -1, 0, 1$ yields

Table 9.3. Numerical values of $K\{A(3,m,l)\} = Z_3^{(l)}(m)$.

m	$l=0$	1	2	3
-3	1	0	0	0
-2	0	0	0	1
-1	1	1	1	1
0	1	2	3	4
1	4	7	9	10
2	10	19	26	30
3	30	56	75	85
4	85	160	216	246
5	246	462	622	707
6	707	1329	1791	2037
7	2037	3828	5157	5864
8	5864	11021	14849	16886
9	16886	31735	42756	48620
10	48620	91376	123111	139997

$$\begin{bmatrix} Z_3(m-2) \\ Z_3(m-1) \\ Z_3(m) \\ Z_3(m+1) \end{bmatrix} = \begin{bmatrix} Z_3^{(0)}(-3) & Z_3^{(1)}(-3) & Z_3^{(2)}(-3) & Z_3^{(3)}(-3) \\ Z_3^{(0)}(-2) & Z_3^{(1)}(-2) & Z_3^{(2)}(-2) & Z_3^{(3)}(-2) \\ Z_3^{(0)}(-1) & Z_3^{(1)}(-1) & Z_3^{(2)}(-1) & Z_3^{(3)}(-1) \\ Z_3^{(0)}(0) & Z_3^{(1)}(0) & Z_3^{(2)}(0) & Z_3^{(3)}(0) \end{bmatrix} \begin{bmatrix} Z_3^{(0)}(m-2) \\ Z_3^{(1)}(m-2) \\ Z_3^{(2)}(m-2) \\ Z_3^{(3)}(m-2) \end{bmatrix} \qquad (9.40)$$

With numerical values for the matrix elements (cf. Table 3):

$$\begin{bmatrix} Z_3(m-2) \\ Z_3(m-1) \\ Z_3(m) \\ Z_3(m+1) \end{bmatrix} = \begin{bmatrix} 1 & 0 & 0 & 0 \\ 0 & 0 & 0 & 1 \\ 1 & 1 & 1 & 1 \\ 1 & 2 & 3 & 4 \end{bmatrix} \begin{bmatrix} Z_3^{(0)}(m-2) \\ Z_3^{(1)}(m-2) \\ Z_3^{(2)}(m-2) \\ Z_3^{(3)}(m-2) \end{bmatrix} \qquad (9.41)$$

This matrix relation may be interpreted as a set of independent linear equations for the unknowns $Z_3^{(i)}(m-2)$. The two first linear equations give readily

$$Z_3^{(0)}(m-2) = Z_3(m-2), \qquad Z_3^{(3)}(m-2) = Z_3(m-1) \qquad (9.42)$$

in consistence with (32). Then the two remaining linear equations, together with (42), yield

$$Z_3(m) - Z_3(m-2) - Z_3(m-1) = Z_3^{(1)}(m-2) + Z_3^{(2)}(m-2) \qquad (9.43)$$

$$Z_3(m+1) - Z_3(m-2) - 4Z_3(m-1) = 2Z_3^{(1)}(m-2) + 3Z_3^{(2)}(m-2) \qquad (9.44)$$

This set was solved for $Z_3^{(1)}(m-2)$ and $Z_3^{(2)}(m-2)$ with the result:

$$Z_3^{(1)}(m-2) = -Z_3(m+1) + 3Z_3(m) + Z_3(m-1) - 2Z_3(m-2) \qquad (9.45)$$

$$Z_3^{(2)}(m-2) = Z_3(m+1) - 2Z_3(m) - 2Z_3(m-1) + Z_3(m-2) \qquad (9.46)$$

We wish the recurrence relation for Z_3, i.e. the linear dependence between the quantities $Z_3(m+j)$. This is achieved by adding one more equation to the set which is represented by (41). The extra equation for $j=2$ reads

$$Z_3(m+2) = 4Z_3^{(0)}(m-2) + 7Z_3^{(1)}(m-2) + 9Z_3^{(2)}(m-2) + 10Z_3^{(3)}(m-2) \qquad (9.47)$$

On inserting the solution for $Z_3^{(l)}(m-2)$ from (42), (45) and (46) it is readily obtained

$$Z_3(m+2) = 2Z_3(m+1) + 3Z_3(m) - Z_3(m-1) - Z_3(m-2) \qquad (9.48)$$

This is the desired recurrence relation and indeed equivalent to the third relation of Table 1.

By suitable manipulations of the solutions (42), (45) and (46), together with the recurrence relation (48), we have attained at the following set of solutions for $Z_3^{(l)}$.

$$Z_3^{(0)}(m) = Z_3(m) \qquad (9.49)$$

$$Z_3^{(1)}(m) = 2Z_3(m) - Z_3(m-2) \qquad (9.50)$$

$$Z_3^{(2)}(m) = 2Z_3(m) + 2Z_3(m-1) - Z_3(m-2) - Z_3(m-3) \qquad (9.51)$$

$$Z_3^{(3)}(m) = 2Z_3(m) + 3Z_3(m-1) - Z_3(m-2) - Z_3(m-3) \qquad (9.52)$$

These linear dependencies imply that also the classes $Z_3^{(l)}$ for every l obey a recurrence relation of the same form as (48).

Here eqn. (49) is valid for all n (not only $n=3$) in a trivial way. By the approach of the next paragraph we will demonstrate that also $Z_n^{(1)}$ takes the same form as (50) for an arbitrary n. On the other hand, eqns. (51) and (52) do not have this wider applicability. In the next paragraph we will also derive some generally valid expressions for $Z_n^{(l)}(m)$ with $l > 1$. Like the above expression (50) for $l=1$ they will contain $Z_n(m)$, $Z_n(m-2)$, $Z_n(m-4)$, ..., etc., which means that the odd and even numbers m are not mixed.

9.9.4 *Alternative Approach*

Formulas for $Z_n^{(l)}(m)$. In order to start with a simple example assume $l=1$ and the fragmentation scheme as shown in Fig. 8. The outcome of this is

$$Z_n^{(1)}(m) = Z_n(m-2) + 2Z_n^{(n-1)}(m-1) \qquad (9.53)$$

The previous scheme of fragmentation (cf. Fig. 4) applied to $l=1$ yields

$$Z_n^{(1)}(m) = Z_n(m) + Z_n^{(n-1)}(m-1) \qquad (9.54)$$

cf. also eqn. (4). On eliminating $Z_n^{(n-1)}(m-1)$ between (53) and (54) one arrives at the form (50), but now with arbitrary n; see Table 4.

This approach may be generalized to higher l values. When a scheme of fragmentation corresponding to Fig. 8 is applied to $Z_n^{(l)}(m)$ l times it is obtained:

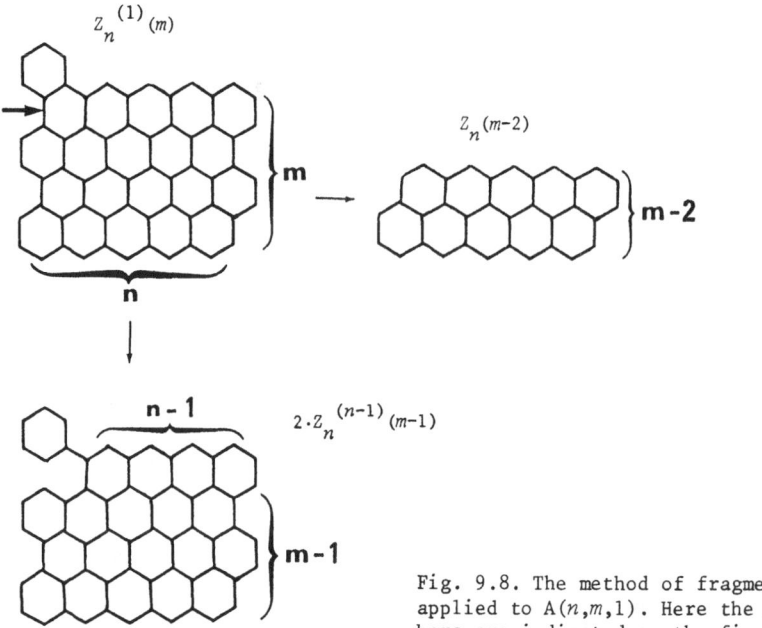

Fig. 9.8. The method of fragmentation applied to $A(n,m,1)$. Here the K numbers are indicated on the figure.

$$Z_n^{(l)}(m) = (l+1)Z_n^{(n-l)}(m-1) + \sum_{i=0}^{l-1} (i+1)Z_n^{(i)}(m-2) \qquad (9.55)$$

The familiar scheme (Fig. 4) gives simply:

$$Z_n^{(l)}(m) = Z_n^{(l-1)}(m) + Z_n^{(n-l)}(m-1) \qquad (9.56)$$

Here $Z_n^{(n-l)}(m-1)$ is eliminated between (55) and (56) with the result:

$$lZ_n^{(l)}(m) = (l+1)Z_n^{(l-1)}(m) - \sum_{i=0}^{l-1} (i+1)Z_n^{(i)}(m-2) \qquad (9.57)$$

This expression is amenable for successive derivation of $Z_n^{(l)}(m)$ expressions for increasing values of l. We do not go into further details here (but see below). The results for l up to 5 are collected in Table 4.

Recurrence Relation for $Z_n(m)$. Finally we give the recurrence relations obtained from the expressions of Table 4. From eqn. (26) or (34) with $j=0$ one has

$$Z_n(m) = \sum_{i=0}^{n} Z_n^{(i)}(m-2) \qquad (9.58)$$

This equation we apply successively for $n = 1, 2, 3, \ldots,$ etc. with the expressions from Table 4, where m is replaced by $m-2$. The alternative recurrence relations ob-

Table 9.4. Expressions of $K\{A(n,m,l)\} = Z_n^{(l)}(m)$ in terms of $K\{A(n, m-j)\} = K\{Z(m-j, n)\} = Z_n(m-j)$.

$Z_n^{(l)}(m)$	Expression
$Z_n^{(0)}(m)$	$Z_n(m)$
$Z_n^{(1)}(m)$	$2Z_n(m) - Z_n(m-2)$
$Z_n^{(2)}(m)$	$3Z_n(m) - 4Z_n(m-2) + Z_n(m-4)$
$Z_n^{(3)}(m)$	$4Z_n(m) - 10Z_n(m-2) + 6Z_n(m-4) - Z_n(m-6)$
$Z_n^{(4)}(m)$	$5Z_n(m) - 20Z_n(m-2) + 21Z_n(m-4) - 8Z_n(m-6) + Z_n(m-8)$
$Z_n^{(5)}(m)$	$6Z_n(m) - 35Z_n(m-2) + 56Z_n(m-4) - 36Z_n(m-6) + 10Z_n(m-8) - Z_n(m-10)$

tained in this way are collected in Table 5. Here (in contrast to Table 1) the m values constantly step by two units; hence the odd and even numbers for this parameter are separated.

The forms of the recurrence relations of Table 5 are also valid for $Z_n^{(l)}$ with arbitrary l values.

General Explicit Formulas. Cyvin SJ, Cyvin, Brunvoll and Gutman (1987) have pursued this analysis and arrived at a general formulation in terms of explicit formulas. Firstly, they derived the following recurrence relation from eqn. (57).

Table 9.5. Alternative recurrence relations for $K\{A(n,m)\} = K\{Z(m,n)\} = Z_n(m)$; see also Table 1.

n	Recurrence relation	
1	$Z_1(m) = 3Z_1(m-2) - Z_1(m-4);$	$m \geq 4$
2	$Z_2(m) = 6Z_2(m-2) - 5Z_2(m-4) + Z_2(m-6);$	$m \geq 6$
3	$Z_3(m) = 10Z_3(m-2) - 15Z_3(m-4) + 7Z_3(m-6) - Z_3(m-8);$	$m \geq 8$
4	$Z_4(m) = 15Z_4(m-2) - 35Z_4(m-4) + 28Z_4(m-6) - 9Z_4(m-8) + Z_4(m-10);$	$m \geq 10$
5	$Z_5(m) = 21Z_5(m-2) - 70Z_5(m-4) + 84Z_5(m-6) - 45Z_5(m-8) + 11Z_5(m-10) - Z_5(m-12);$	$m \geq 12$

$$Z_n^{(l+1)}(m) = 2Z_n^{(l)}(m) - Z_n^{(l-1)}(m) - Z_n^{(l)}(m-2); \qquad l \geq 1 \qquad (9.59)$$

Especially for $l=0$:

$$Z_n^{(1)}(m) = 2Z_n^{(0)}(m) - Z_n^{(0)}(m-2) \qquad (9.60)$$

By studying the repeated application of the recurrence properties it was obtained

$$Z_n^{(l)}(m) = \sum_{i=0}^{l} (-1)^i \binom{l+i+1}{2i+1} Z_n(m-2i) \qquad (9.61)$$

and finally:

$$Z_n(m+2) = \sum_{i=0}^{n} (-1)^i \binom{n+i+2}{2i+2} Z_n(m-2i) \qquad (9.62)$$

Eqns. (61) and (62) are the general forms of the relations of Tables 4 and 5, respectively.

CHAPTER 10

REGULAR THREE-, FOUR- AND FIVE-TIER STRIPS

10.1 PREVIOUS WORK

Studies of the number of Kekulé structures (K) of benzenoid classes here referred to as t-tier strips started by the work of Gordon and Davison (1952). The 3-
tier strips are relatively simple and well known (Gordon and Davison 1952; Yen 1971;
Cyvin 1986d). A systematic study of 4-tier strips has been performed more recently
by Cyvin (1986d). Before that some studies of 5-tier strips had been published
(Gordon and Davison 1952; Yen 1971; Ohkami and Hosoya 1983). The first systematic
treatment of 5-tier strips is due to Cyvin SJ, Cyvin and Gutman (1985).

10.2 DEFINITIONS

10.2.1 *Regular t-Tier Strip*

A t-tier strip is a benzenoid consisting of t tier condensed linear chains
(rows), conventionally drawn horizontally. In order to be a regular t-tier strip it
should possess the following two properties.

(a) The bottom and top row should have the same length in terms of the number
of hexagons, say n.

(b) The vertical sides (rims) should each consist of a connected chain of t
hexagons, of which any *LA*-sequence (cf. Paragraph 2.3.1) is permitted.

A precise definition of the rims is warranted here. It is assumed that the
rows are drawn (as usual) horizontally. The left rim consists of the first (or
extreme left) hexagon from each row. The right rim consists of the last (or extreme
right) hexagon from each row.

It is of interest to consider classes of regular t-tier strips so that the
benzenoids of a certain class, say $S(n)$, contain one parameters (n).

The definition sorts out a limited number of classes for every t value, and
yet it embraces the conventional classes as parallelograms, hexagons, chevrons, etc.
For the sake of brevity we will occasionally refer to (t-tier) strips, tacitly assuming that they are regular according to the above definition.

10.2.2 *Straight and Skew Strips: the Top-Bottom Shift*

In a straight t-tier strip the top and bottom rows are vertically aligned;
otherwise we speak about a skew strip. The displacement of the top row in relation
to the bottom row is measured by the top-bottom shift, say sh, in terms of the num-

ber of hexagons. This number (sh) is defined as positive or zero, and it may have
integer or half-integer values. For a given t value sh is maximum for the parallelogram, viz. $\frac{1}{2}(t-1)$. For odd numbers t one has the possibilities $sh = 0, 1, 2, \ldots$;
for even t numbers $sh = \frac{1}{2}, \frac{3}{2}, \frac{5}{2}, \ldots$.

10.3 CLASSIFICATION OF REGULAR t-TIER STRIPS

10.3.1 *Series of t-Tier Strips*

General. Based on the top-bottom shift (sh) the t-tier strips are classified into
series, conventionally designated by roman numerals. One finds the number of series
equal to $[\frac{1}{2}(t+1)]$; $[x]$ is the integer $\leq x$. Each series is characterized by a leading class of hexagon-shaped benzenoids, viz. $O(k,m,n)$, where $k+m = t+1$. The class
of parallelograms, $L(t,n)$, always constitutes a series alone; it should be identified
with the degenerated hexagon class $O(1,t,n)$. Otherwise a series consists of the leading hexagon class and classes of sub-graphs of this hexagon.

Survey of the Series of t-Tier Strips with t up to 5. The 1-tier and 2-
tier strips are trivial cases. In each case there is one class only, the linear
single chain, $L(n)$, and the parallelogram with two rows, $L(2,n)$, respectively. Table
1 gives a survey of the series of t-tier strips with $t = 1, 2, \ldots, 5$. The numbers
of classes within each series are given in parentheses.

Table 10.1. Series of regular t-tier strips.

t	$m+k$	Series	Leading hexagon class	sh	(Number of classes)
1	2	I	$O(1,1,n)$	0	(1)
2	3	I	$O(1,2,n)$	1/2	(1)
3	4	I	$O(2,2,n)$	0	(3)
		II	$O(1,3,n)$	1	(1)
4	5	I	$O(2,3,n)$	1/2	(6)
		II	$O(1,4,n)$	3/2	(1)
5	6	I	$O(3,3,n)$	0	(16)
		II	$O(2,4,n)$	1	(10)
		III	$O(1,5,n)$	2	(1)

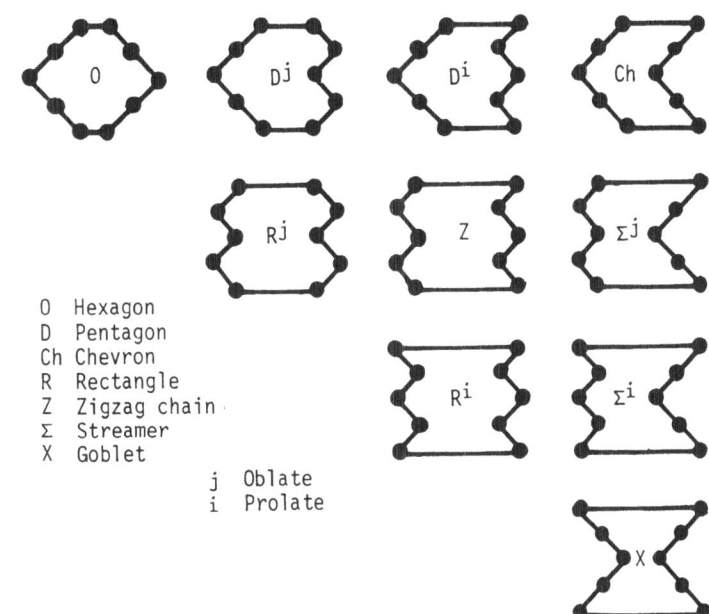

O   Hexagon
D   Pentagon
Ch Chevron
R   Rectangle
Z   Zigzag chain
Σ   Streamer
X   Goblet

j   Oblate
i   Prolate

Fig. 10.1. Main types of straight regular strips.

10.3.2 *Types of Straight Strips*

Hexagon, Chevron and Goblet. Among the straight (regular) strips some main
types of benzenoids are distinguished. We consider first the cases where both rims
contain only one kink each; they have the LA-sequence $L^p A L^p$, where p is a positive
integer, and $t = 2p+1$. Notice that t is always an odd number for straight strips. We
have excluded the trivial cases of one row ($t=1$), viz. $L(n)$. The considered types of
classes may all be classified into hexagons, chevrons and goblets. The hexagons and
goblets are dihedral (D_{2h}), while the chevrons are mirror-symmetrical (C_{2v}). Illustrations of these types are found in Fig. 1. Here the rims are drawn as dualist
graphs (cf. Paragraph 2.2.4).

Oblate and Prolate Pentagons, Rectangles and Streamers. In pentagons and
streamers the configuration $L^p A L^p$ is maintained for one of the rims, while the other
one is indented. In Fig. 1 the cases with regular indentations are depicted; they
are represented by $LA^{2p-1}L$. In these cases one may distinguish between the indentation outwards (j) or inwards (i). The designations oblate and prolate pertain to j
and i, respectively. In an oblate rectangle (cf. Fig. 1) both rims are indented
outwards; in a prolate rectangle inwards.

Zigzag Chain. Finally we characterize a zigzag chain as having both rims indented, one outwards and one inwards (cf. Fig. 1).

Notation. Figure 1 summarizes the symbols adopted for the benzenoid types in question. The hexagons, pentagons, chevrons, streamers and goblets are identified by indices following the same system: $O(m,m,n) = O(m,n)$, $D^j(m,m,n) = D^j(m,n)$, ..., $X(m,m,n) = X(m,n)$, where $2m-1 = t$. The rectangles are identifed similarly by $R^j(m,n)$ and $R^i(m,n)$; only two indices are used invariably. Finally one has the notation $Z(t,n)$ for the t-tier zigzag chain. These notations are consistent with the usage other places in this book.

10.3.3 *Multiple Chain*

An n-tuple chain consists of n condensed identical single chains. The concept of *LA*-sequence is applicable to multiple chains. Here the kinks must follow (in contrast to single chains) in an alternating manner as ... -left-right-left- ... . We use the notation M_n followed by the *LA*-sequence in parentheses. The multiple chains fall under the definition of regular t-tier strips, straight or skew. They reveal themselves as having all rows of the same length (n hexagons). Previously encountered examples are the parallelograms, $M_n(L^t)$, chevrons, $M_n(L^p A L^q)$, where $p+q+1 = t$, and the zigzag chains, $M_n(LA^{t-2}L)$; cf. Chapters 8 and 9.

10.3.4 *General Classification*

Two-Segment Rim. A two-segment rim is defined as a two-segment chain, which is characterized by the *LA*-sequence $L^p A L^q$, where p and q are positive integers, and $p+q = t-1$ ($t > 2$). We will define a protruding (two-segment) rim and an intruding rim. If the angularly annelated (*A*-mode) hexagon of a left two-segment rim is the extreme left hexagon, or if it is the extreme right hexagon in a right rim, the rim is protruding. Otherwise the two-segment rim is intruding. This terminology is convenient for a precise definition of the main classes of regular t-tier strips.

Six Main Types. The five designations hexagon, pentagon, chevron, streamer and goblet are generally applicable to t-tier strips, straight or skew. In consistence with Paragraph 10.3.2 we recognize both rims as two-segment rims in hexagons, chevrons and goblets. In pentagons and streamers only one of the rims is of the two-segment type.

More precise definitions: A hexagon has two protruding two-segment rims. A goblet has two intruding two-segment rims. A chevron has one protruding and one intruding two-segment rim. A pentagon has exactly one two-segment rim, which is protruding. A streamer has exactly one two-segment rim, which is intruding.

When the two segments in $L^p A L^q$ have unequal lengths ($p \neq q$) the hexagons and goblets are centrosymmetrical (C_{2h}), while the pentagons, chevrons and streamers are unsymmetrical (C_s).

The definitions of oblate and prolate types (Paragraph 10.3.2) are not adequate for t-tier skew strips. Nevertheless is it practical to use these designations for some low values of t; we will see some examples for $t=5$ in the following.

The designation tower is devised in order to cover all types of t-tier strips which do not belong to any of the five types treated above. All multiple chains others than chevrons (but including parallelograms and zigzag chains) are examples of towers according to this definition. Rectangles are also towers.

In practice it seems not natural to designate parallelograms as towers. Also otherwise the concept of towers can be avoided for small values of t. Instead we may use the special designations as rectangle, zigzag chain, other multiple chains, hexagon without corners, etc. Furthermore, many t-tier strips are essentially disconnected (cf. Paragraph 2.3.2) and may be specified (although not always unambiguously) by means of their units. Below we will see how the concept of towers is avoided through $t=5$.

Systematic Notation. In a systematic notation for t-tier strips the types are identified by the symbols O, D, Ch, Σ and X (see above), in addition to H for towers. Three indices are added in parentheses according to the parameters (k,m,n) of the leading hexagon of the particular series. Subscripts are applied to distinguish different classes belonging to the same series and main type.

10.3.5 *Special Classification up to $t=5$*

Classes for $t=1$, 2 and 3. The trivial cases of $t=1$ and 2 are mentioned in Paragraph 10.3.1.

The regular strips of $t=3$ contain four classes divided into two series: (I) Hexagon $O(2,2,n)$, Chevron $Ch(2,2,n)$, Tower $H(2,2,n) = R^j(2,n)$ and (II) Tower $H(1,3,n) = L(3,n)$.

Classes for $t=4$. For $t=4$ the strips are all skew since t is an even integer. Series I exhibits exactly one example each of the six main types of benzenoid classes; cf. Fig. 2 (schematic representations with the aid of dualist graphs). The zigzag chain, $Z(4,n)$, and the parallelogram, $L(4,n)$, should in the systematic notation (Paragraph 10.3.4) be termed $H(2,3,n)$ and $H(1,4,n)$, respectively.

Classes for $t=5$. The 5-tier strips form three series (cf. Table 1); series I contains straight strips, the other skew.

Among the straight strips (Series I) there is one more class of pentagons and streamers (cf. Fig. 3) in addition to those (oblate and prolate) shown in Fig. 1. Similarly we recognize some classes (Fig. 3) similar to the rectangles of Fig. 1. The classes D and R of Fig. 3 may suitably be called intermediate pentagons and rectangles, respectively. The classes R' and Σ (Fig. 3) consist of essentially dis-

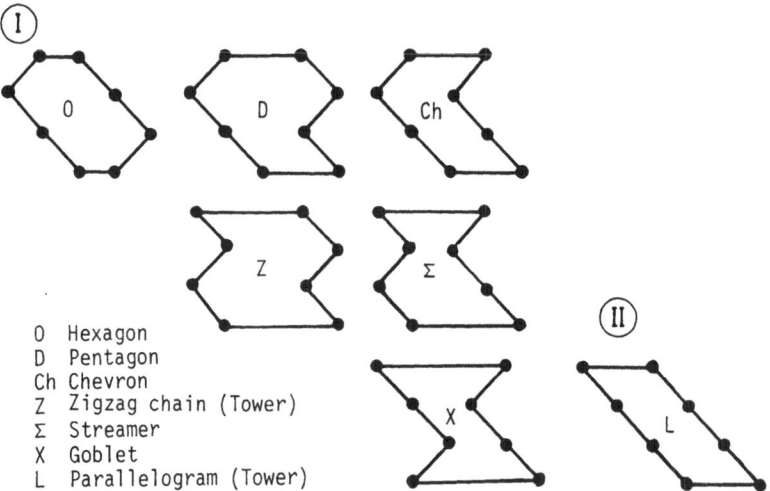

Fig. 10.2. All classes of regular 4-tier strips; series I and II; all of them are skew strips.

O Hexagon
D Pentagon
Ch Chevron
Z Zigzag chain (Tower)
Σ Streamer
X Goblet
L Parallelogram (Tower)

connected benzenoids.

In Series II of 5-tier strips there are two pentagons and two streamers; they may be termed "oblate" and "prolate", although these designations are not precisely defined for skew strips.

The above designations for 5-tier strips include in general some terms which do not belong to the systematic notation (Paragraph 10.3.4). The concept of towers has been avoided. Table 2 summarizes the systematic and practical notation for the existing classes of 5-tier strips.

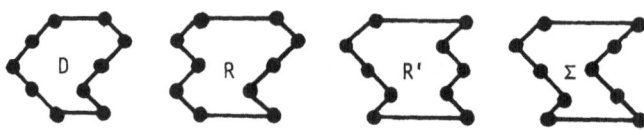

Fig. 10.3. Intermediate types of pentagon (D), rectangles (R, R') and streamer (Σ) among Series I of regular 5-tier strips.

Table 10.2. Terminology for the classes of regular 5-tier strips.

(Series) No.	Systematic notation	Practical notation and designation
(I) 1.	$O(3,3,n)$	$O(3,n)$ Hexagon (dihedral)
2.	$D_1(3,3,n)$	$D^j(3,n)$ Oblate pentagon
3.	$D_2(3,3,n)$	$D(3,n)$ Intermediate pentagon
4.	$D_3(3,3,n)$	$D^i(3,n)$ Prolate pentagon
5.	$Ch(3,3,n)$	$Ch(3,n)$ Chevron (mirror-symmetrical)
6.	$H_{11}(3,3,n)$	$R^j(3,n)$ Oblate rectangle
7.	$H_{12}(3,3,n)$	$R(3,n)$ Intermediate rectangle
8.	$H_{13}(3,3,n)$	$Z(5,n)$ Zigzag chain
9.	$\Sigma_1(3,3,n)$	$\Sigma^j(3,n)$ Oblate streamer; also $L(2,n)\cdot L(2,n)$ as Series II, No. 8
10.	$H_{221}(3,3,n)$	$L(n)\cdot O(2,n)$ Essentially disconnected
11.	$H_{222}(3,3,n)$	$M_n(LALAL)$ Multiple chain
12.	$H_{23}(3,3,n)$	$L(n)\cdot Ch(2,n)$ Essentially disconnected
13.	$\Sigma_2(3,3,n)$	$L(n)\cdot L(2,n)$ Essentially disconnected; see also Series II, No. 9
14.	$H_{33}(3,3,n)$	$R^i(3,n)$ Prolate rectangle
15.	$\Sigma_3(3,3,n)$	$\Sigma^i(3,n)$ Prolate streamer; also $L(n)\cdot L(n)$ as Series II, No. 10
16.	$X(3,3,n)$	$X(3,n)$ Goblet (dihedral)
(II) 1.	$O(2,4,n)$	$O(2,4,n)$ Hexagon (centrosymmetrical)
2.	$D_1(2,4,n)$	$D^j(2,4,n)$ "Oblate" pentagon
3.	$D_2(2,4,n)$	$D^i(2,4,n)$ "Prolate" pentagon
4.	$Ch(2,4,n)$	$Ch(2,4,n)$ Chevron (asymmetrical)
5.	$H_{11}(2,4,n)$	$Ob(2,4,n)$ Hexagon without two corners
6.	$H_{12}(2,4,n)$	$M_n(LLAAL)$ Multiple chain
7.	$\Sigma_1(2,4,n)$	$\Sigma^j(2,4,n)$ "Oblate" streamer
8.	$H_{22}(2,4,n)$	$L(2,n)\cdot L(2,n)$ Essentially disconnected; see also Series I, No. 9
9.	$\Sigma_2(2,4,n)$	$\Sigma^i(2,4,n)$ "Prolate" streamer; also $L(n)\cdot L(2,n)$ as Series I, No. 13
10.	$X(2,4,n)$	$X(2,4,n)$ Goblet (centrosymmetrical); also $L(n)\cdot L(n)$ as Series I, No. 15
(III)	$H(1,5,n)$	$L(5,n)$ Parallelogram

10.4 EXAMPLES OF NON-REGULAR t-TIER STRIPS

Cyvin SJ, Cyvin and Gutman (1985) gave some examples of strips which are not regular according to the definition of Paragraph 10.2.1. Either the property (a), the property (b) or both could be violated.

The below figure shows two 5-tier strips where the property (a) of Paragraph 10.2.1 is violated, but not (b).

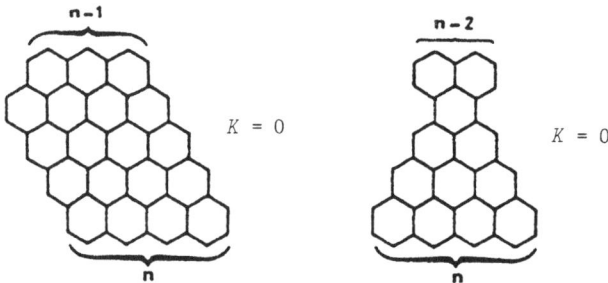

When (b) is valid all peaks are necessarily found in the top row and all valleys in the bottom row. Hence a non-regular strip of this category, i.e. when only (a) is violated, is an obvious non-Kekuléan benzenoid.

Figure 4 shows a class of 5-tier strips, which contains two parameters (k,n). In general (unless $k=n=2$) these strips are non-regular. The special case of $k=n$ is depicted separately. In these benzenoid systems, when $k=n > 2$ (see the left-hand

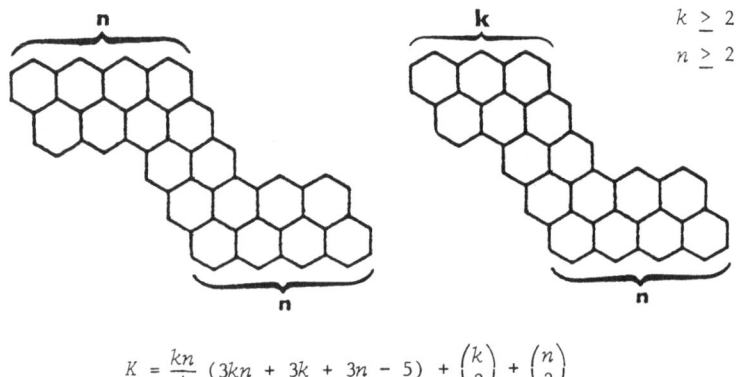

$$K = \frac{kn}{4}\,(3kn + 3k + 3n - 5) + \binom{k}{2} + \binom{n}{2}$$

Fig. 10.4. A benzenoid class of 5-tier strips. The middle part of 10 hexagons, viz. L(2,5), is supposed to be unaffected by the two parameter values. For $k=n=2$ the system degenerates to only this parallelogram. The K formula is given. Its special case for $k=n$ was derived by Cyvin SJ, Cyvin and Gutman (1985).

drawing of Fig. 4) only the condition (b) is violated. When $k \neq n$, and $k > 2$ or $n > 2$ (right-hand drawing), both (a) and (b) are violated. Nevertheless all members of this benzenoid class are Kekuléan. It is, however, not difficult to construct examples of non-Kekuléans where both (a) and (b) are violated.

Below we show an example of a benzenoid system which is not a 5-tier strip at all.

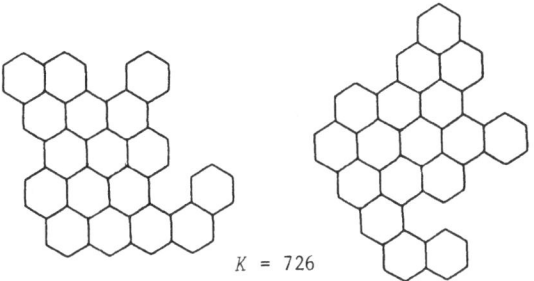

$K = 726$

A re-orientation, which implies a re-defining of the rows (cf. the right-hand drawing above) shows that the system may be interpreted as a non-regular 7-tier strip. In the same way the systems of Fig. 4 may be interpreted as regular $(k+n+2)$-tier strips.

In this connection we wish to clarify something of the nature of ribbons. The definition of a class of ribbons as put forward in Section 8.4 places these benzenoids among non-regular m-tier strips. However, every individual ribbon $V(k,m,n)$ may be interpreted as a regular $(m+n-1)$-tier strip belonging to a class of pentagons, say $D(m,n,n')$, where $n' = 1$. For example, the first three members for $k=2$, $m=3$, $n=4$ are:

$$V(2,3,4) = D(3,4,1) \qquad D(3,4,2) \qquad D(3,4,3)$$

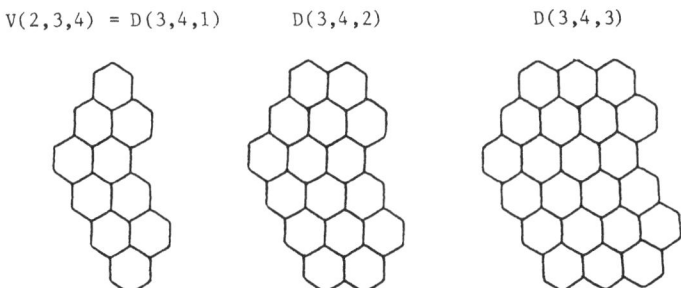

Only $D(3,4,1)$ is a ribbon. Finally we point out that the triangle-shaped benzenoids introduced in Paragraph 8.5.5 (see especially Fig. 8.14), which obviously are non-regular m-tier strips, also may be interpreted as regular $(2m-1)$-tier strips. In fact the prolate and oblate triangles coincide with the corresponding prolate and oblate pentagons for $n=1$, respectively: $T^i(m) = D^i(m,1)$, $T^j(m) = D^j(m,1)$.

10.5 DICTIONARY OF K FORMULAS FOR REGULAR 3-, 4- AND 5-TIER STRIPS

10.5.1 *General Features*

When a regular t-tier strip is oriented in the conventional way (cf. Paragraph 10.2.1) it has exactly n peaks and n valleys. Hence $\Delta=0$ (cf. Paragraph 2.2.6).

Usually all members of a class of a (regular) t-tier strip, $S(n)$, are Keku-léan. Then the combinatorial formulas of $K\{S(n)\}$ are polynomials in n. Furthermore, $n=0$ gives

$$K\{S(0)\} = 1 \tag{10.1}$$

corresponding to the trivial case of no hexagons.

A t-tier strip may be a concealed non-Kekuléan (cf. Paragraph 2.3.2). If this is the case for one member of a class it must be the case for all members;

$$K\{S_o(n)\} = 0 \tag{10.2}$$

for all n.

10.5.2 *Tables*

A comprehensive list is given in Tables 3-5. The formulas are given both in terms of binomial coefficients and in a factored polynomial form. The tables include the number of hexagons, h. Numerical values of h and K are given for $n = 6$, 5 and 4 when $t = 3$, 4 and 5, respectively.

Table 10.3. Classes of regular 3-tier strips. Formulas for h and K.

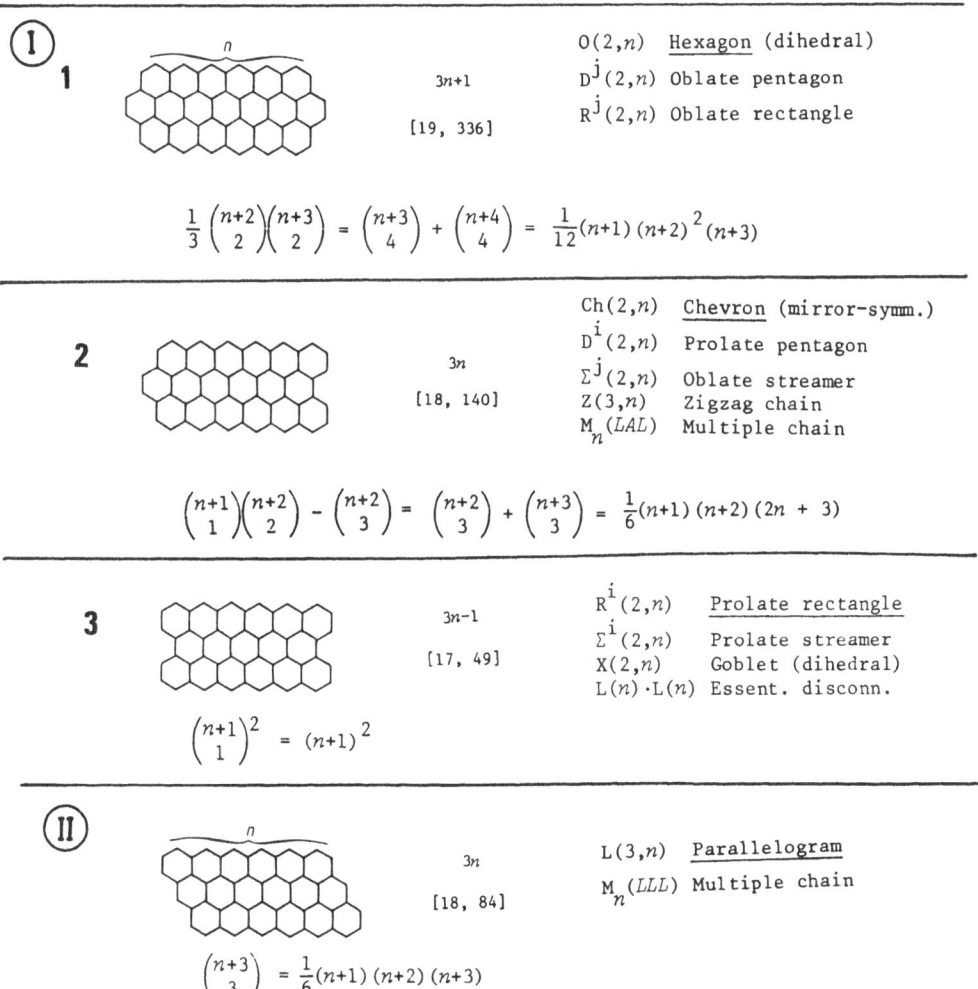

①
1

$3n+1$

[19, 336]

O$(2,n)$ Hexagon (dihedral)
D$^j(2,n)$ Oblate pentagon
R$^j(2,n)$ Oblate rectangle

$$\frac{1}{3}\binom{n+2}{2}\binom{n+3}{2} = \binom{n+3}{4} + \binom{n+4}{4} = \frac{1}{12}(n+1)(n+2)^2(n+3)$$

2

$3n$

[18, 140]

Ch$(2,n)$ Chevron (mirror-symm.)
D$^i(2,n)$ Prolate pentagon
$\Sigma^j(2,n)$ Oblate streamer
Z$(3,n)$ Zigzag chain
M$_n(LAL)$ Multiple chain

$$\binom{n+1}{1}\binom{n+2}{2} - \binom{n+2}{3} = \binom{n+2}{3} + \binom{n+3}{3} = \frac{1}{6}(n+1)(n+2)(2n+3)$$

3

$3n-1$

[17, 49]

R$^i(2,n)$ Prolate rectangle
$\Sigma^i(2,n)$ Prolate streamer
X$(2,n)$ Goblet (dihedral)
L$(n)\cdot L(n)$ Essent. disconn.

$$\binom{n+1}{1}^2 = (n+1)^2$$

②

$3n$

[18, 84]

L$(3,n)$ Parallelogram
M$_n(LLL)$ Multiple chain

$$\binom{n+3}{3} = \frac{1}{6}(n+1)(n+2)(n+3)$$

Table 10.4. Classes of regular 4-tier strips. Formulas for h and K.

(I)

1

O(2,3,n) Hexagon (centrosymm.)

$$\frac{1}{4} \binom{n+3}{3}\binom{n+4}{3} = \binom{n+3}{3}^2 - \binom{n+3}{2}\binom{n+3}{4}$$

$$= \frac{1}{144}(n+1)(n+2)^2(n+3)^2(n+4)$$

$4n+2$ [22, 1176]

2

D(2,3,n) Pentagon

$$\frac{1}{4} \binom{n+2}{1}^2\binom{n+3}{3} = \binom{n+2}{2}\binom{n+3}{3} - \binom{n+2}{1}\binom{n+3}{4}$$

$$= \frac{1}{24}(n+1)(n+2)^3(n+3)$$

$4n+1$ [21, 686]

3

Ch(2,3,n) Chevron (unsymm.)
$M_n(LLAL)$ Multiple chain

$$\binom{n+1}{1}\binom{n+3}{3} - \binom{n+3}{4} = \binom{n+4}{4} + 2\binom{n+3}{4}$$

$$= \frac{1}{24}(n+1)(n+2)(n+3)(3n+4)$$

$4n$ [20, 266]

4

Z(4,n) Zigzag chain
$M_n(LAAL)$ Multiple chain

$$\binom{n+2}{2}^2 - \binom{n+3}{4} = \frac{1}{24}(n+1)(n+2)(5n^2+15n+12)$$

$4n$ [20, 371]

5

$4n-1$

[19, 126]

$\Sigma(2,3,n)$ Streamer
$L(n) \cdot L(2,n)$ Essent. disconn.

$$\binom{n+1}{1}\binom{n+2}{2} = \frac{1}{2}(n+1)^2(n+2)$$

6

$4n-2$

[18, 36]

X(2,3,n) Goblet (centrosymm.)
$L(n) \cdot L(n)$ Essent. disconn.

$$\binom{n+1}{1}^2 = (n+1)^2$$

(II)

$4n$

[20, 126]

L(4,n) Parallelogram
$M_n(LLLL)$ Multiple chain

$$\binom{n+4}{4} = \frac{1}{24}(n+1)(n+2)(n+3)(n+4)$$

Table 10.5. Classes of regular 5-tier strips. Formulas for h and K.

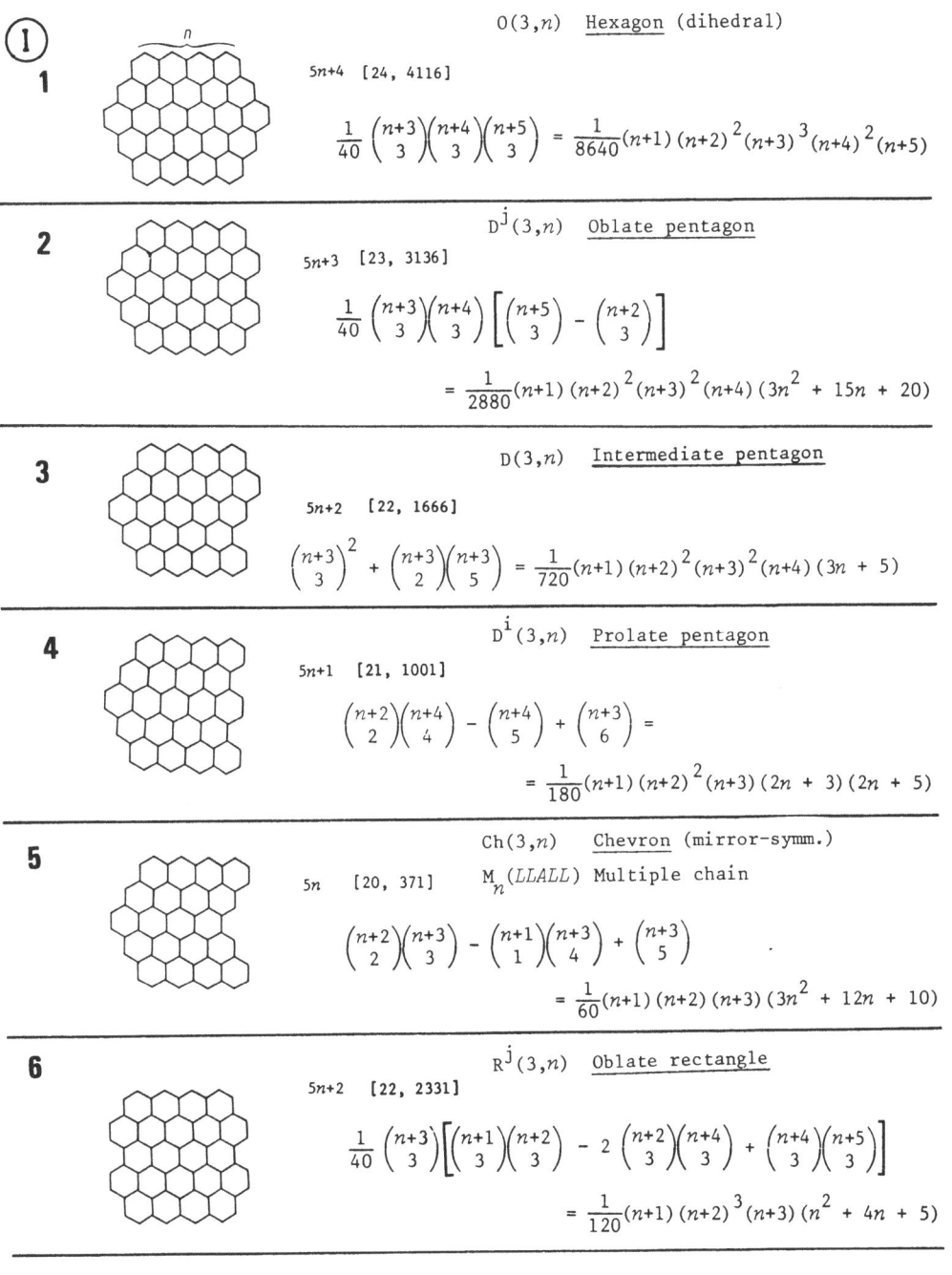

①

1

O(3,n) Hexagon (dihedral)

$5n+4$ [24, 4116]

$$\frac{1}{40}\binom{n+3}{3}\binom{n+4}{3}\binom{n+5}{3} = \frac{1}{8640}(n+1)(n+2)^2(n+3)^3(n+4)^2(n+5)$$

2

$D^j(3,n)$ Oblate pentagon

$5n+3$ [23, 3136]

$$\frac{1}{40}\binom{n+3}{3}\binom{n+4}{3}\left[\binom{n+5}{3} - \binom{n+2}{3}\right]$$

$$= \frac{1}{2880}(n+1)(n+2)^2(n+3)^2(n+4)(3n^2 + 15n + 20)$$

3

D(3,n) Intermediate pentagon

$5n+2$ [22, 1666]

$$\binom{n+3}{3}^2 + \binom{n+3}{2}\binom{n+3}{5} = \frac{1}{720}(n+1)(n+2)^2(n+3)^2(n+4)(3n + 5)$$

4

$D^i(3,n)$ Prolate pentagon

$5n+1$ [21, 1001]

$$\binom{n+2}{2}\binom{n+4}{4} - \binom{n+4}{5} + \binom{n+3}{6} =$$

$$= \frac{1}{180}(n+1)(n+2)^2(n+3)(2n + 3)(2n + 5)$$

5

Ch(3,n) Chevron (mirror-symm.)

$M_n(LLALL)$ Multiple chain

$5n$ [20, 371]

$$\binom{n+2}{2}\binom{n+3}{3} - \binom{n+1}{1}\binom{n+3}{4} + \binom{n+3}{5}$$

$$= \frac{1}{60}(n+1)(n+2)(n+3)(3n^2 + 12n + 10)$$

6

$R^j(3,n)$ Oblate rectangle

$5n+2$ [22, 2331]

$$\frac{1}{40}\binom{n+3}{3}\left[\binom{n+1}{3}\binom{n+2}{3} - 2\binom{n+2}{3}\binom{n+4}{3} + \binom{n+4}{3}\binom{n+5}{3}\right]$$

$$= \frac{1}{120}(n+1)(n+2)^3(n+3)(n^2 + 4n + 5)$$

7

R(3,n) Intermediate rectangle

$5n+1$ [21, 1176]

$$\binom{n+2}{2}\binom{n+4}{4} + \binom{n+2}{1}\binom{n+3}{5} = \frac{1}{240}(n+1)(n+2)^2(n+3)(7n^2 + 23n + 20)$$

8

Z(5,n) Zigzag chain

$M_n(LAAAL)$ Multiple chain

$5n$ [20, 671]

$$\binom{n+2}{2}\binom{n+2}{3} + \binom{n+1}{1}\binom{n+4}{4} + \binom{n+3}{5}$$

$$= \frac{1}{30}(n+1)(n+2)(2n+3)(2n^2 + 6n + 5)$$

---

9 $\Sigma^j(3,n)$ Oblate streamer

$L(2,n)\cdot L(2,n)$ Essent. disconn.

$5n-1$ [19, 225]

$$\binom{n+2}{2}^2 = \frac{1}{4}(n+1)^2(n+2)^2$$

10 L(n)·O(2,n) Essent. disconn.

$5n$ [20, 525]

$$\binom{n+2}{2}\binom{n+3}{3}$$

$$= \frac{1}{12}(n+1)^2(n+2)^2(n+3)$$

11

$M_n(LALAL)$ Multiple chain

$5n$ [20, 546]

$$\binom{n+2}{2}\binom{n+3}{3} + \binom{n+3}{5} = \frac{1}{120}(n+1)(n+2)(n+3)(11n^2 + 29n + 20)$$

---

12 L(n)·Ch(2,n) Essent. disconn.

$5n-1$ [19, 275]

$$2\binom{n+2}{2}^2 - \binom{n+1}{1}\binom{n+3}{3}$$

$$= \frac{1}{6}(n+1)^2(n+2)(2n+3)$$

13 L(n)·L(2,n) Essent. disconn.

$5n-2$ [18, 75]

$$\binom{n+1}{1}\binom{n+2}{2} = \frac{1}{2}(n+1)^2(n+2)$$

---

14 $R^i(3,n)$ Prolate rectangle

$L(n)\cdot L(n)\cdot L(n)$ Essent. disconn.

$5n-2$ [18, 125]

$$\binom{n+1}{1}^3 = (n+1)^3$$

15 $\Sigma^i(3,n)$ Prolate streamer

$L(n)\cdot L(n)$ Essent. disconn.

$5n-3$ [17, 25]

$$\binom{n+1}{1}^2 = (n+1)^2$$

16

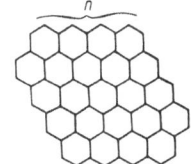

$5n-4$ [16, 0]

$X(3,n)$ Goblet (dihedral)

0

(II)

1

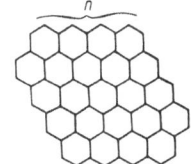

n

$5n+3$ [23, 1764]

$O(2,4,n)$ Hexagon (centrosymm.)

$$\frac{1}{5}\binom{n+4}{4}\binom{n+5}{4} = \frac{1}{2880}(n+1)(n+2)^2(n+3)^2(n+4)^2(n+5)$$

2

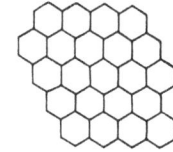

$5n+2$ [22, 1274]

$D^j(2,4,n)$ Oblate pentagon

$$\frac{1}{5}\binom{n+4}{4}\left[\binom{n+5}{4} - \binom{n+3}{4}\right] = \frac{1}{720}(n+1)(n+2)^2(n+3)^2(n+4)(2n+5)$$

3

$5n+1$ [21, 714]

$D^i(2,4,n)$ Prolate pentagon

$$\binom{n+2}{2}\binom{n+4}{4} - \binom{n+2}{1}\binom{n+4}{5} = \frac{1}{240}(n+1)(n+2)^2(n+3)(n+4)(3n+5)$$

4

$5n$ [20, 294]

$Ch(2,4,n)$ Chevron (unsymm.)

$M_n(LLLAL)$ Multiple chain

$$\binom{n+1}{1}\binom{n+4}{4} - \binom{n+4}{5} = \frac{1}{120}(n+1)(n+2)(n+3)(n+4)(4n+5)$$

5

$5n+1$ [21, 889]

$Ob(2,4,n)$ Hexagon without two corners

$$\frac{1}{5}\left[\binom{n+2}{4}\binom{n+3}{4} - 2\binom{n+3}{4}\binom{n+4}{4} + \binom{n+4}{4}\binom{n+5}{4}\right]$$

$$= \frac{1}{360}(n+1)(n+2)^2(n+3)(7n^2+28n+30)$$

6  $5n$ [20, 469]

$M_n(LLAAL)$ Multiple chain

$$\binom{n+2}{2}\binom{n+3}{3} - \binom{n+4}{5} = \frac{1}{120}(n+1)(n+2)(n+3)(9n^2 + 26n + 20)$$

---

7 $\Sigma^j(2,4,n)$ Oblate streamer

$L(n) \cdot L(3,n)$ Essent. disconn.

    $5n-1$ [19, 175]

$$\binom{n+1}{1}\binom{n+3}{3}$$
$$= \frac{1}{6}(n+1)^2(n+2)(n+3)$$

8 $L(2,n) \cdot L(2,n)$ Essent. disconn.

    $5n-1$ [19, 225]

$$\binom{n+2}{2}^2 = \frac{1}{4}(n+1)^2(n+2)^2$$

---

9 $\Sigma^i(2,4,n)$ Prolate streamer

$L(n) \cdot L(2,n)$ Essent. disconn.

    $5n-2$ [18, 75]

$$\binom{n+1}{1}\binom{n+2}{2} = \frac{1}{2}(n+1)^2(n+2)$$

10 $X(2,4,n)$ Goblet (centrosymm.)

$L(n) \cdot L(n)$ Essent. disconn.

    $5n-3$ [17, 25]

$$\binom{n+1}{1}^2 = (n+1)^2$$

(III) n

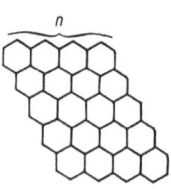 $5n$ [20, 126]

$L(5,n)$ Parallelogram

$M_n(LLLLL)$ Multiple chain

$$\binom{n+5}{5} = \frac{1}{120}(n+1)(n+2)(n+3)(n+4)(n+5)$$

10.6 METHODS OF DERIVATION OF K FORMULAS FOR t-TIER STRIPS

10.6.1 *Schematic Survey*

Several methods are available for the derivation of combinatorial K formulas for classes of t-tier strips. A schematic survey is given below. The methods are going to be illustrated by examples selected from 4-tier and 5-tier strips.

(1) Application of more general combinatorial formulas;
(2) stripping;
(3) chopping;
(4) addition of algorithm numerals.

Both stripping (point 2) and chopping (point 3) are direct applications of the fragmentation method. The method is supported by algebraic computations. In both cases one tends to take advantage of essentially disconnected benzenoids. In the latter case (point 3) this is in fact a part of the method.

The evaluation of sums is frequently encountered and is a part of the methods under points (2-4). Different techniques are available: (i) Direct expansion in terms of sums of powers. It was found advantageous to employ sums of $(i+1)^p$, where p is a positive integer. The method may be demonstrated already under point (1) by the application of a chevron formula. (ii) Summation of expressions with binomial coefficients. (iii) Identification of a sum with the K number of a benzenoid class, for which the formula is known. (iv) Refined application of chevron formulas. − In practice it is not possible to make a sharp division between all the different techniques which are listed.

10.6.2 *Application of More General Combinatorial Formulas*

Among the actual combinatorial formulas are those for $L(m,n)$ and for the three-parameter classes $O(k,m,n)$ and $Ch(k,m,n)$. Also hexagons without one or two corners are of interest; cf. Chapter 8.

Consider first the 4-tier pentagon $D(2,3,n)$, as a hexagon without corner, viz. $Oa(2,3,n)$; cf. Table 4(I-2). By means of eqn. (8.2) one obtains

$$K\{D(2,3,n)\} = K\{Oa(2,3,n)\} = K\{O(2,3,n)\} - K\{O(2,\ 3,\ n-1)\}; \quad n \geq 1 \qquad (10.3)$$

and

$$K\{D(2,3,n)\} = \frac{1}{4} \binom{n+3}{3}\binom{n+4}{3} - \frac{1}{4}\binom{n+2}{3}\binom{n+3}{3} \qquad (10.4)$$

which is easily identified with the expressions of Table 4(I-2). Here we have used eqn. (i) of CHART 8-I. The result in one of the forms of Table 4 would be obtained directly by means of eqn. (ii) of CHART 8-I.

In the next example consider the 5-tier mirror-symmetrical chevron, viz.

$Ch(3,n) = Ch(3,3,n)$. By means of eqn. (i) of CHART 8-II it is obtained

$$K\{Ch(3,3,n)\} = \sum_{i=0}^{n} \binom{i+2}{2}^2 = \frac{1}{4} \sum_{i=0}^{n} (i+1)^4 + \frac{1}{2} \sum_{i=0}^{n} (i+1)^3 + \frac{1}{4} \sum_{i=0}^{n} (i+1)^2$$

$$= \frac{1}{120}(n+1)(n+2)(2n+3)(3n^2+9n+5) + \frac{1}{8}(n+1)^2(n+2)^2 + \frac{1}{24}(n+1)(n+2)(2n+3) \quad (10.5)$$

By elementary calculations the polynomial reduces to the answer of Table 5(I-5). This direct expansion in terms of sums of powers is rather tedious. The problem is solved considerably easier by means of eqn. (ii) of CHART 8-II, which gives directly the binomial-coefficient form of Table 5(I-5):

$$K\{Ch(3,3,n)\} = \binom{n}{0}\binom{n+3}{5} - \binom{n+1}{1}\binom{n+3}{4} + \binom{n+2}{2}\binom{n+3}{3} \quad (10.6)$$

The treatment of essentially disconnected benzenoids often falls into the category considered here. For example in the case of I-12 of Table 5 we have

$$K\{L(n) \cdot Ch(2,n)\} = K\{L(n)\} \cdot K\{Ch(2,n)\} = (n+1)\left[(n+1)\binom{n+2}{2} - \binom{n+2}{3}\right] \quad (10.7)$$

where the formula from Table 3(I-2) was utilized. The expression (7) reduces to the formula of Table 5(I-12).

10.6.3 *Stripping*

Introduction. The method of fragmentation is employed. The idea is to attain at the benzenoid $S(n-1)$ when starting from $S(n)$, i.e. the same kind only with the n parameter lowered by unity. The procedure leads to a recurrence relation and subsequent summation formula.

Inter-Related Classes. On basis of the single fragmentation procedure (see below) connections of a certain kind are obtained between the K numbers of some of the classes. Cyvin SJ, Cyvin and Gutman (1985) have listed all such connections for 5-tier strips. A complete listing for 3-, 4- and 5-tier strips is found in Table 6.

Single Fragmentation. In the simplest case of stripping the goal of producing $S(n-1)$ from $S(n)$ is achieved by one single fragmentation. The method was already encountered under chevrons; cf. Fig. 8.2 and the accompanying text, especially eqn. (8.19) therein. The application to $D(2,3,n)$ is shown in Fig. 5. It is deduced (compare with Table 6):

$$K\{D(2,3,n)\} = K\{D(2, 3, n-1)\} + K\{Z(4,n)\}; \qquad n \geq 1 \quad (10.8)$$

which is a recurrence relation. By means of the initial condition $K\{D(2,3,0)\} = K\{Z(4,0)\} = 1$ the following summation formula is obtained.

Table 10.6. Connections between K numbers of inter-related classes S_1 and S_2;

$$K\{S_2(n)\} = K\{S_1(n)\} - K\{S_1(n-1)\} \; ; \quad n \geq 1 , \qquad K\{S_1(n)\} = \sum_{i=0}^{n} K\{S_2(i)\}$$

t	$S_1(n)$	$S_2(n)$	t	$S_1(n)$	$S_2(n)$	$S_1(n)$	$S_2(n)$
3	$O(2,n)$	$Ch(2,n)$	5	$O(3,n)$	$D^j(3,n)$	$O(2,4,n)$	$D^j(2,4,n)$
	$Ch(2,n)$	$R^i(2,n)$		$D^j(3,n)$	$R^j(3,n)$	$D^j(2,4,n)$	$Ob(2,4,n)$
4	$O(2,3,n)$	$D(2,3,n)$		$D(3,n)$	$R(3,n)$		
	$D(2,3,n)$	$Z(4,n)$		$D^i(3,n)$	$Z(5,n)$	$D^i(2,4,n)$	$M_n(LLAAL)$
	$Ch(2,3,n)$	$\Sigma(2,3,n)$		$Ch(3,n)$	$\Sigma^j(3,n)$	$Ch(2,4,n)$	$\Sigma^j(2,4,n)$

$$K\{D(2,3,n)\} = \sum_{i=0}^{n} K\{Z(4,i)\} = \sum_{i=0}^{n}\left[\binom{i+2}{2}^2 - \binom{i+3}{4}\right] = \sum_{i=0}^{n}\binom{i+2}{2}^2 - \sum_{i=1}^{n}\binom{i+3}{4} \qquad (10.9)$$

Here it has been made use of a formula from Table 4(I-4). Eqn. (9) is also valid for $n=0$. The last summation is obtained straightforwardly as

$$\sum_{i=1}^{n}\binom{i+3}{4} = \binom{n+4}{5} \qquad (10.10)$$

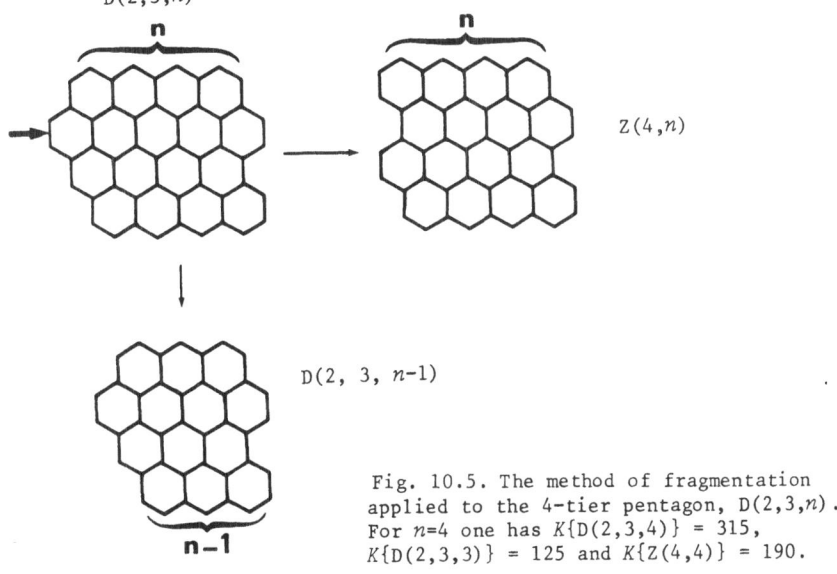

D(2,3,n)

Z(4,n)

D(2, 3, $n-1$)

Fig. 10.5. The method of fragmentation applied to the 4-tier pentagon, D(2,3,n). For $n=4$ one has $K\{D(2,3,4)\} = 315$, $K\{D(2,3,3)\} = 125$ and $K\{Z(4,4)\} = 190$.

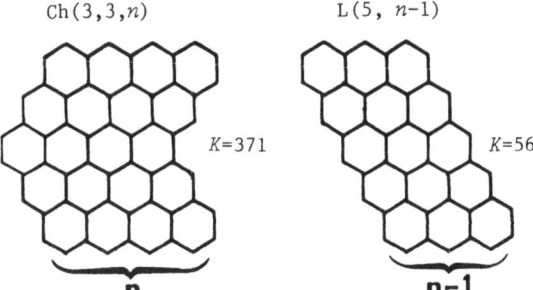

Fig. 10.6. Two of the benzenoids of eqn. (12). The given K numbers apply to $n=4$. The difference between them is the K number of D(2,3,4); cf. Fig. 10.5.

It is an example of point (ii) of Paragraph 10.9.1. The expression (10) may be identified with $K\{L(5, n-1)\}$ for $n \geq 1$. Also the first summation on the right-hand side of (10) may be identified with the K number of a class of benzenoids, namely chevrons. It is an exemplification of point (iv). Write the sum as

$$\sum_{i=0}^{n} \binom{i+2}{2}^2 = \sum_{i=0}^{n} \binom{3+i-1}{i}^2 \tag{10.11}$$

On comparison with formula (i) from CHART 8-II this is seen to be identical with $K\{Ch(3,3,n)\}$. Consequently

$$K\{D(2,3,n)\} = K\{Ch(3,3,n)\} - K\{L(5, n-1)\}; \qquad n \geq 1 \tag{10.12}$$

The benzenoids of this curious relation are found in Figs. 5 and 6. A combinatorial formula for $K\{D(2,3,n)\}$ is now easily found from (12) on application of formula (ii) from CHART 8-II; cf. also Table 5(I-5) and eqn. (6).

$$K\{D(2,3,n)\} = \binom{n+3}{5} - (n+1)\binom{n+3}{4} + \binom{n+2}{2}\binom{n+3}{3} - \binom{n+4}{5} \tag{10.13}$$

This equation is also valid for $n=0$. It reduces to

$$K\{D(2,3,n)\} = \binom{n+2}{2}\binom{n+3}{3} - (n+2)\binom{n+3}{4} \tag{10.14}$$

which is an alternative form of eqn. (4).

Multiple Fragmentation. The example of Fig. 7 shows how the step from $S(n)$ to $S(n-1)$ may be achieved through a double fragmentation. At the same time the exploitation of essentially disconnected benzenoids is demonstrated. It is deduced

$$K\{Z(4,n)\} = K\{Z(4, n-1)\} + K\{L(2,n)\} \cdot K\{L(n)\} + K\{Ch(2, 2, n-1)\}; \qquad n \geq 1 \tag{10.15}$$

By means of formula (ii) from CHART 8-II it was attained at:

$$K\{Z(4,n)\} - K\{Z(4, n-1)\} = \binom{n+2}{2}(n+1) - \binom{n+1}{3} + n\binom{n+1}{2}$$

$$= 2(n+1)\binom{n+2}{2} - (n+1)^2 - \binom{n+2}{3}; \qquad n \geq 1 \qquad (10.16)$$

Consequently, by means of $K\{Z(4,0)\} = 1$:

$$K\{Z(4,n)\} = \sum_{i=0}^{n}\left[2(i+1)\binom{i+2}{2} - (i+1)^2 - \binom{i+2}{3}\right] = 2K\{Ch(2,3,n)\} - K\{Ch(2,2,n)\} - \binom{n+3}{4}$$

$$= 2\left[(n+1)\binom{n+3}{3} - \binom{n+3}{4}\right] - \left[(n+1)\binom{n+2}{2} - \binom{n+2}{3}\right] - \binom{n+3}{4}$$

$$= 2(n+3)\binom{n+3}{3} - (n+2)\binom{n+2}{2} - 3\binom{n+4}{4} \qquad (10.17)$$

This derivation follows the same principles as those of eqns. (9)-(14). Eqn. (17) is

Z(4,*n*)

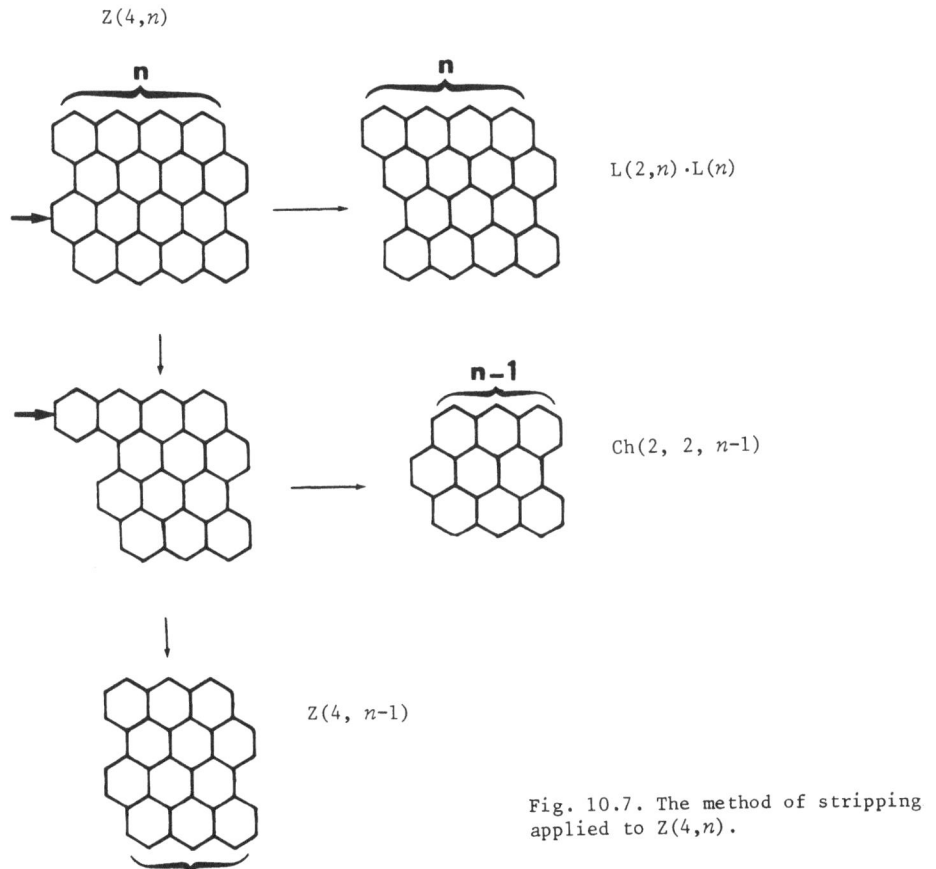

Fig. 10.7. The method of stripping applied to Z(4,*n*).

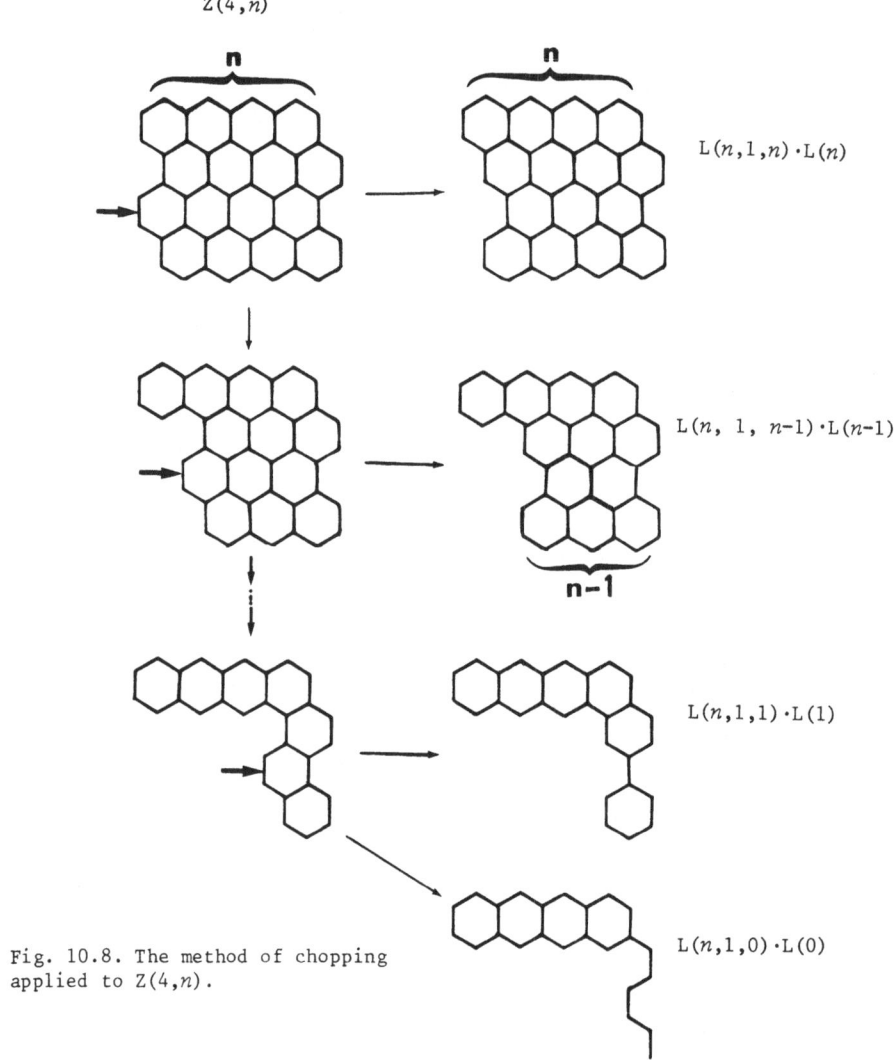

Fig. 10.8. The method of chopping applied to Z(4,n).

equivalent to the different forms given previously; cf. Chapter 9 and Table 4(I-4).

In the following we show instances where a summation is identified with a K number for a benzenoid other than chevron. A double fragmentation of the 5-tier intermediate rectangle, R(3,n); cf. Table 5(I-7) - has given

$$K\{R(3,n)\} - K\{R(3,\ n-1)\} = K\{Z(5,n)\} + K\{D(2,\ 3,\ n-1)\}; \qquad n \geq 1 \qquad (10.18)$$

Consequently

$$K\{R(3,n)\} = \sum_{i=0}^{n} K\{Z(5,i)\} + \sum_{i=1}^{n} K\{D(2,\ 3,\ i-1)\} \qquad (10.19)$$

The two summations are identified with the K numbers of different benzenoid classes according to Table 6:

$$K\{R(3,n)\} = K\{Di(3,3,n)\} + K\{O(2, 3, n-1)\}; \qquad n \geq 1 \qquad (10.20)$$

10.6.4 *Chopping*

The method of chopping is a systematic application of the fragmentation method, which leads to an addition of K numbers for essentially disconnected (or disconnected) benzenoid systems. Here we shall demonstrate a variant of the summation technique under point (iv) of Paragraph 10.6.1, which involves a difference between K numbers of two chevron classes. Consider the 4-tier zigzag chain, $Z(4,n)$, as in the preceding example. Let the first step of the fragmentation be identical with the one of Fig. 7. Then continue along the same row throughout. The resulting pattern is shown in Fig. 8 and is reflected in algebraic form by the equation

$$K\{Z(4,n)\} = \sum_{i=0}^{n} K\{L(n,1,i)\} \cdot K\{L(i)\} \qquad (10.21)$$

This is actually an example of eqn. (9.27) with $m=4$, $p=1$, $q=0$. By means of eqn. (9.13) or CHART 9-I, when realizing that $L(n,1,i) = A(n,1,i)$, one obtains

$$K\{Z(4,n)\} = \sum_{i=0}^{n} \left[(n+1)(i+1) - \binom{i+1}{2}\right](i+1) = (n+1)\sum_{i=0}^{n}(i+1)^2 - \sum_{i=1}^{n}\binom{i+1}{2}(i+1) \quad (10.22)$$

Here the first summation on the right-hand side may be treated as above; cf. (11);

$$\sum_{i=0}^{n}(i+1)^2 = \sum_{i=0}^{n}\binom{2+i-1}{i}^2 = K\{Ch(2,2,n)\} \qquad (10.23)$$

In a similar way the last summation of (22) is

$$\sum_{i=1}^{n}\binom{2+i-1}{i-1}\binom{2+i-1}{i} = K\{Ch(3,2,n)\} - K\{Ch(2,2,n)\}$$

$$= \binom{n+2}{4} - n\binom{n+2}{3} + \binom{n+1}{2}\binom{n+2}{2} \qquad (10.24)$$

in accord with eqns. (8.17) and (8.18). Here it is convenient to insert into eqn. (22) the difference between chevron K numbers rather than the final form of (24). Together with (23) it yields

$$K\{Z(4,n)\} = (n+2)K\{Ch(2,2,n)\} - K\{Ch(2,3,n)\} \qquad (10.25)$$

where the K numbers again may be worked out by eqn. (ii) of CHART 8-II.

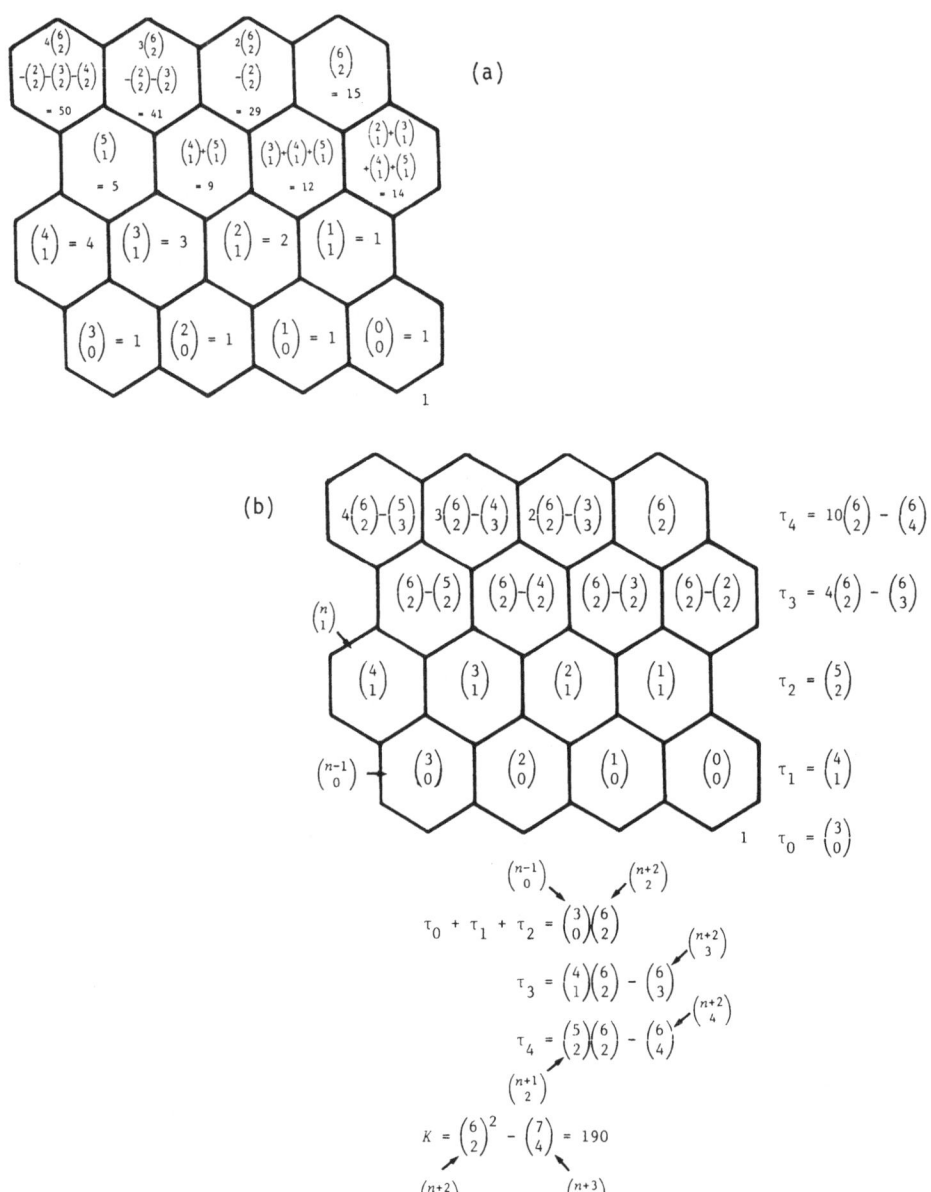

Fig. 10.9. The zigzag chain, $Z(4,n)$, with algorithm numerals: (a) integers and some expressions with binomial coefficients; (b) modified expressions in the two top rows. τ_0 is the external unity; τ_1, ..., τ_4 are the sums of numerals in the four respective rows. $K = \tau_0 + \ldots + \tau_4$ is the total sum of numerals, equal to the number of Kekulé structures.

10.6.5 *Addition of Algorithm Numerals*

The title method was demonstrated in Chapter 8 for chevrons (see Paragraph 8.3.4). Below we show briefly an application to the 4-tier zigzag chain. It is referred to Fig. 9, where $Z(4,n)$ with $n=4$ is depicted. The addition of numerals starts in the general case (arbitrary n) by

$$\tau_0 + \tau_1 + \tau_2 = \binom{n-1}{0} + \binom{n}{1} + \binom{n+1}{2} = \binom{n+2}{2} = \binom{n-1}{0}\binom{n+2}{2} \qquad (10.26)$$

Furthermore,

$$\tau_3 = \binom{n}{1}\binom{n+2}{2} - \binom{n+2}{3}, \qquad \tau_4 = \binom{n+1}{2}\binom{n+2}{2} - \binom{n+2}{4} \qquad (10.27)$$

Hence

$$K\{Z(4,n)\} = \sum_{i=0}^{4} \tau_i = \left[\binom{n-1}{0} + \binom{n}{1} + \binom{n+1}{2}\right]\binom{n+2}{2} - \left[\binom{n+2}{3} + \binom{n+2}{4}\right]$$

$$= \binom{n+2}{2}^2 - \binom{n+3}{4} \qquad (10.28)$$

The formula is identical with previous results.

10.7 THE 4-TIER ZIGZAG CHAIN

The benzenoid class $Z(4,n)$ has been used extensively to illustrate the different methods of this chapter. In Table 7 the formula of $K\{Z(4,n)\}$ is given in a number of different representations.

The benzenoid $Z(4,n)$ is a hexagon without two corners, $Ob(2,3,n)$; cf. Section 8.2. That gives rise to the form (a) of Table 7. Eqns. (b) and (c) are identical with (9.17) and (9.16), respectively. Also the addition of algorithm numerals (Paragraph 10.6.5) leads naturally to the form (c). Many different representations in terms of binomial coefficients are possible. Eqn. (17) is one of them; it is given as (d) in Table 7. A simpler form is found under (e), while (f) contains only one kind of binomial coefficients. If allowance is made for multiples of n in the binomial coefficients one may produce representations like (g). The forms (h) and (i) were obtained by the fully computerized method (cf. Section 4.10). The factored and unfactored polynomials are given as (j) and (k), respectively. Eqn. (ℓ) is given in terms of the powers of $(n+1)$. Finally in eqn. (m) we made the characteristic term $(2n + 3)$ visible; it is a common factor for $K\{Z(m,n)\}$ with odd integers $m \geq 3$, but not here (cf. Section 9.7).

Table 10.7. The formula of $K\{Z(4,n)\}$ in different representations.

$Z(4,n)$

Form	$K\{Z(4,n)\}$

(a) $\quad \frac{1}{4}\left[\binom{n+1}{3}\binom{n+2}{3} - 2\binom{n+2}{3}\binom{n+3}{3} + \binom{n+3}{3}\binom{n+4}{3}\right]$

(b) $\quad (n+1)^2\binom{n+2}{2} - 2(n+1)\binom{n+2}{3} + \binom{n+2}{4}$

(c) $\quad \binom{n+2}{2}^2 - \binom{n+3}{4}$

(d) $\quad 2(n+3)\binom{n+3}{3} - (n+2)\binom{n+2}{2} - 3\binom{n+4}{4}$

(e) $\quad (n+2)\binom{n+2}{3} + \binom{n+4}{4}$

(f) $\quad \frac{1}{6}\binom{n+2}{2}\left[5\binom{n+2}{2} + 1\right]$

(g) $\quad \frac{1}{2}\left[\frac{1}{6}\binom{2n+3}{2}\binom{2n+4}{2} - \binom{n+2}{2}^2\right]$

(h) $\quad \frac{1}{12}(n+1)(n+2)\left[5\binom{n}{2} + 10n + 6\right]$

(i) $\quad 5\binom{n}{4} + 15\binom{n}{3} + 16\binom{n}{2} + 7n + 1$

(j) $\quad \frac{1}{24}(n+1)(n+2)(5n^2 + 15n + 12)$

(k) $\quad \frac{1}{24}(5n^4 + 30n^3 + 67n^2 + 66n + 24)$

(ℓ) $\quad \frac{1}{24}(n+1)[5(n+1)^3 + 10(n+1)^2 + 7(n+1) + 2]$

(m) $\quad \frac{1}{24}(n+1)(n+2)[2(2n + 3)^2 - 3(n+1)(n+2)]$

11.1 INTRODUCTION

This is a continuation of the account of some classes of basic benzenoids; for the previous parts, see Chapters 8 and 9.

The main classes considered here are: pentagons (prolate, problate and oblate), triangles (prolate, problate and oblate), streamers and goblets.

11.2 PENTAGONS

11.2.1 *Definitions*

A class of pentagon-shaped benzenoids or pentagons has the main symbol D; cf. Chapter 10. Also according to the definitions in that chapter the prolate and oblate pentagons belong to straight regular strips. One of the rims (conventionally the left) is a protruding two-segment chain, while the other rim (conventionally the right) is indented either inwards (prolate) or outwards (oblate). When each segment of the two-segment chain holds m hexagons the prolate and oblate pentagons are denoted by $D^i(m,n)$ and $D^j(m,n)$, respectively. They are $(2m-1)$-tier strips. The cases of 5-tier strips ($m=3$) are found as No. I-4 (D^i) and No. I-2 (D^j) in Table 10.5. Figure 1 shows $D^i(4,3)$ and $D^j(3,3)$.

In addition, the "problate" pentagon is defined as a system in-between $D^i(m+1,\ n)$ and $D^j(m,n)$. It is designated D (without superscript), but all three

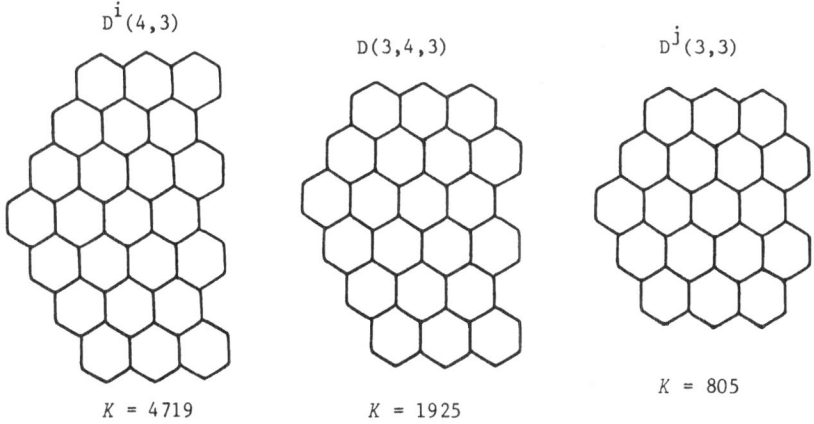

$$D^i(4,3)$$
$$D(3,4,3)$$
$$D^j(3,3)$$

$$K = 4719 \qquad K = 1925 \qquad K = 805$$

Fig. 11.1. Examples of a prolate (D^i), problate (D) and oblate (D^j) pentagon. K numbers are given.

CHART 11-I. Prolate pentagon

$$D^i(m,n)$$

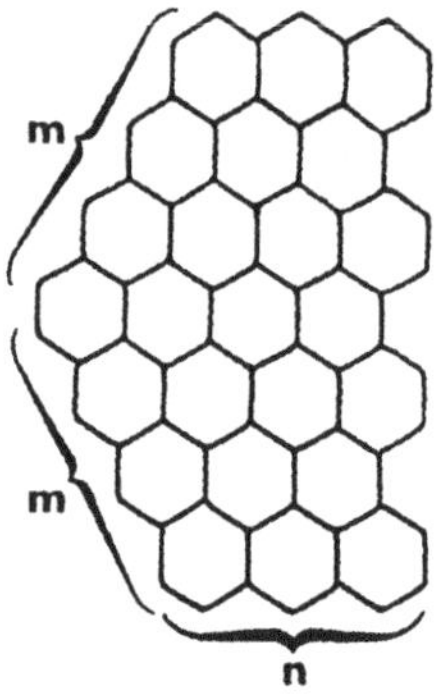

$$^aK\{D^i(m,n)\} = \prod_{i=1}^{n} \frac{\binom{2m+2i}{m}}{\binom{m+2i-1}{m}} \qquad (n > 0) \qquad \text{(i)}$$

$$^aK\{D^i(m,n)\} = \begin{cases} \prod_{i=1}^{\frac{m}{2}} \dfrac{\binom{m+2n+2i}{m+1}}{\binom{m+2i}{m+1}} ; & m = 2, 4, 6, \ldots. \\[2em] \prod_{i=1}^{\frac{m+1}{2}} \dfrac{\binom{m+2n+2i-1}{m}}{\binom{m+2i-1}{m}} ; & m = 1, 3, 5, \ldots. \end{cases} \qquad \text{(ii)}$$

[a] Cyvin SJ, Cyvin BN (1987). J Mol Struct (Theochem) 152: 347

parameters are included, the two first of them (m, $m+1$) always differing by unity. It means that the two linear segments of the not indented rim differ in their lenghts by one hexagon. D(m, $m+1$, n) is a skew regular $2m$-tier strip; cf. Fig. 1.

Non-degenerate ($m > 1$) prolate pentagons are mirror-symmetrical (C_{2v}). The same is the case for oblate pentagons with $m > 2$. Non-degenerate problate pentagons are unsymmetrical (C_s).

11.2.2 Prolate Pentagon

Previous Work and General Formulas. The prolate pentagon $D^i(2,n)$ coincides with the 3-tier chevron, Ch(2,n); cf. Chapters 8 and 10. The K formula for the class of 5-tier prolate pentagons, $D^i(3,n)$, was given by Cyvin SJ, Cyvin and Gutman (1985); see also Chapter 10. Cyvin SJ and Cyvin (1986b) derived the K formula for $D^i(4,n)$ and conjectured the corresponding formula for $D^i(5,n)$. The general formulas of $K\{D^i(m,n)\}$ (see CHART I) were presented by Cyvin SJ and Cyvin (1987a). Both forms (i) and (ii) are useful for different purposes since the parameters m and n are not equivalent; they can not be permuted.

In the trivial case of m or n equal to zero one has $K\{D^i(0,n)\}=K\{D^i(m,0)\}=1$.

For $m=1$ and $m=2$ the K formulas may be written:

$$K\{D^i(1,n)\} = n + 1, \qquad K\{D^i(2,n)\} = \frac{(n+1)(n+2)(2n+3)}{1\cdot 2\cdot 3}$$

The general formulas (for $m > 2$) in the same style read:

$$K\{D^i(m,n)\}$$

$$= \left(\frac{2n+m+1}{m+1}\right)^{\frac{m}{2}} \prod_{i=1}^{\frac{m}{2}} \left[\frac{(n+i)(n+m-i+1)}{i(m-i+1)}\right]^i \prod_{i=1}^{\frac{m}{2}-1} \left[\frac{(2n+2i+1)(2n+2m-2i+1)}{(2i+1)(2m-2i+1)}\right]^i ;$$

$$m = 4, 6, 8, \ldots ,$$

$$K\{D^i(m,n)\} = \left(\frac{2n+m+1}{m+1}\right)^{\frac{m+1}{2}} \prod_{i=1}^{\frac{m-1}{2}} \left[\frac{(n+i)(n+m-i+1)(2n+2i+1)(2n+2m-2i+1)}{i(m-i+1)(2i+1)(2m-2i+1)}\right]^i ;$$

$$m = 3, 5, 7, \ldots \qquad (11.1)$$

Applications of (1) for some few m values are given below.

$$K\{D^i(3,n)\} = \frac{(n+1)(n+2)^2(n+3)(2n+3)(2n+5)}{1\cdot 2^2\cdot 3\cdot 3\cdot 5}$$

$$K\{D^i(4,n)\} = \frac{(n+1)(n+2)^2(n+3)^2(n+4)(2n+3)(2n+5)^2(2n+7)}{1\cdot 2^2\cdot 3^2\cdot 4\cdot 3\cdot 5^2\cdot 7}$$

$$K\{D^i(5,n)\} = \frac{(n+1)(n+2)^2(n+3)^3(n+4)^2(n+5)(2n+3)(2n+5)^2(2n+7)^2(2n+9)}{1\cdot 2^2\cdot 3^3\cdot 4^2\cdot 5\cdot 3\cdot 5^2\cdot 7^2\cdot 9}$$

Numerical K values for prolate pentagons are found in Table 1.

Table 11.1. Numerical values of $K\{D^i(m,n)\}$.

m \ n	1	2	3	4	5	6	7	8	9	10
1	2	3	4	5	6	7	8	9	10	11
2	5	14	30	55	91	140	204	285	385	506
3	14	84	330	1001	2548	5712	11628	21945	38962	65780
4	42	594	4719	26026	111384	395352	1215126	3331251	8321170	19240650
5	132	4719	81796	884884	6852768	41314284	204951252	869562265		
6	429	40898	1643356	37119160	553361016					
7	1430	379236	37119160	1844536720						
8	4862	3711916	922268360							
9	16796	37975756								
10	58786	403127256								

$D^i(m,n)$

Quotient Recurrence Relations. Below we give the recurrence relations in terms of quotients as derived from the formulas of CHART I.

$$\frac{K\{D^i(m,n)\}}{K\{D^i(m-1,\ n)\}} = \frac{\binom{2m+2n}{m}}{\binom{2m}{m}} = \prod_{i=1}^{m} \frac{m+2n+i}{m+i}\ ; \qquad m \geq 1 \qquad (11.2)$$

$$\frac{K\{D^i(m,n)\}}{K\{D^i(m,\ n-1)\}} = \frac{\binom{2m+2n}{m}}{\binom{m+2n-1}{m}} = \prod_{i=1}^{m} \frac{m+2n+i}{2n+i-1}\ ; \qquad n \geq 1 \qquad (11.3)$$

The product form of (3) is not adequate for $m=0$, in which case the quotient is equal to 1.

Prolate Pentagon Without Apex or Without Corner. The protruding (L_3-mode) hexagon, which connects the two segments of the not indented rim, is referred to as the apex. The prolate pentagon without apex (cf. CHART II) is denoted $Da^i(m,n)$. For this benzenoid system one has (cf. Fig. 2)

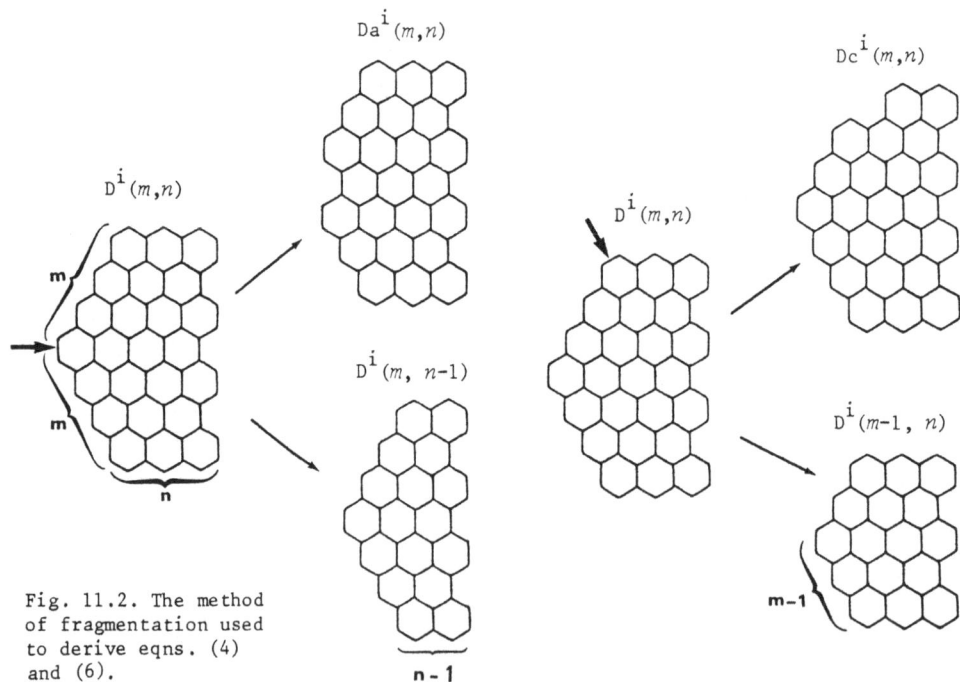

$Da^i(m,n)$

$Dc^i(m,n)$

$D^i(m,n)$

$D^i(m,n)$

$D^i(m,\ n-1)$

$D^i(m-1,\ n)$

m

m

n

$m-1$

$n-1$

Fig. 11.2. The method of fragmentation used to derive eqns. (4) and (6).

$$K\{Da^i(m,n)\} = K\{D^i(m,n)\} - K\{D^i(m,\ n-1)\}; \qquad n \geq 1 \qquad (11.4)$$

On combining with the preceding formulas it was attained at:

$$K\{Da^i(m,n)\} = \left[1 - \frac{\binom{m+2n-1}{m}}{\binom{2m+2n}{m}} \right] K\{D^i(m,n)\} \qquad (11.5)$$

Results for $K\{Da^i(m,n)\}$ with some fixed values of m are collected in CHART II.

A pentagon-shaped benzenoid has also L_3-mode hexagons at the ends of the two-segment rim. These are referred to as corners. A prolate pentagon without corner,

CHART 11-II. Prolate pentagon without apex, and prolate pentagon without corner

$Da^i(m,n)$ $Dc^i(m,n)$

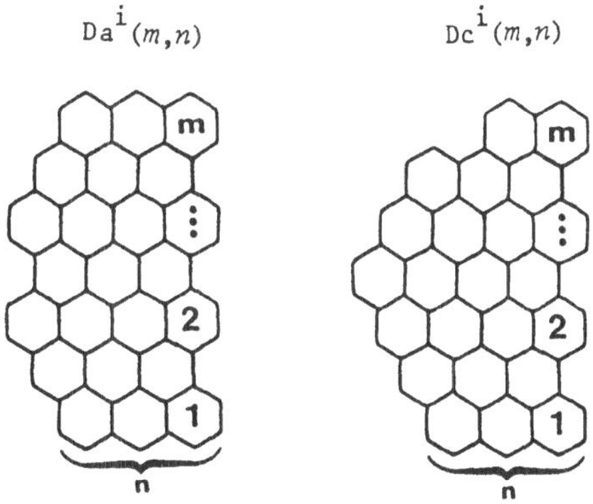

$K\{Da^i(1,n)\} = 1$

[a]$K\{Da^i(2,n)\} = (n+1)^2$

[b]$K\{Da^i(3,n)\} = \frac{1}{30}(n+1)(n+2)(2n+3)(2n^2+6n+5)$

[c]$K\{Da^i(4,n)\} = \frac{1}{7560}(n+1)(n+2)^3(n+3)(2n+3)(2n+5)(4n^2+16n+21)$

$K\{Da^i(5,n)\} = \frac{1}{19051200}(n+1)(n+2)^2(n+3)^2(n+4)(2n+3)(2n+5)^2(2n+7)(4n^4+40n^3$
$$+ 165n^2 + 325n + 252)$$

$K\{Dc^i(1,n)\} = n$

[d]$K\{Dc^i(2,n)\} = \frac{1}{6}n(n+1)(2n+7)$

$K\{Dc^i(3,n)\} = \frac{1}{180}n(n+1)(n+2)(2n+3)(2n^2+15n+37)$

$K\{Dc^i(4,n)\} = \frac{1}{75600}n(n+1)(n+2)^2(n+3)(2n+3)(2n+5)(2n+13)(2n^2+13n+41)$

$K\{Dc^i(5,n)\} = \frac{1}{285768000}n(n+1)(n+2)^2(n+3)^2(n+4)(2n+3)(2n+5)^2(2n+7)(4n^4$
$$+ 80n^3 + 635n^2 + 2500n + 4881)$$

[a]$Da^i(2,n) = R^i(2,n)$

[b]$Da^i(3,n) = Z(5,n)$

[c]Cyvin SJ, Cyvin BN (1986). Match 21: 295; correction: Cyvin BN, Cyvin SJ (1987). Match 22: 157

[d]$Dc^i(2,n) = Cc(2,2,n)$

$Dc^i(m,n)$, is shown in CHART II. One has (cf. Fig. 2)

$$K\{Dc^i(m,n)\} = K\{D^i(m,n)\} - K\{D^i(m-1, n)\}; \qquad m \geq 1 \qquad (11.6)$$

Similarly to eqn. (5) one obtains:

$$K\{Dc^i(m,n)\} = \left[1 - \frac{\binom{2m}{m}}{\binom{2m+2n}{m}} \right] K\{D^i(m,n)\} \qquad (11.7)$$

CHART 11-III. Problate pentagon

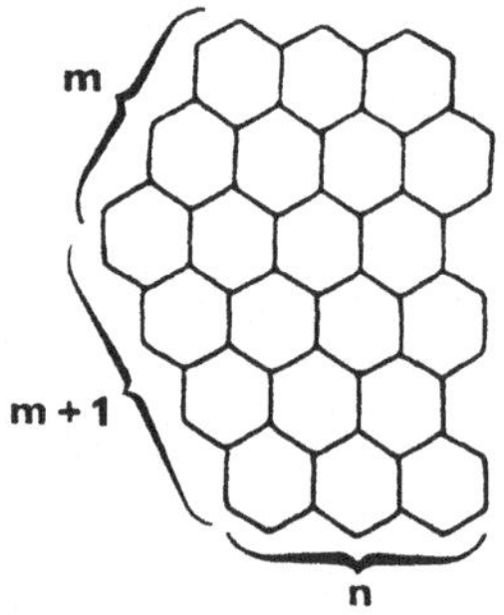

m

$D(m, m+1, n)$

m + 1

n

[a]$K\{D(1,2,n)\} = \frac{1}{2} (n+1)(n+2)$

[b]$K\{D(2,3,n)\} = \frac{1}{24} (n+1)(n+2)^3(n+3)$

$K\{D(3,4,n)\} = \frac{1}{4320} (n+1)(n+2)^3(n+3)^3(n+4)(2n + 5)$

$K\{D(4,5,n)\} = \frac{1}{7257600} (n+1)(n+2)^3(n+3)^4(n+4)^3(n+5)(2n + 5)(2n + 7)$

---

[a]$D(1,2,n) = L(2,n)$

[b]Cyvin SJ (1986). Monatsh Chem 117: 33

11.2.3 *Problate Pentagon*

A problate pentagon is actually a two-parameter system although written as $D(m, m+1, n)$. In CHART III the combinatorial formulas for some fixed values of m are given. For the trivial case of $D(0,1,n)$ one has $K=1$.

We may define a problate pentagon without apex (Da) and two types of problate pentagons without corner (Dc and Dc') as shown in Fig. 3. In Dc and Dc' the corner is deleted from the short or long segment, respectively, of the two-segment rim. For these systems one has:

$$K\{Da(m, m+1, n)\} = K\{D(m, m+1, n)\} - K\{D(m, m+1, n-1)\}; \qquad n \geq 1 \qquad (11.8)$$

$$K\{Dc(m, m+1, n)\} = K\{D(m, m+1, n)\} - K\{D^i(m,n)\}; \qquad (11.9)$$

$$K\{Dc'(m, m+1, n)\} = K\{D(m, m+1, n)\} - K\{D(m-1, m, n)\}; \qquad m \geq 1 \qquad (11.10)$$

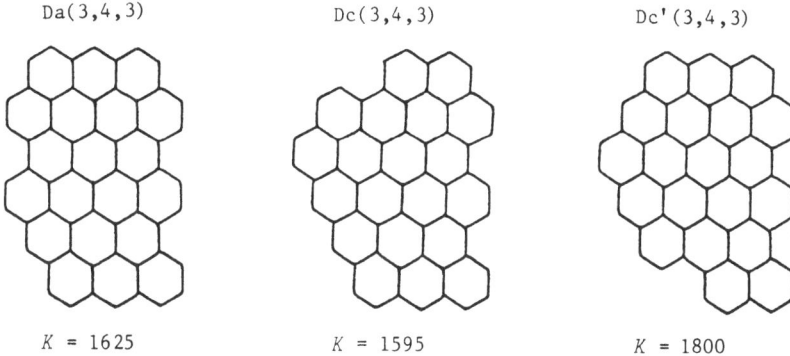

Da(3,4,3)   Dc(3,4,3)   Dc'(3,4,3)

$K = 1625$ $K = 1595$ $K = 1800$

Fig. 11.3. Problate pentagon without apex (Da) and two problate pentagons without corner (Dc and Dc').

11.2.4 *Oblate Pentagon*

Previous Work. The oblate pentagon $D^j(2,n)$ coincides with the 3-tier hexagon, $O(2,n)$; cf. Chapters 8 and 10. The formula for $K\{D^j(3,n)\}$, the number of Kekulé structures for an oblate 5-tier pentagon, was given by Cyvin SJ, Cyvin and Gutman (1985); see also Chapter 10. The K formula for $D^j(4,n)$ was derived by Cyvin BN, Cyvin and Brunvoll (1986) using the fully computerized method (cf. Paragraph 4.10.6).

Combinatorial K Formulas for $D^j(m,n)$. CHART IV shows the combinatorial formulas of $K\{D^j(m,n)\}$ for some fixed values of m. For the trivial case of $m=0$:

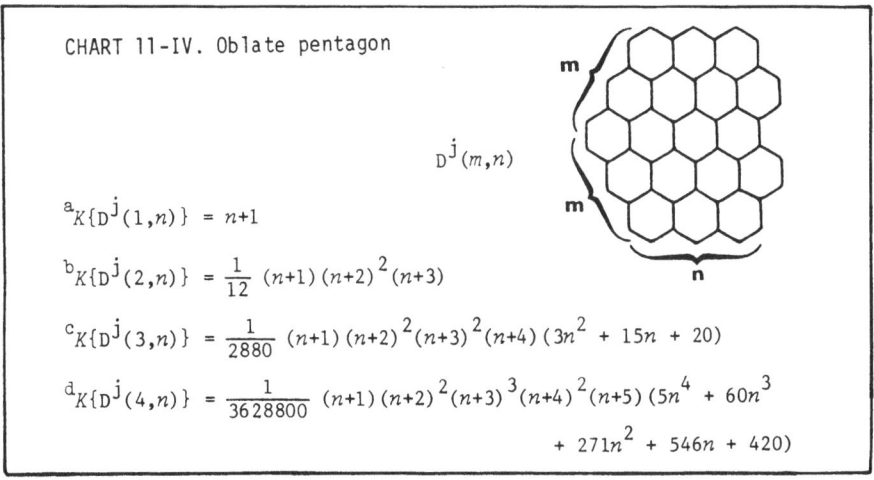

CHART 11-IV. Oblate pentagon

$$D^j(m,n)$$

$$^a K\{D^j(1,n)\} = n+1$$

$$^b K\{D^j(2,n)\} = \frac{1}{12}(n+1)(n+2)^2(n+3)$$

$$^c K\{D^j(3,n)\} = \frac{1}{2880}(n+1)(n+2)^2(n+3)^2(n+4)(3n^2 + 15n + 20)$$

$$^d K\{D^j(4,n)\} = \frac{1}{3628800}(n+1)(n+2)^2(n+3)^3(n+4)^2(n+5)(5n^4 + 60n^3 + 271n^2 + 546n + 420)$$

[a] $D^j(1,n) = L(n)$
[b] Gordon M, Davison WHT (1952). J Chem Phys 20: 428; also $O(2,n)$
[c] Cyvin SJ, Cyvin BN, Gutman I (1985). Z Naturforsch 40a: 1253
[d] Cyvin BN, Cyvin SJ, Brunvoll J (1986). Match 21: 291

Table 11.2. Numerical values of $K\{D^j(m,n)\}$.

m \ n	1	2	3	4	5	6	7	8	9	10
1	2	3	4	5	6	7	8	9	10	11
2	6	20	50	105	196	336	540	825	1210	1716
3	19	155	805	3136	9996	27468	67320	150645	313027	611611
4	62	1315	15218	118188	690480	3256308	12991770	45316557	141547978	403129727
5	207	11957	326053	5355756	60864012	522261531				
6	704	114972	7736092	282497568						
7	2431	1157650	199806100							
8	8502	12115220								
9	30056	131015260								
10	107236									

$D^j(m,n)$

$$K\{D^j(0,n)\} = 1.$$

In Table 2 some numerical K values for oblate pentagons are listed.

Oblate Pentagon Without Apex or Without Corner. The definitions of an oblate pentagon without apex (Da^j) and an oblate pentagon without corner (Dc^j) are apparent from CHART V. By the method of fragmentation it is easily found:

$$K\{Da^j(m,n)\} = K\{D^j(m,n)\} - K\{D^j(m,\ n-1)\}; \qquad n \geq 1 \qquad (11.11)$$

$$K\{Dc^j(m,n)\} = K\{D^j(m,n)\} - K\{D(m-1,\ n)\}; \qquad m \geq 1 \qquad (11.12)$$

CHART V shows the combinatorial formulas for $K\{Da^j(m,n)\}$ and $K\{Dc^j(m,n)\}$ with some fixed values of m.

11.2.5 *Mirror-Symmetrical Seven-Tier Pentagons*

Introduction and Listing of Formulas. Many other types of pentagons besides the prolate, problate and oblate pentagons are covered by the present definition (Paragraph 11.2.1). Some additional examples are found among the five-tier strips listed in Table 10.5. In the present paragraph we deal with the mirror-symmetrical pentagons among straight seven-tier strips and the corresponding pentagons without apices (defined in analogy with the preceding definitions of Da^i and Da^j). In addition to the prolate (D^i) and oblate (D^j) pentagons there are four pentagons of the title category; here they are denoted D^k, D^l, D^m and D^n.

In CHART VI the derived combinatorial formulas for the four classes of pentagons in question are given, and also for the corresponding pentagons without apices.

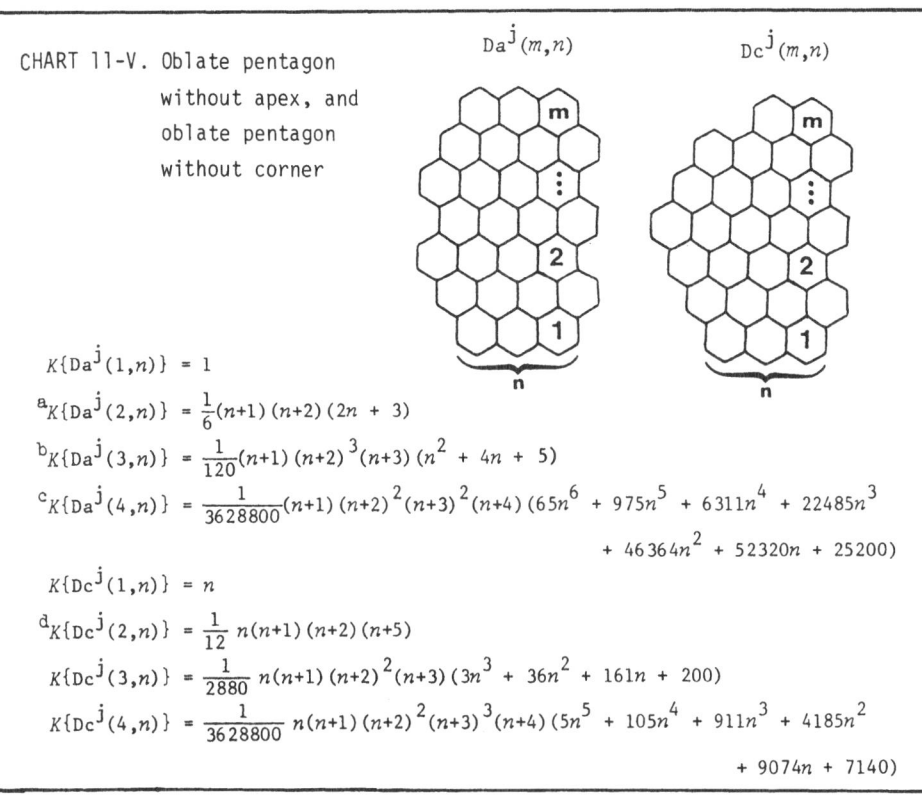

CHART 11-V. Oblate pentagon
without apex, and
oblate pentagon
without corner

$Da^j(m,n)$ $Dc^j(m,n)$

$K\{Da^j(1,n)\} = 1$

[a]$K\{Da^j(2,n)\} = \frac{1}{6}(n+1)(n+2)(2n+3)$

[b]$K\{Da^j(3,n)\} = \frac{1}{120}(n+1)(n+2)^3(n+3)(n^2+4n+5)$

[c]$K\{Da^j(4,n)\} = \frac{1}{3628800}(n+1)(n+2)^2(n+3)^2(n+4)(65n^6+975n^5+6311n^4+22485n^3$

$+ 46364n^2 + 52320n + 25200)$

$K\{Dc^j(1,n)\} = n$

[d]$K\{Dc^j(2,n)\} = \frac{1}{12} n(n+1)(n+2)(n+5)$

$K\{Dc^j(3,n)\} = \frac{1}{2880} n(n+1)(n+2)^2(n+3)(3n^3+36n^2+161n+200)$

$K\{Dc^j(4,n)\} = \frac{1}{3628800} n(n+1)(n+2)^2(n+3)^3(n+4)(5n^5+105n^4+911n^3+4185n^2$

$+ 9074n + 7140)$

[a] Gordon M, Davison WHT (1952). J Chem Phys 20: 428; also Ch(2,n)
[b] Cyvin SJ, Cyvin BN, Gutman I (1985). Z Naturforsch 40a: 1253; also $R^j(3,n)$
[c] Cyvin BN, Cyvin SJ (1987). Match 22: 157
[d] Cyvin SJ (1986). Monatsh Chem 117: 33; $Dc^j(2,n) = Oa(2,n,2)$

Derivation of the K Formulas for D^k *and* Da^k. As an example we will show the
derivation of $K\{D^k(4,n)\}$ and $K\{Da^k(4,n)\}$. These K numbers are inter-related by

$$K\{Da(4,n)\} = K\{D^k(4,n)\} - K\{D^k(4, n-1)\}; \qquad n \geq 1 \qquad (11.13)$$

and

$$K\{D^k(4,n)\} = \sum_{i=0}^{n} K\{Da^k(4,i)\} \qquad (11.14)$$

We start with the derivation of $K\{Da^k(4,n)\}$ according to the method of chopping; cf.
Paragraph 10.6.4. The present example is somewhat more advanced than the one of the
mentioned paragraph (Fig. 10.8).

The fragmentation scheme shown in Fig. 4 leads to the following fundamental
relation.

CHART 11-VI. Some mirror-symmetrical seven-tier pentagons and the corresponding pentagons without apex

$D^k(4,n)$ $D^l(4,n)$ $D^m(4,n)$ $D^n(4,n)$

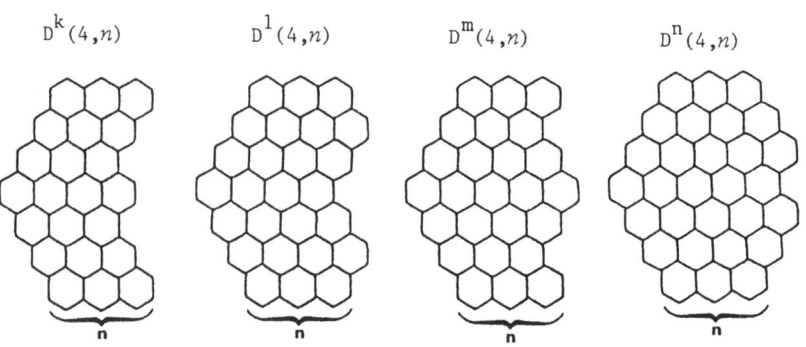

$^aK\{D^k(4,n)\} = \dfrac{1}{20160}(n+1)(n+2)^2(n+3)(n+4)(45n^3 + 320n^2 + 671n + 420)$

$^aK\{D^l(4,n)\} = \dfrac{1}{907200}(n+1)(n+2)^2(n+3)^2(n+4)^2(n+5)(5n^4 + 60n^3 + 259n^2 + 474n + 315)$

$^aK\{D^m(4,n)\} = \dfrac{1}{604800}(n+1)(n+2)^2(n+3)^3(n+4)^2(n+5)(2n + 5)(9n + 14)$

$^bK\{D^n(4,n)\} = \dfrac{1}{108864000}(n+1)(n+2)^2(n+3)^3(n+4)^3(n+5)^2(n+6)(2n + 7)(n^2 + 7n + 15)$

$Da^k(4,n)$ $Da^l(4,n)$ $Da^m(4,n)$ $Da^n(4,n)$

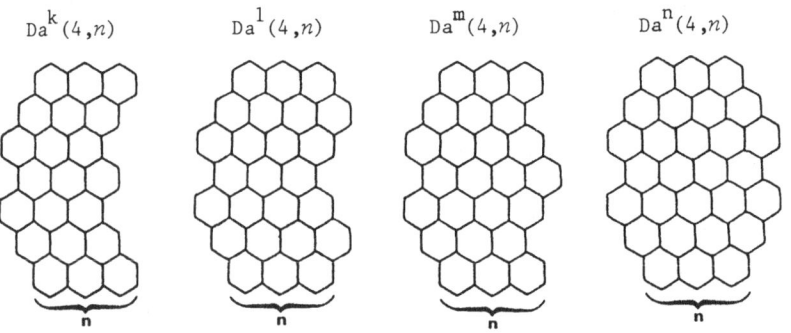

$^{a,c}K\{Da^k(4,n)\} = \dfrac{1}{2520}(n+1)(n+2)(n+3)(45n^4 + 325n^3 + 852n^2 + 983n + 420)$

$^aK\{Da^l(4,n)\} = \dfrac{1}{30240}(n+1)(n+2)^3(n+3)^3(n+4)(2n + 5)(n^2 + 5n + 7)$

$^aK\{Da^m(4,n)\} = \dfrac{1}{302400}(n+1)(n+2)^2(n+3)^2(n+4)(99n^4 + 815n^3 + 2616n^2 + 3820n + 2100)$

$^dK\{Da^n(4,n)\} = \dfrac{1}{3628800}(n+1)(n+2)^2(n+3)^4(n+4)^2(n+5)(n^2 + 6n + 10)(n^2 + 6n + 14)$

[a]Cyvin BN, Cyvin SJ (1987). Match 22: 157
[b]$D^n(4,n) = Oa(4,n)$
[c]$Da^k(4,n) = M_n(L^2A^3L^2)$
[d]$Da^n(4,n) = Ob(4,n)$

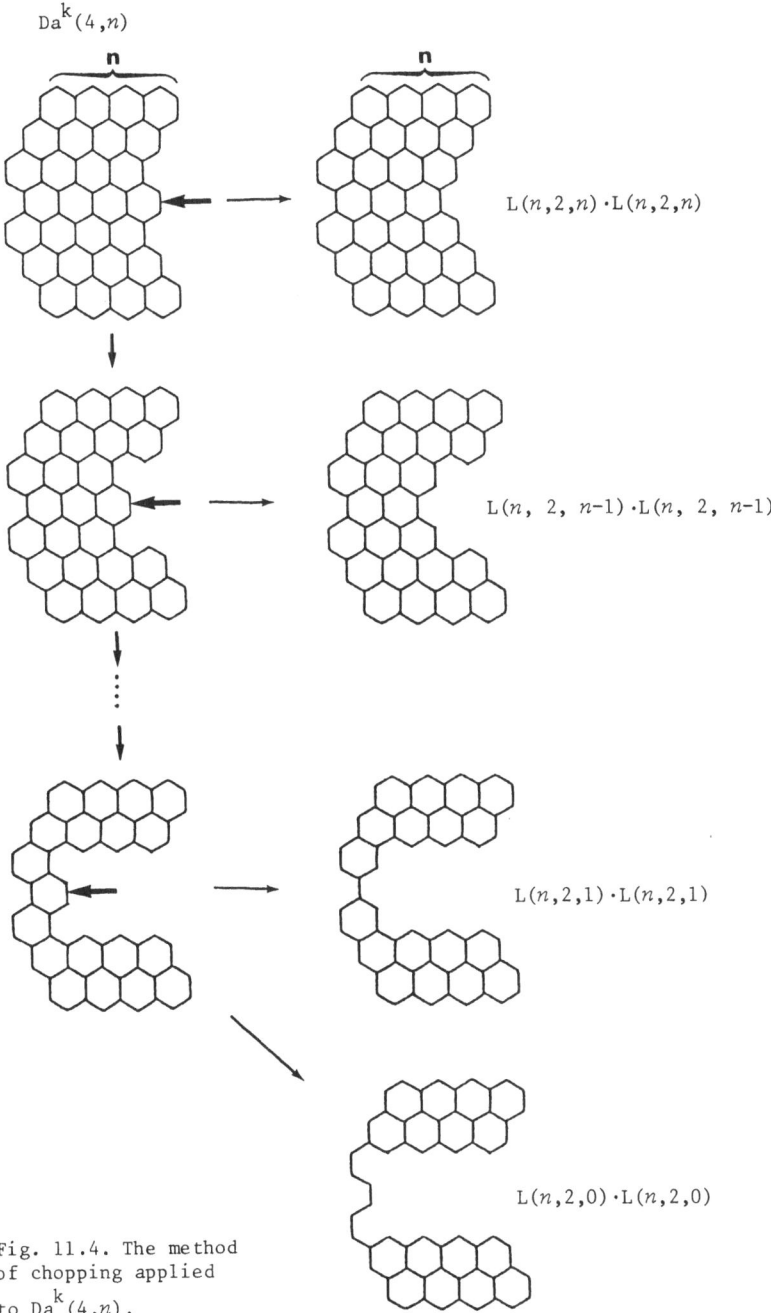

Fig. 11.4. The method
of chopping applied
to $Da^k(4,n)$.

$$K\{Da^{k}(4,n)\} = \sum_{i=0}^{n}[K\{L(n,2,i)\}]^{2} = \sum_{i=0}^{n}\left[\binom{n+3}{3} - \binom{n-i+2}{3}\right]^{2} \tag{11.15}$$

where we have applied eqn. (i) of CHART 8-V. On substituting the summation index by $j = n-i$ and taking the terms in the reverse order it is obtained

$$K\{Da^{k}(4,n)\} = \sum_{j=0}^{n}\left[\binom{n+3}{3} - \binom{j+2}{3}\right]^{2}$$

$$= \binom{n+3}{3}^{2}\sum_{j=0}^{n}1 - 2\binom{n+3}{3}\sum_{j=1}^{n}\binom{j+2}{3} + \sum_{j=1}^{n}\binom{j+2}{3}^{2} \tag{11.16}$$

Here only the last summation presents some difficulties. It may be expanded according to the refined applications of chevron formulas, as was exemplified at several
occasions in Section 10.6. Here we obtain

$$\sum_{j=1}^{n}\binom{j+2}{3}^{2} = \sum_{i=0}^{n-1}\binom{i+3}{i}^{2} = \sum_{i=0}^{n}\binom{4+i-1}{i}^{2} - \binom{n+3}{3}^{2} = K\{Ch(4,n)\} - \binom{n+3}{3}^{2} \tag{11.17}$$

The chevron formula (ii) of CHART 8-II was applied, and the result was simplified to

$$\sum_{j=1}^{n}\binom{j+2}{3}^{2} = \binom{n+3}{3}\binom{n+3}{4} - \binom{n+3}{2}\binom{n+4}{5} + (n+3)\binom{n+5}{6} - \binom{n+6}{7} \tag{11.18}$$

Eqn. (16) was finally expanded into the following form with the aid of (18).

$$K\{Da^{k}(4,n)\} = (n+2)\binom{n+3}{3}^{2} - \binom{n+3}{3}\binom{n+4}{4} - \binom{n+3}{2}\binom{n+4}{5} + (n+3)\binom{n+5}{6} - \binom{n+6}{7} \tag{11.19}$$

This formula is equivalent to the polynomial form of CHART VI.

The formula of $K\{D^{k}(4,n)\}$ was derived from eqn. (14) after inserting the result for $K\{Da^{k}(4,n)\}$ from CHART VI. The direct expansion of summations was employed,
as also is exemplified in Paragraph 10.6.2. An elementary, but tedious calculation
leads to

$$K\{D^{k}(4,n)\} = \frac{1}{2520}\left[45\sum_{i=0}^{n}(i+1)^{7} + 280\sum_{i=0}^{n}(i+1)^{6} + 672\sum_{i=0}^{n}(i+1)^{5}\right.$$

$$\left. + 805\sum_{i=0}^{n}(i+1)^{4} + 525\sum_{i=0}^{n}(i+1)^{3} + 175\sum_{i=0}^{n}(i+1)^{2} + 18\sum_{i=0}^{n}(i+1)\right] \tag{11.20}$$

and finally the answer which was entered into CHART VI.

Cyvin BN and Cyvin (1987) have given the numerical values of K for the pentagons $D^{k}(4,n)$, $D^{1}(4,n)$ and $D^{m}(4,n)$ up to $n=10$, along with the corresponding data for
the pentagons without apices.

11.2.6 *Auxiliary Benzenoid Classes*

Some auxiliary classes shall be defined in connection with the pentagons. They are in many ways analogous to the zigzag chains augmented by a row; cf. Section 9.3.

Let $C(n,m,l)$ denote a problate or oblate pentagon augmented by a row of l hexagons where $0 \leq l \leq n$, as shown in Fig. 5. Notice that a problate pentagon is augmented at the shorter segment (cf. the left-hand part of Fig. 5). Here the pentagon to be augmented is an m-tier strip, which (by definition) is obtained as a special case for $l=0$:

$$C(n,\; 2p,\; 0) = D(p,\; p+1,\; n), \qquad C(n,\; 2p+1,\; 0) = D^j(p+1,\; n)$$

It is seen that we have to distinguish between even and odd numbers m, viz. $2p$ and $2p+1$, respectively. For $l=n$ a prolate or problate pentagon is obtained in the following way

$$C(n,\; 2p,\; n) = D^i(p+1,\; n), \qquad C(n,\; 2p+1,\; n) = D(p+1,\; p+2,\; n)$$

It is noted that the prolate pentagon without corner, Dc^i (cf. CHART II), and the problate pentagon without corner denoted Dc' (cf. Fig. 3) are special members of the auxiliary classes (for $l = n-1$).

In CHART VII a set of combinatorial formulas of $K\{C(n,m,l)\}$ for m up to 6 are listed.

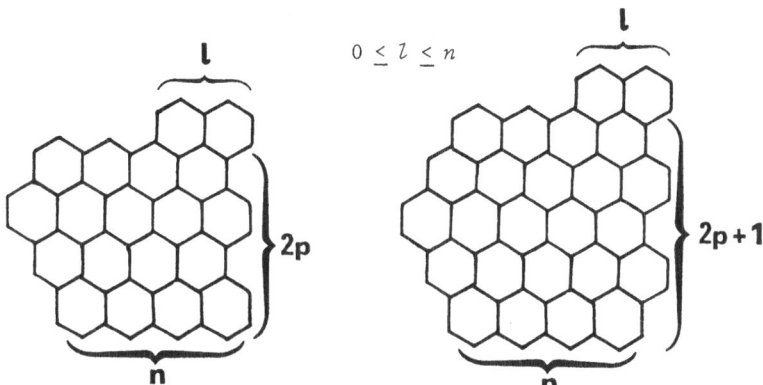

Fig. 11.5. The definition of $C(n,m,l)$. The depicted examples are $C(4,4,2)$ (K = 811; left) and $C(4,5,2)$ (K = 7623; right). They have an even m (= $2p$) and odd m (= $2p+1$), respectively.

CHART 11-VII. Problate or oblate pentagon augmented with a row

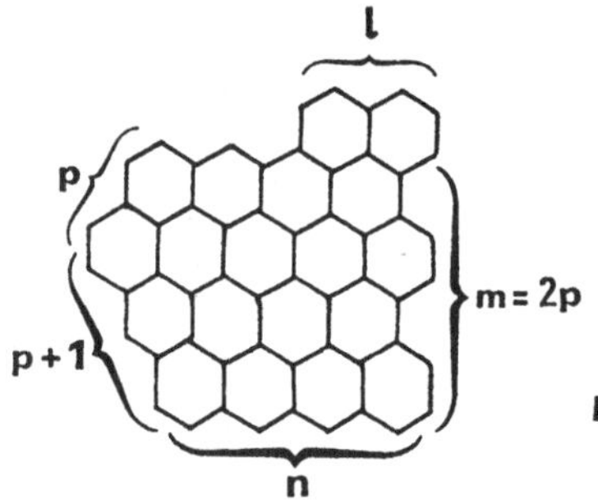

$C(n,m,l)$

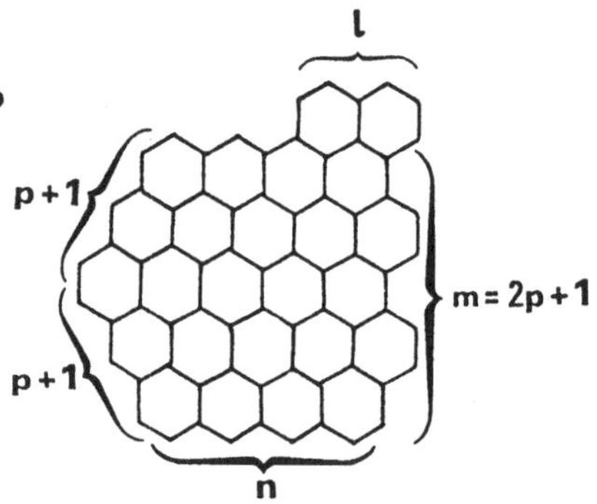

$$^aK\{C(n,0,l)\} = l+1$$

$$^bK\{C(n,1,l)\} = (n+1)(l+1) - \binom{l+1}{2}$$

$$^cK\{C(n,2,l)\} = \binom{n+2}{2}(l+1) - \binom{l+2}{3}$$

$$K\{C(n,3,l)\} = \left[\binom{n+2}{2}^2 - (n+2)\binom{n+2}{3}\right](l+1) - \binom{n+2}{2}\binom{l+2}{3} + (n+2)\binom{l+2}{4}$$

$$K\{C(n,4,l)\} = \left[\binom{n+2}{2}\binom{n+3}{3} - (n+2)\binom{n+3}{4}\right](l+1) - \binom{n+3}{3}\binom{l+2}{3} + (n+2)\binom{l+3}{5}$$

$$K\{C(n,5,l)\} = \left\{\binom{n+2}{2}\left[\binom{n+3}{3}^2 - \binom{n+3}{2}\binom{n+3}{4}\right] - \binom{n+3}{4}\left[(n+2)\binom{n+3}{3} - \binom{n+3}{4}\right]\right.$$
$$\left. + \binom{n+3}{5}\left[(n+2)\binom{n+3}{2} - \binom{n+3}{3}\right]\right\}(l+1)$$
$$- \left[\binom{n+3}{3}^2 - \binom{n+3}{2}\binom{n+3}{4}\right]\binom{l+2}{3} + \left[(n+2)\binom{n+3}{3} - \binom{n+3}{4}\right]\binom{l+3}{5}$$
$$- \left[(n+2)\binom{n+3}{2} - \binom{n+3}{3}\right]\binom{l+3}{6}$$

$$K\{C(n,6,l)\} = \left\{\binom{n+2}{2}\left[\binom{n+3}{3}\binom{n+4}{4} - \binom{n+3}{2}\binom{n+4}{5}\right] - \binom{n+3}{4}\left[(n+2)\binom{n+4}{4} - \binom{n+4}{5}\right]\right.$$
$$\left. + \binom{n+4}{6}\left[(n+2)\binom{n+3}{2} - \binom{n+3}{3}\right]\right\}(l+1)$$
$$- \left[\binom{n+3}{3}\binom{n+4}{4} - \binom{n+3}{2}\binom{n+4}{5}\right]\binom{l+2}{3} + \left[(n+2)\binom{n+4}{4} - \binom{n+4}{5}\right]\binom{l+3}{5}$$
$$- \left[(n+2)\binom{n+3}{2} - \binom{n+3}{3}\right]\binom{l+4}{7}$$

$$^aC(n,0,l) = L(l)$$
$$^bC(n,1,l) = L(n,1,l)$$
$$^cC(n,2,l) = A(n,2,l)$$

11.3 TRIANGLES

11.3.1 *Definitions*

The prolate, problate and oblate triangle-shaped benzenoids or triangles are special cases of the corresponding pentagons where the last parameter equals to unity (cf. also the end of Section 10.4);

$$T^i(m) = D^i(m,1), \qquad T(m,\ m+1) = D(m,\ m+1,\ 1), \qquad T^j(m) = D^j(m,1)$$

For the K numbers we introduce the abbreviations:

$$T_m{}^i = K\{T^i(m)\}, \qquad T_m = K\{T(m,\ m+1)\}, \qquad T_m{}^j = K\{T^j(m)\} \qquad (11.21)$$

11.3.2 *Previous Work and General Formulas*

The K numbers (21) are coupled through the following recurrence relations (Cyvin 1986c).

$$T_m{}^i = T_{m-1}{}^i + T_{m-1} \ ; \qquad m \geq 1 \qquad (11.22)$$

$$T_m = T_m{}^j + T_{m-1} \ ; \qquad m \geq 1 \qquad (11.23)$$

In spite of the known initial conditions, viz. $T_0{}^i = T_0 = T_0{}^j = 1$, the two equations are not sufficient to deduce successively the K numbers for increasing m values. A necessary additional piece of information is contained in the following equations, which were derived by Cyvin and Gutman (1986b) during their application of an algorithm to prolate triangles; see also Paragraph 8.5.5.

$$T_0{}^i = \binom{0}{0} = 1\ , \qquad T_1{}^i = \binom{2}{1} = 2\ , \qquad T_m{}^i = \binom{2m}{m} - \sum_{i=0}^{m-2} T_i{}^i \binom{2m-2-2i}{m-i} \ ; \qquad m \geq 2 \qquad (11.24)$$

With the aid of the relations (22) and (23) it was obtained (Cyvin 1986c) from (24):

$$T_0 = \binom{1}{0} = 1\ , \qquad T_1 = \binom{3}{1} = 3\ , \qquad T_m = \binom{2m+1}{m} - \sum_{i=0}^{m-2} T_i \binom{2m-2-2i}{m-i} \ ; \qquad m \geq 2 \qquad (11.25)$$

and:

$$T_0{}^j = \binom{0}{0} = 1\ , \qquad T_1{}^j = \binom{2}{1} = 2\ , \qquad T_2{}^j = \binom{4}{2} = 6\ ,$$

$$T_m{}^j = \binom{2m}{m} - \sum_{i=0}^{m-3} T_i \binom{2m-3-2i}{m-i} \ ; \qquad m \geq 3 \qquad (11.26)$$

where

$$T_i = \sum_{k=0}^{i} T_k{}^j \qquad (11.27)$$

Numerical values of K for the three classes of triangles are given in Table 3.

An explicit formula for $T_m{}^i$ is obtained when n is put equal to 1 in eqn. (i) of CHART I. The result is found in CHART VIII. Similar formulas for T_m and $T_m{}^j$ were derived from the explicit $T_m{}^i$ formula and the connections (22) and (23); see CHART VIII.

A recurrence relation in terms of a quotient for the prolate triangle reads:

$$\frac{T_m{}^i}{T_{m-1}{}^i} = \frac{\dbinom{2m+2}{m}}{\dbinom{2m}{m}} = \frac{2(2m+1)}{m+2} \; ; \; m \geq 1 \quad (11.28)$$

11.3.3 *Triangles Without Apex*

The triangles without apices are naturally defined by $Ta^i(m) = Da^i(m,1)$, $Ta(m, m+1) = Da(m, m+1, 1)$ and $Ta^j(m) = Da^j(m,1)$.

Every triangle-shaped benzenoid has exactly one Kekulé structure where a single bond is assigned to the protruding edge of the apex (marked by arrows in Fig. 6). This feature has the following consequences for the numbers of Kekulé structures for triangles without apices:

$$K\{Ta^j(m)\} = T_m{}^i - 1 \qquad (11.29)$$

$$K\{Ta(m, m+1)\} = T_m - 1 \qquad (11.30)$$

$$K\{Ta^j(m)\} = T_m{}^j - 1 \qquad (11.31)$$

Table 11.3. Numerical values of K for the benzenoids $T^i(m)$, $T(m, m+1)$ and $T^j(m)$.

m	$T_m{}^i$	T_m	$T_m{}^j$
0	1	1	1
1	2	3	2
2	5	9	6
3	14	28	19
4	42	90	62
5	132	297	207
6	429	1001	704
7	1430	3432	2431
8	4862	11934	8502
9	16796	41990	30056
10	58786	149226	107236
11	208012	534888	385662
12	742900	1931540	1396652
13	2674440	7020405	5088865
14	9694845	25662825	18642420
15	35357670	94287120	68624295
16	129644790	347993910	253706790

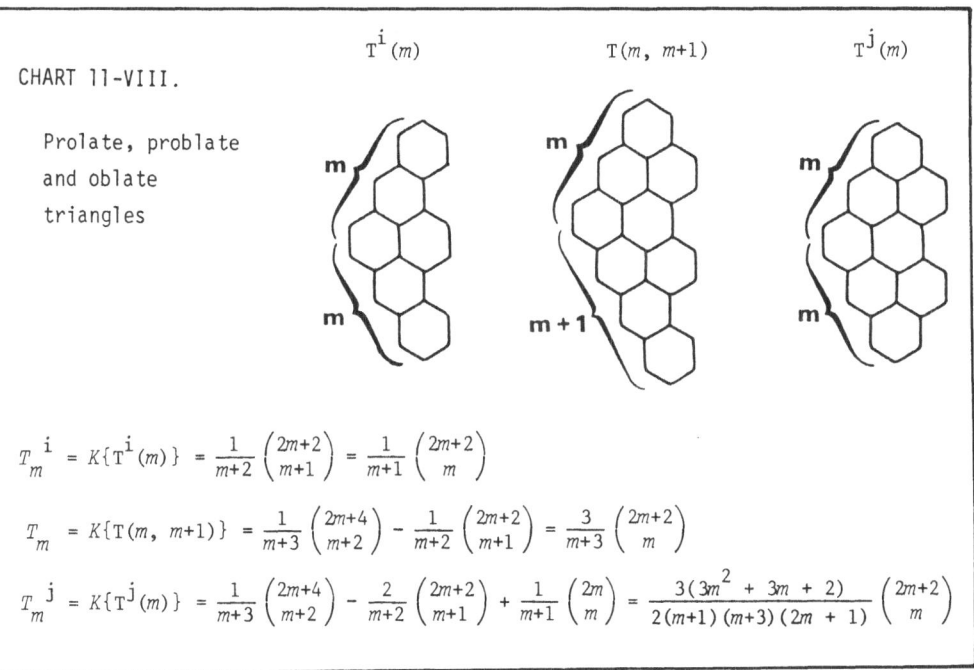

CHART 11-VIII.

Prolate, problate and oblate triangles

$T^i(m)$ $T(m, m+1)$ $T^j(m)$

$$T_m^{\ i} = K\{T^i(m)\} = \frac{1}{m+2}\binom{2m+2}{m+1} = \frac{1}{m+1}\binom{2m+2}{m}$$

$$T_m = K\{T(m, m+1)\} = \frac{1}{m+3}\binom{2m+4}{m+2} - \frac{1}{m+2}\binom{2m+2}{m+1} = \frac{3}{m+3}\binom{2m+2}{m}$$

$$T_m^{\ j} = K\{T^j(m)\} = \frac{1}{m+3}\binom{2m+4}{m+2} - \frac{2}{m+2}\binom{2m+2}{m+1} + \frac{1}{m+1}\binom{2m}{m} = \frac{3(3m^2 + 3m + 2)}{2(m+1)(m+3)(2m + 1)}\binom{2m+2}{m}$$

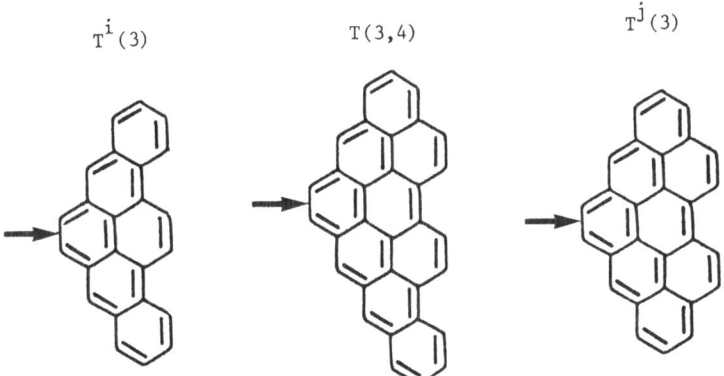

$T^i(3)$ $T(3,4)$ $T^j(3)$

Fig. 11.6. Unique Kekulé structures in a prolate (T^i), problate (T) and oblate (T^j) triangle.

11.4 STREAMERS AND GOBLETS

11.4.1 *General*

A streamer is a regular strip with one intruding two-segment chain, while a goblet has two intruding two-segment chains; cf. Chapter 10 for precise definitions. Among the 5-tier strips (see Table 10.5) we find streamers as Nos. I-9, I-13, I-15, II-7 and II-9; Nos. I-16 and II-10 represent goblets.

All streamers and goblets are either essentially disconnected or concealed non-Kekuléan benzenoids. This fact simplifies the problem of enumeration of the Kekulé structures.

11.4.2 *Streamer*

A streamer is a three-parameter system, viz. $\Sigma(k,m,n) = \Sigma(m,k,n)$. If $k=m$ it may be denoted by $\Sigma(m,n)$.

Figure 6 shows the existing classes of mirror-symmetrical 7-tier streamers. They are analogous to the corresponding six pentagons (cf. Section 11.2.5) and also designated correspondingly with regard to the superscripts: Σ^i and Σ^j are the prolate and oblate streamer, respectively (see the middle column of Fig. 6), and in addition

Fig. 11.6. The six mirror-symmetrical 7-tier streamers, depicted for $n=3$.

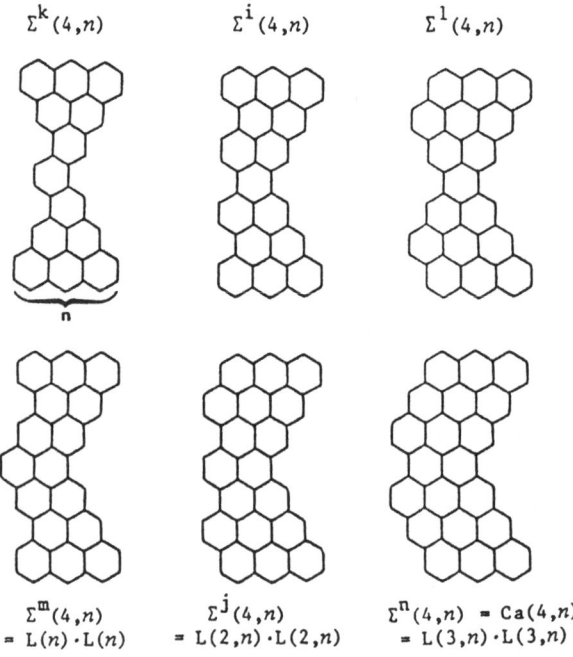

$\Sigma^k(4,n)$ $\Sigma^i(4,n)$ $\Sigma^l(4,n)$

$\Sigma^m(4,n)$
$= L(n) \cdot L(n)$

$\Sigma^j(4,n)$
$= L(2,n) \cdot L(2,n)$

$\Sigma^n(4,n) = Ca(4,n)$
$= L(3,n) \cdot L(3,n)$

one has Σ^k, Σ^l, Σ^m and Σ^n to be compared with the systems of CHART VI. Out of these six streamers three are concealed non-Kekuléans and three essentially disconnected. The combinatorial formulas for the numbers of Kekulé structures read (Cyvin BN and Cyvin 1987):

$$K\{\Sigma^k(4,n)\} = 0, \qquad K\{\Sigma^i(4,n)\} = 0, \qquad K\{\Sigma^l(4,n)\} = 0 \qquad (11.32)$$

$$K\{\Sigma^m(4,n)\} = (n+1)^2, \qquad K\{\Sigma^j(4,n)\} = \frac{1}{4}(n+1)^2(n+2)^2,$$

$$K\{\Sigma^n(4,n)\} = \frac{1}{36}(n+1)^2(n+2)^2(n+3)^2 \qquad (11.33)$$

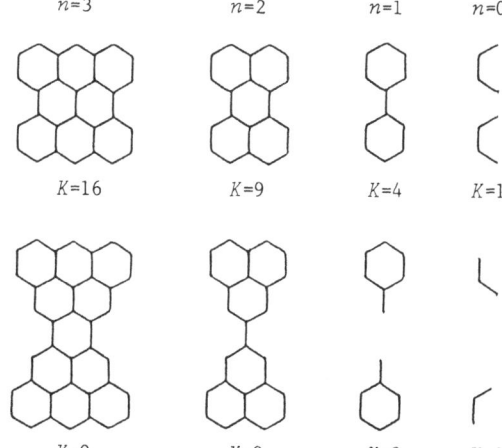

Fig. 11.7. The goblets X(2,n) and X(3,n) for $0 \leq n \leq 3$. Degenerate systems are involved.

11.4.3 Goblet

A goblet, X(k,m,n) = X(m,k,n), is a non-degenerate regular strip when $\min(k,m) \geq 2$ and $n \geq \min(k,m)$. A non-degenerate goblet is dihedral (D_{2h}) when $k{=}m$, and centrosymmetrical (C_{2h}) when $k \neq m$. In the former case ($k{=}m$) the system may be designated X(m,n).

If $\min(k,m) = 2$, the goblet is essentially disconnected, and then

$$K\{X(2,m,n)\} = (n+1)^2 \qquad (11.34)$$

These systems include *perylene* (Fig. 1.1) and *zethrene* (Fig. 2.4).

When $\min(k,m) \geq 3$, the goblet is a concealed non-Kekuléan.

Figure 7 illustrates the classes of the 3-tier and 5-tier dihedral goblets; it shows how the benzenoids degenerate to disconnected systems when n decreases.

CHAPTER 12

CLASSES OF BASIC BENZENOIDS (IV): RECTANGLES

12.1 DEFINITIONS

    A class of rectangle-shaped benzenoids or rectangles, which has the main sym-
bol R, consists of straight regular strips; cf. Chapter 10. Both rims are (single)
zigzag chains. We distinguish between a prolate rectangle, $R^i(m,n)$, and an oblate
rectangle, $R^j(m,n)$, which possess an inward or outward indentation, respectively;
cf. Fig. 1. They are $(2m-1)$-tier strips. The cases of 5-tier strips $(m=3)$ are found
as No. I-14 (R^i) and No. I-6 (R^j) in Table 10.5. Figure 1 shows $R^i(4,4)$ and $R^j(3,3)$,
a 7-tier and a 5-tier regular strip, respectively. An intermediate system is identified as a non-regular 6-tier strip (see Fig. 1); it is an obvious non-Kekuléan.

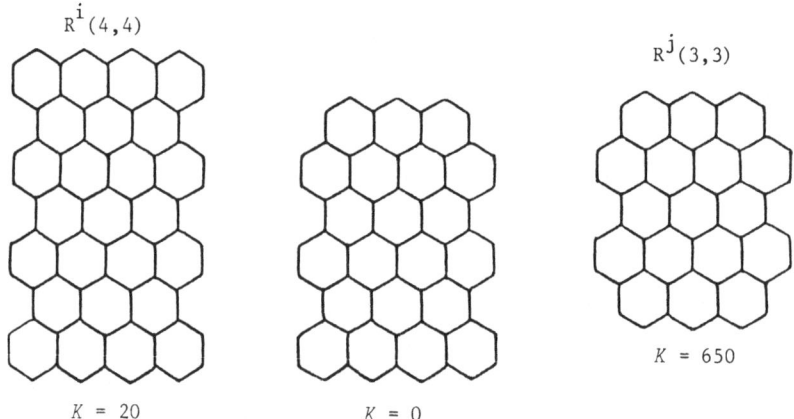

$R^i(4,4)$

$R^j(3,3)$

$K = 650$

$K = 20$ $K = 0$

Fig. 12.1. Examples of a prolate rectangle (R^i), oblate rectangle (R^j),
and an intermediate form, which is an obvious non-Kekuléan. K numbers
are given.

12.2 PROLATE RECTANGLE

    Gordon and Davison (1952) gave the formula for K, the number of Kekulé struc-
tures, of the 3-tier prolate rectangle, $R^i(2,n)$. The systematic studies of 3-tier
(Cyvin 1986d) and 5-tier (Cyvin SJ, Cyvin and Gutman 1985) strips include the appropriate prolate rectangles. A general K formula for $R^i(m,n)$ is due to Yen (1971).
Later it was re-derived in different ways by others (Jiang 1980; Ohkami and Hosoya
1983).

CHART 12-I. Prolate rectangle

$R^i(m,n)$

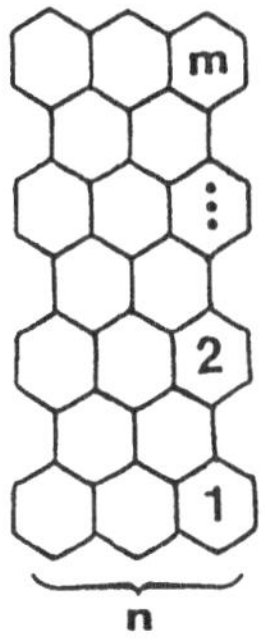

$^a K\{R^i(m,n)\} = (n+1)^m$

^{a}Yen TF (1971). Theor Chim Acta 20: 399

A prolate rectangle is an essentially disconnected benzenoid: $R^i(m,n) = L(n) \cdot L(n) \cdots L(n)$ [m times]. The combinatorial K formula is given in CHART I. It applies to, as special cases, the *polyphenylenes* ($n=1$; see Fig. 4.5), *polyrylenes* ($n=2$) and *polyanthenes* ($n=3$).

12.3 OBLATE RECTANGLE

12.3.1 *Previous Work*

The problem to determine K for $R^j(2,n)$, which coincides with the hexagon $O(3,n)$, was solved by the early investigators (Gordon and Davison 1952; Yen 1971), and also considered later (Cyvin 1986d). The K formula for $R^j(3,n)$, the 5-tier oblate rectangle, has also been derived by several investigators in different ways (Gordon and Davison 1952; Yen 1971; Ohkami and Hosoya 1983; Cyvin SJ, Cyvin and Gutman 1985). The cases of $m=4$ (Cyvin SJ, Cyvin and Bergan 1986) and $m=5$ (Cyvin 1986a) were solved much later, and recently for $m=6$ (Chen 1986b), $m=7$ (Chen 1986c), and finally for $m=8$ (Cyvin SJ, Cyvin and Chen 1987).

So far the studies of K formulas for oblate rectangles, $R^j(m,n)$, with fixed values of m have been reviewed. Gutman (1985b) attacked the problem of K enumeration for $R^j(m,n)$ with fixed values of n. He solved this problem for $n=1$ and $n=2$ by introducing classes of auxiliary benzenoids and treating systems of linearly coupled recurrence relations. Explicit K formulas were achieved. The paper was followed by several corresponding studies of oblate rectangles and related classes (Chen 1986a; Cyvin SJ, Cyvin and Bergan 1986; Su 1986; Zhang and Chen 1986a; Chen, Cyvin and Cyvin 1987). The related classes consist of oblate rectangles with modifications at one (not-indented) end and occasionally at both ends. It is relevant to mention two additional works: Cyvin SJ, Chen and Cyvin (1987); Cyvin SJ, Cyvin, Brunvoll, Chen and Su (1987).

12.3.2 *Compilation of Some Results*

Combinatorial K Formulas for $R^j(m,n)$ *with Fixed Values of m.* CHART II
shows the available formulas of $K\{R^j(m,n)\}$, where $m = 1, 2, \ldots, 8$.
   Cyvin SJ, Cyvin and Bergan (1986) showed that $K\{R^j(m,n)\}$ for a given m is a

CHART 12-II. Oblate rectangle with
fixed values of m

$R^j(m,n)$

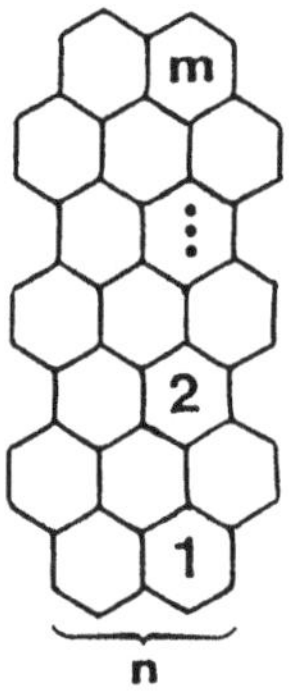

[a]$K\{R^j(1,n)\} = n+1$

[b]$K\{R^j(2,n)\} = \frac{1}{12}(n+1)(n+2)^2(n+3)$

[b]$K\{R^j(3,n)\} = \frac{1}{120}(n+1)(n+2)^3(n+3)(n^2 + 4n + 5)$

[c]$K\{R^j(4,n)\} = \frac{1}{20160}(n+1)(n+2)^4(n+3)(17n^4 + 136n^3 + 439n^2 + 668n + 420)$

[d]$K\{R^j(5,n)\} = \frac{1}{362880}(n+1)(n+2)^5(n+3)(31n^6 + 372n^5 + 1942n^4 + 5616n^3 + 9511n^2$
$$+ 8988n + 3780)$$

[e]$K\{R^j(6,n)\} = \frac{1}{79833600}(n+1)(n+2)^6(n+3)(691n^8 + 11056n^7 + 79788n^6 + 338320n^5$
$$+ 921759n^4 + 1654264n^3 + 1915562n^2 + 1315560n + 415800)$$

[f]$K\{R^j(7,n)\} = \frac{1}{6227020800}(n+1)(n+2)^7(n+3)(5461n^{10} + 109220n^9 + 1006407n^8$
$$+ 5617392n^7 + 210022809n^6 + 55133100n^5 + 102705053n^4 + 134421928n^3$$
$$+ 118632870n^2 + 64047960n + 16216200)$$

[g]$K\{R^j(8,n)\} = \frac{1}{10461394944000}(n+1)(n+2)^8(n+3)(929569n^{12} + 22309656n^{11}$
$$+ 250158485n^{10} + 1731086820n^9 + 8229767127n^8 + 28315930608n^7$$
$$+ 72322500575n^6 + 138258580980n^5 + 1965594445604n^4 + 203012336736n^3$$
$$+ 144957849840n^2 + 64500408000n + 13621608000)$$

[a]$R^j(1,n) = L(n)$

[b]Gordon M, Davison WHT (1952). J Chem Phys 20: 428; also O(3,n)

[c]Cyvin SJ, Cyvin BN, Bergan JL (1986). Match 19: 189

[d]Cyvin SJ (1986). Match 19: 213

[e]Chen RS (1986). Match 21: 259; correction[g]

[f]Chen RS (1986). Match 21: 277

[g]Cyvin SJ, Cyvin BN, Chen RS (1987). Match 22: 151

polynomial in n of $(3m-2)$-th degree. Chen (1986b) proved that it has the linear factors $(n+1)(n+2)^{m}(n+3)$ for $m > 1$.

Practical applications of the formulas of CHART II are inconvenient when m is large enough. For $m=6$ in particular, an intermediate result from Chen (1986b) obtained by the fully computerized method (cf. Section 4.10) may be more convenient;

$$K\{R^{j}(6,n)\} = (n+1)(n+2)^{6}(n+3)\left[\frac{691}{1980}\binom{n}{8} + \frac{691}{360}\binom{n}{7} + \frac{4507}{1008}\binom{n}{6} + \frac{3847}{672}\binom{n}{5}\right.$$
$$\left. + \frac{14683}{3360}\binom{n}{4} + \frac{129}{64}\binom{n}{3} + \frac{521}{960}\binom{n}{2} + \frac{5}{64}n + \frac{1}{192}\right] \quad (12.1)$$

The formula for $n=8$ in CHART II is of course the most impractical one because of the large integers involved. Cyvin SJ, Cyvin and Chen (1987) have reported an intermediate result, where the number of digits in the integers does not exceed 10.

Table 12.1. Numerical values of $K\{R^{j}(m,n)\}$.

m\n	1	2	3	4	5	6	7	8	9	10
1	2	3	4	5	6	7	8	9	10	11
2	6	20	50	105	196	336	540	825	1210	1716
3	18	136	650	2331	6860	17472	39852	83325	162382	298584
4	54	928	8500	52137	242158	916992	2969946	8501625	22020064	52509600
5	162	6336	111250	1167291	8557164	48179200	221578092	868388125		
6	486	43264	1456250	26137809	302425158					
7	1458	295424	19062500	585284211						
8	4374	2017280	249531250							
9	13122	13774848								
10	39336	94060544								

$R^{j}(m,n)$

Table 1 gives the numerical K values for $R^{j}(m,n)$ for some values of $m \leq 10$ and $n \leq 10$.

K Numbers for $R^{j}(m,n)$ with Fixed Values of n. Table 2 shows the recurrence relations for $K\{R^{j}(m,n)\}$ with fixed values of n up to 10.

For some of the lowest n values explicit combinatorial K formulas are available; see CHART III.

12.3.3 Limit Values Involving K Numbers

We get the following limit values of the quantities of $(\ln K)/m$ from the formulas of CHART III.

$$\lim_{m \to \infty} \frac{\ln K\{R^{j}(m,1)\}}{m} = \ln 3 \quad (12.2)$$

$$\lim_{m \to \infty} \frac{\ln K\{R^j(m,2)\}}{m} = \ln(4 + \sqrt{8}) \tag{12.3}$$

$$\lim_{m \to \infty} \frac{\ln K\{R^j(m,3)\}}{m} = \ln\left(\frac{15 + 5\sqrt{5}}{2}\right) \tag{12.4}$$

Table 12.2. Recurrence relations for oblate rectangles with fixed values of n: $R_n(m) = K\{R^j(m,n)\}$.*

n	Recurrence relation	
1	[a]$R_1(m) = 3R_1(m-1)$;	$m > 1$
2	[a]$R_2(m) = 8R_2(m-1) - 8R_2(m-2)$;	$m > 2$
3	[b,c]$R_3(m) = 15R_3(m-1) - 25R_3(m-2)$;	$m > 2$
4	[c,d]$R_4(m) = 27R_4(m-1) - 108R_4(m-2) + 108R_4(m-3)$;	$m > 3$
5	[e]$R_5(m) = 42R_5(m-1) - 245R_5(m-2) + 343R_5(m-3)$;	$m > 3$
6	[f]$R_6(m) = 64R_6(m-1) - 640R_6(m-2) + 2048R_6(m-3) - 2048R_6(m-4)$;	$m > 4$
7	[g]$R_7(m) = 90R_7(m-1) - 1215R_7(m-2) + 5103R_7(m-3) - 6561R_7(m-4)$;	$m > 4$
8	[g]$R_8(m) = 125R_8(m-1) - 2500R_8(m-2) + 17500R_8(m-3) - 50000R_8(m-4)$ $+ 50000R_8(m-2)$;	$m > 5$
9	[g]$R_9(m) = 165R_9(m-1) - 4235R_9(m-2) + 37268R_9(m-3) - 131769R_9(m-4)$ $+ 161051R_9(m-5)$;	$m > 5$
10	[g]$R_{10}(m) = 216R_{10}(m-1) - 7560R_{10}(m-2) + 96768R_{10}(m-3)$ $- 559872R_{10}(m-4) + 1492992R_{10}(m-5) - 142992R_{10}(m-6)$;	$m > 6$

*The same relations also apply to sets of modified classes (see the text).

[a]Gutman I (1985). Match 17: 3

[b]Chen RS (1986). J Xinjiang Univ 3(2): 13

[c]Cyvin SJ, Cyvin BN, Bergan JL (1986). Match 19: 189

[d]Su LX (1986). Match 20: 229

[e]Chen RS, Cyvin SJ, Cyvin BN (1987). Match 22: 111

[f]Cyvin SJ, Chen RS, Cyvin BN (1987). Match 22: 129

[g]see Paragraph 12.6.3

$$\lim_{m \to \infty} \frac{\ln K\{R^j(m,4)\}}{m} = \ln(12 + 6\sqrt{3}) \tag{12.5}$$

12.4 AUXILIARY BENZENOID CLASSES

Definitions. The auxiliary classes $B(n, 2m-2, t)$ are defined for $m \geq 1$ and $-n \leq t \leq n$.

First we define $B(n, 2m-2, l)$ for $l > 0$. The classes consist of $(2m - 1)$-tier strips; cf. Fig. 2. For $0 < l < n$ they may be interpreted as $R^j(m,n)$ rectangles of which the top row is incomplete; they are irregular strips. The "completed" rectangle emerges when $l=n$. We have by definition

$$B(n, 2m-2, n) = R^j(m,n)$$

The system for $l=0$, viz. $B(n, 2m-2, 0)$, is defined as a regular $(2m - 2)$-tier strip as shown in Fig. 2.

CHART 12-III. Oblate rectangle with fixed values of n

$R^j(m,n)$

$m \geq 1$

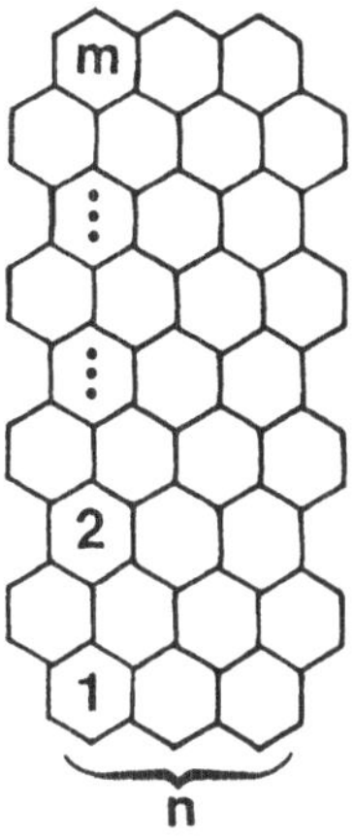

$$^a K\{R^j(m,1)\} = 2 \cdot 3^{m-1}$$

$$^a K\{R^j(m,2)\} = \frac{1}{16}\left[(4 + \sqrt{8})^{m+1} + (4 - \sqrt{8})^{m+1}\right]$$

$$^{b,c} K\{R^j(m,3)\} = \frac{2}{\sqrt{5}}\left[(\sqrt{5} + 2)\left(\frac{15 + 5\sqrt{5}}{2}\right)^{m-1} + (\sqrt{5} - 2)\left(\frac{15 - 5\sqrt{5}}{2}\right)^{m-1}\right]$$

$$^d K\{R^j(m,4)\} = \frac{1}{27}\left[3^{m+1} + \frac{1}{4}(12 + 6\sqrt{3})^{m+1} + \frac{1}{4}(12 - 6\sqrt{3})^{m+1}\right]$$

[a] Gutman I (1985). Match 17: 3
[b] Chen RS (1986). J Xinjiang Univ 3(2): 13
[c] Cyvin SJ, Cyvin BN, Bergan JL (1986). Match 19: 189
[d] Su LX (1986). Match 20: 229

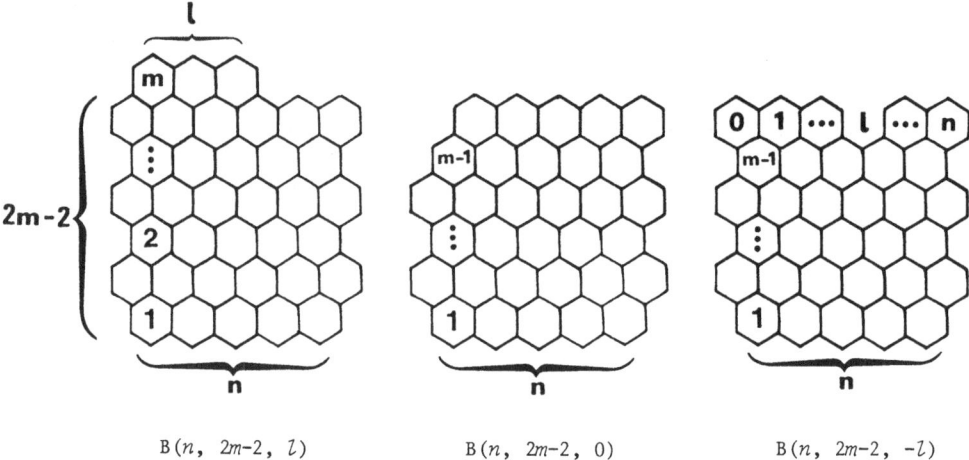

$$B(n, \ 2m\text{-}2, \ l) \qquad\qquad B(n, \ 2m\text{-}2, \ 0) \qquad\qquad B(n, \ 2m\text{-}2, \ -l)$$

Fig. 12.2. Definition of the auxiliary classes $B(n, \ 2m\text{-}2, \ t)$.

Finally we define auxiliary benzenoid classes associated to the incomplete rectangles as $B(n, \ 2m\text{-}2, \ -l)$, where $0 \le l \le n$. The definition is explained by Fig. 2. It is noted that the definitions of $B(n, \ 2m\text{-}2, \ l)$ and $B(n, \ 2m\text{-}2, \ -l)$ coincide for $l=0$.

For the sake of brevity we introduce the symbol

$$m' = 2m - 2 \tag{12.6}$$

Here $m' \ge 0$. The parameter l is restricted to $0 \le l \le n$ throughout.

Special Cases. For the degenerate case of $m=1$ ($m' = 0$) one has $B(n,0,l) = L(l)$; hence

$$K\{B(n,0,l)\} = l+1 \tag{12.7}$$

We will find it expedient to define for the associated classes:

$$K\{B(n, \ 0, \ -l)\} = 1 \tag{12.8}$$

independently of l (and n).

Eqn. (7) applied to $l=n$ gives

$$K\{B(n,0,n)\} = n+1 \tag{12.9}$$

which also is a special case of

$$K\{B(n, \ 2m\text{-}2, \ n)\} = K\{R^j(m,n)\} \tag{12.10}$$

For $n=0$ one has necessarily also $l=0$, and

$$K\{B(0, m', 0)\} = 1 \qquad (12.11)$$

corresponding to the trivial case of no hexagons.

Some Relations. It is immediately clear that the associated classes obey the symmetry condition $B(n, m', -l) = B(n, m', l-n)$; hence

$$K\{B(n, m', -l)\} = K\{B(n, m', l-n)\} \qquad (12.12)$$

CHART 12-IV. Incomplete

oblate rectangle

$$l \qquad B(n, 2m-2, l)$$

$$2m-2$$

$$n$$

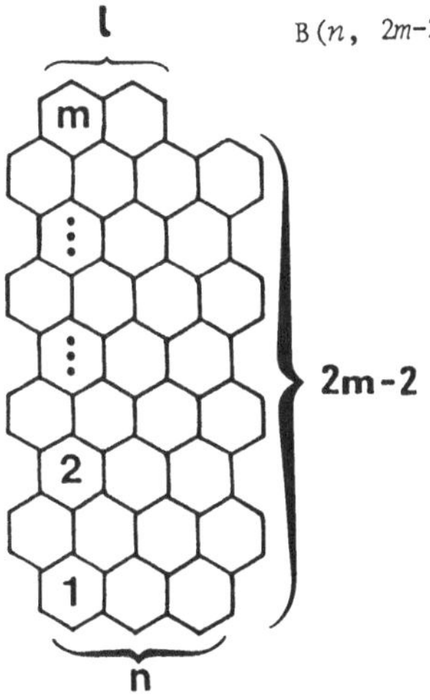

$$^a R_n^{(l)}(1) = K\{B(n,0,l)\} = l+1$$

$$^b R_n^{(l)}(2) = K\{B(n,2,l)\} = \binom{n+2}{2}\binom{l+2}{2}$$

$$- (n+2)\binom{l+2}{3}$$

$$^{c,d} R_n^{(l)}(3) = K\{B(n,4,l)\} = \left[\binom{n+2}{2}\binom{n+3}{3} - (n+2)\binom{n+3}{4}\right]\binom{l+2}{2}$$

$$- (n+2)\binom{n+2}{2}\binom{l+3}{4} + (n+2)^2\binom{l+3}{5}$$

$$^d R_n^{(l)}(4) = K\{B(n,6,l)\}$$

$$= \left\{\binom{n+3}{3}\left[\binom{n+2}{2}\binom{n+3}{3} - (n+2)\binom{n+3}{4}\right] - (n+2)\binom{n+2}{2}\binom{n+4}{5}\right\}\binom{l+2}{2}$$

$$- (n+2)\left[\binom{n+2}{2}\binom{n+3}{3} - (n+2)\binom{n+3}{4}\right]\binom{l+3}{4} + (n+2)^2\binom{n+2}{2}\binom{l+4}{6} - (n+2)^3\binom{l+4}{7}$$

[a] $B(n,0,l) = L(l)$
[b] Cyvin SJ (1987). Monatsh Chem (in press)
[c] Cyvin SJ (1986). Match 19: 213
[d] Cyvin SJ, Cyvin BN, Brunvoll J, Chen RS, Su LX (1987). Match 22: 141

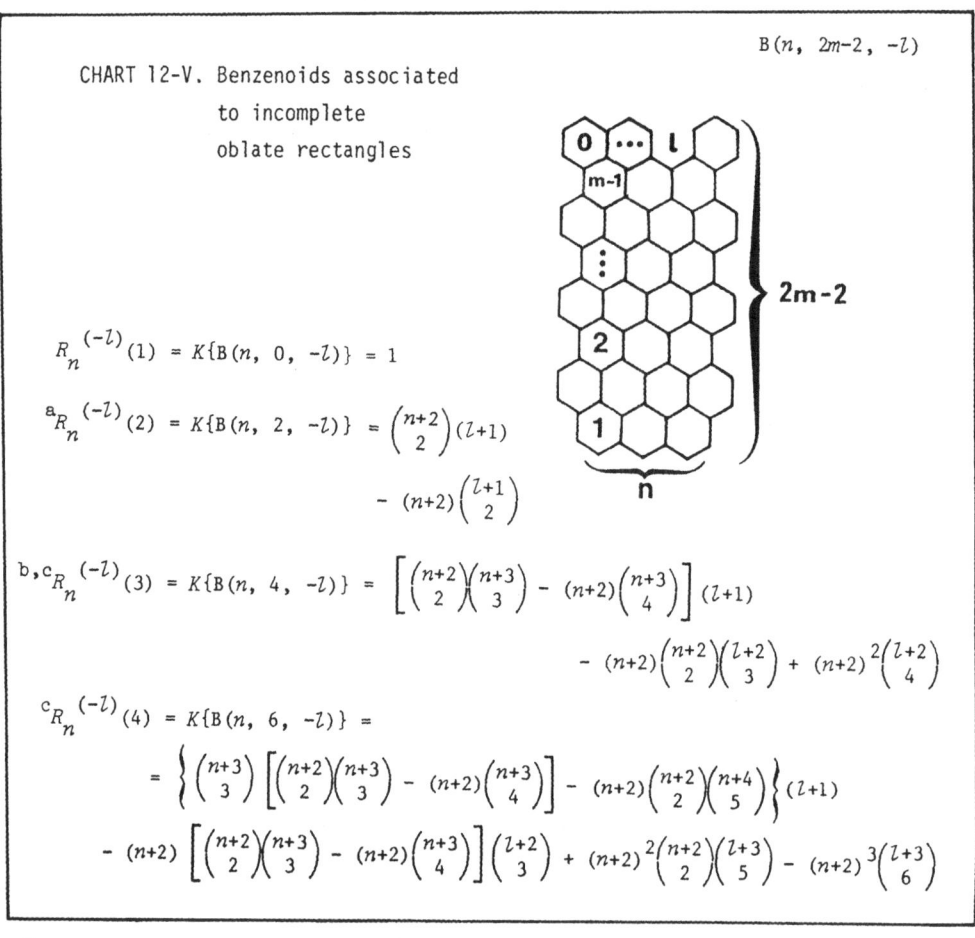

CHART 12-V. Benzenoids associated
to incomplete
oblate rectangles

$B(n, 2m-2, -l)$

$$R_n^{(-l)}(1) = K\{B(n, 0, -l)\} = 1$$

$${}^aR_n^{(-l)}(2) = K\{B(n, 2, -l)\} = \binom{n+2}{2}(l+1)$$

$$- (n+2)\binom{l+1}{2}$$

$${}^{b,c}R_n^{(-l)}(3) = K\{B(n, 4, -l)\} = \left[\binom{n+2}{2}\binom{n+3}{3} - (n+2)\binom{n+3}{4}\right](l+1)$$

$$- (n+2)\binom{n+2}{2}\binom{l+2}{3} + (n+2)^2\binom{l+2}{4}$$

$${}^cR_n^{(-l)}(4) = K\{B(n, 6, -l)\} =$$

$$= \left\{\binom{n+3}{3}\left[\binom{n+2}{2}\binom{n+3}{3} - (n+2)\binom{n+3}{4}\right] - (n+2)\binom{n+2}{2}\binom{n+4}{5}\right\}(l+1)$$

$$- (n+2)\left[\binom{n+2}{2}\binom{n+3}{3} - (n+2)\binom{n+3}{4}\right]\binom{l+2}{3} + (n+2)^2\binom{n+2}{2}\binom{l+3}{5} - (n+2)^3\binom{l+3}{6}$$

[a]Cyvin SJ, Cyvin BN, Bergan JL (1986). Match 19: 189

[b]Cyvin SJ, Chen RS, Cyvin BN (1987). Match 22: 129

[c]Cyvin SJ, Cyvin BN, Brunvoll J, Chen RS, Su LX (1987). Match 22: 141

A fundamental relation, which connects the K numbers for incomplete rectangles and their associated classes, reads

$$K\{B(n, m', -l)\} = K\{B(n, m', l)\} - K\{B(n, m', l-1)\}; \qquad l \geq 1 \qquad (12.13)$$

Hence also:

$$K\{B(n, m', l)\} = \sum_{i=0}^{l} K\{B(n, m', -i)\} \qquad (12.14)$$

Eqns. (13) and (14) are also valid for $m' = 0$ by virtue of the definition (8).

Listing of Formulas and K Numerical Values. CHART IV gives the combinatorial K formulas of $K\{B(n, 2m-2, l)\}$ for $m = 1, 2, 3, 4$. The corresponding formulas of $K\{B(n, 2m-2, -l)\}$ are found in CHART V.

The formulas of CHART V are available in expanded forms for $m=2$ (Chen 1986b) and $m=3$ (Cyvin SJ, Chen and Cyvin 1987). Below they are extended to $m=4$.

$$K\{B(n, 2, -l)\} = \frac{1}{2}(n+2)(l+1)(n-l+1) \tag{12.15}$$

$$K\{B(n, 4, -l)\} = \frac{1}{24}(n+2)(l+1)(n-l+1)[(n-l)(n-l+2)(n + 3l + 11)$$

$$+ (l+2)(l+3)(4n - 3l + 8) - 6(n+2)(n+3)] \tag{12.16}$$

$$K\{B(n, 6, -l)\} = \frac{1}{720}(n+1)(n+2)^4(n+3)(l+1)[3(n^2 + 4n + 5) - 5l(l+2)]$$

$$+ \frac{1}{720}(n+2)^3(l-1)l(l+1)(l+2)(l+3)(3n - l + 5) \tag{12.17}$$

For $l=0$ the two classes (CHART IV and CHART V) coincide, and the K formulas become polynomials in n. They are listed in CHART VI; here the range of m is extended to $m=8$.

Numerical values of $K\{B(n, 2m-2, 0)\}$ are listed in Table 3. Additional useful tabulations of K numbers for the auxiliary classes considered here are available (Chen, Cyvin and Cyvin 1987).

12.5 MODIFIED OBLATE RECTANGLES

12.5.1 *Introduction*

Here we give a number of relations, including combinatorial K formulas for benzenoid classes derived from oblate rectangles, $R^j(m,n)$ with fixed values of n. The oblate rectangles are modified by adding (fusion or condensation) or deleting one or more hexagons at one of the not-indented ends, and occasionally at both ends. All the auxiliary classes considered above (Section 12.4) fall into the category of modified oblate rectangles.

The existing material (cf. Paragraph 12.3.1) is too voluminous to be surveyed in details here.

Furthermore, a number of limit values of the quantities $(\ln K)/m$ may be deduced from the available formulas. Whenever the benzenoid class in question obeys one of the four first recurrence relations of Table 2, the limit value may be identified with the appropriate value of eqns. (2)-(5). Otherwise the limit values are not to be treated in the remainder of this section.

We shall not specify any boundaries of m (as is done in Table 2, for instance), which only means that we allow for nominal K values (cf. Section 6.4).

Table 12.3. Numerical values of $K\{B(n, 2m-2, 0)\} = R_n^{(0)}(m)$.

n / m	1	2	3	4	5	6	7	8	9	10
1	1	1	1	1	1	1	1	1	1	1
2	3	6	10	15	21	28	36	45	55	66
3	9	40	125	315	686	1344	2430	4125	6655	10296
4	27	272	1625	6993	24010	69888	179334	416625	893101	1791504
5	81	1856	21250	156411	847553	3667968	13364757	42508125	121110352	315057600
6	243	12672	278125	3501873	29950074	192716800	997101414			
7	729	86528	3640625	78413427						
8	2187	590848	47656250							
9	6561	4034560	623828125							
10	19683	27549696								

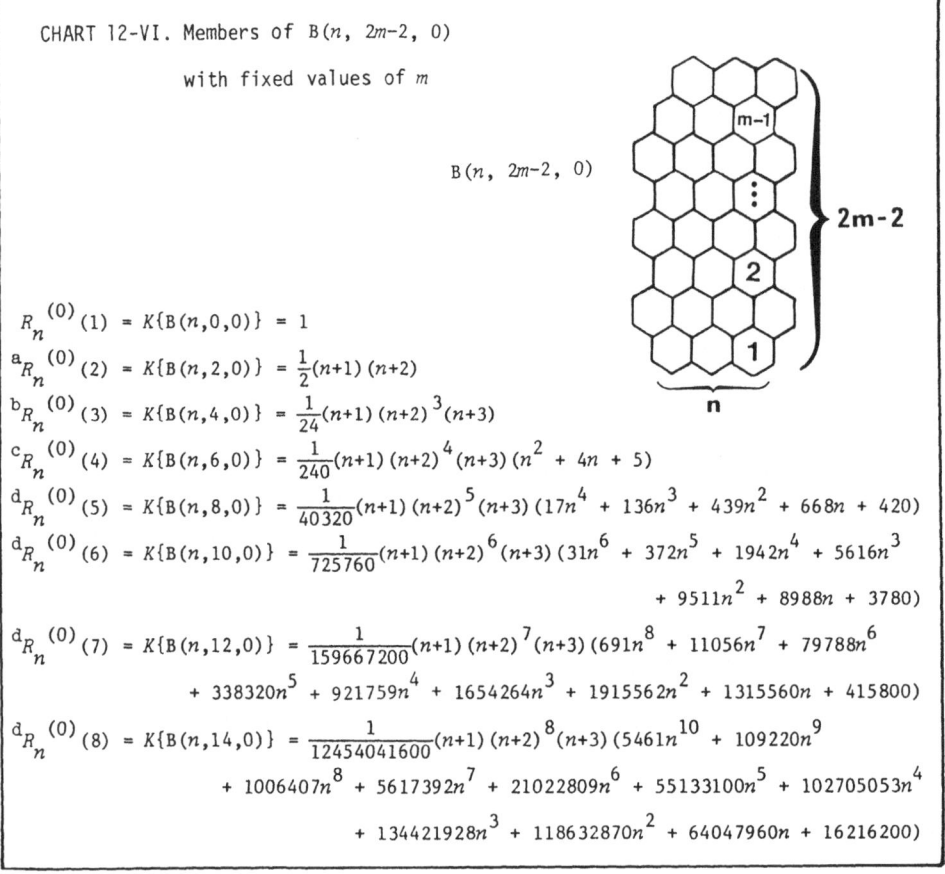

CHART 12-VI. Members of $B(n, 2m-2, 0)$

with fixed values of m

$B(n, 2m-2, 0)$

$R_n^{(0)}(1) = K\{B(n,0,0)\} = 1$

[a] $R_n^{(0)}(2) = K\{B(n,2,0)\} = \frac{1}{2}(n+1)(n+2)$

[b] $R_n^{(0)}(3) = K\{B(n,4,0)\} = \frac{1}{24}(n+1)(n+2)^3(n+3)$

[c] $R_n^{(0)}(4) = K\{B(n,6,0)\} = \frac{1}{240}(n+1)(n+2)^4(n+3)(n^2 + 4n + 5)$

[d] $R_n^{(0)}(5) = K\{B(n,8,0)\} = \frac{1}{40320}(n+1)(n+2)^5(n+3)(17n^4 + 136n^3 + 439n^2 + 668n + 420)$

[d] $R_n^{(0)}(6) = K\{B(n,10,0)\} = \frac{1}{725760}(n+1)(n+2)^6(n+3)(31n^6 + 372n^5 + 1942n^4 + 5616n^3$
$+ 9511n^2 + 8988n + 3780)$

[d] $R_n^{(0)}(7) = K\{B(n,12,0)\} = \frac{1}{159667200}(n+1)(n+2)^7(n+3)(691n^8 + 11056n^7 + 79788n^6$
$+ 338320n^5 + 921759n^4 + 1654264n^3 + 1915562n^2 + 1315560n + 415800)$

[d] $R_n^{(0)}(8) = K\{B(n,14,0)\} = \frac{1}{12454041600}(n+1)(n+2)^8(n+3)(5461n^{10} + 109220n^9$
$+ 1006407n^8 + 5617392n^7 + 21022809n^6 + 55133100n^5 + 102705053n^4$
$+ 134421928n^3 + 118632870n^2 + 64047960n + 16216200)$

[a] $B(n,2,0) = L(2,n)$

[b] Cyvin SJ (1986). Monatsh Chem 117: 33; also $D(2,3,n)$

[c] Cyvin SJ, Cyvin BN, Bergan JL (1986). Match 19: 189

[d] Cyvin SJ, Cyvin BN, Brunvoll J, Chen RS, Su LX (1987). Match 22: 141

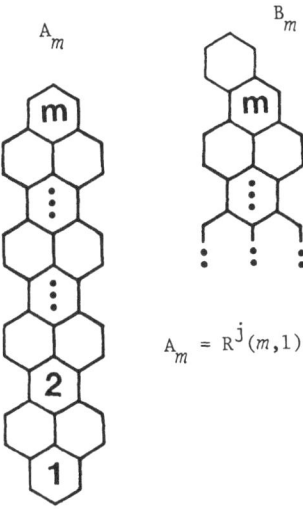

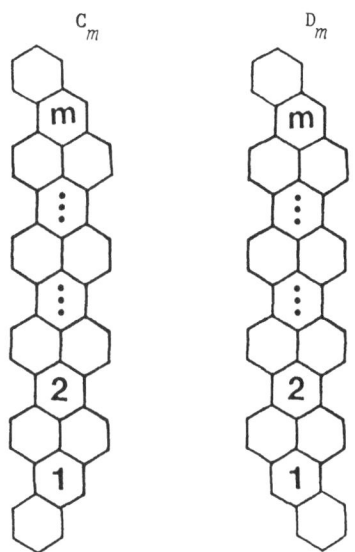

Fig. 12.3. The class of
oblate rectangles with
$n=1$ (A_m) and a related
class (B_m).

Fig. 12.4. Classes of mirror-
symmetrical (C_m) and centrosymmetrical (D_m) benzenoids
obtained by additions of hexagons to $R^j(m,1)$.

12.5.2 *Modifications of* $R^j(m,1)$

Gutman (1985b) considered the two classes A_m and B_m of Fig. 3. They both obey
the first, simple recurrence relation of Table 2.

Throughout this section we apply the abbreviations of the type $A_m = K\{A_m\}$, B_m
$= K\{B_m\}$, etc. Then

$$B_m = \frac{3}{2} A_m = 3^m \qquad (12.18)$$

The explicit form of $A_m = K\{R^j(m,1)\}$ is given in CHART III.

Figure 4 defines the classes C_m and D_m obtained by modifications of the oblate
rectangle with $n=1$ at both ends. We find the following recurrence properties of the
K numbers.

$$C_m = B_m + C_{m-1}, \qquad D_m = B_m + D_{m-1} \qquad (12.19)$$

It follows

$$C_m = 2 + \sum_{i=1}^{m} 3^i, \qquad D_m = 1 + \sum_{i=1}^{m} 3^i \qquad (12.20)$$

where we have employed the initial conditions $C_1 = 5$ and $D_1 = 4$, which imply $C_0 = 2$ and $D_0 = 1$. Consequently the explicit formulas read

$$C_m = \frac{1}{2}(3^{m+1} + 1) \, , \qquad D_m = \frac{1}{2}(3^{m+1} - 1) \qquad (12.21)$$

12.5.3 *Modifications of* $R^j(m,2)$

Figure 5 defines five benzenoid classes including the oblate rectangle with $n=2$, viz. E_m. Gutman (1985b) considered all these classes except I_m and solved the K enumeration problem. All these five classes obey the second recurrence relation of Table 2. Here we give the expressions found for the different K numbers in terms of the K numbers for the oblate rectangle itself (E_m and E_{m-1}).

$$\begin{array}{ll}
F_m = E_m - 2E_{m-1}, & G_m = 2E_m, \\
H_m = 4E_m - 8E_{m-1}, & I_m = 2E_m - 2E_{m-1}
\end{array} \qquad (12.22)$$

Numerical values can be obtained by means of the data in Table 1. In this way also

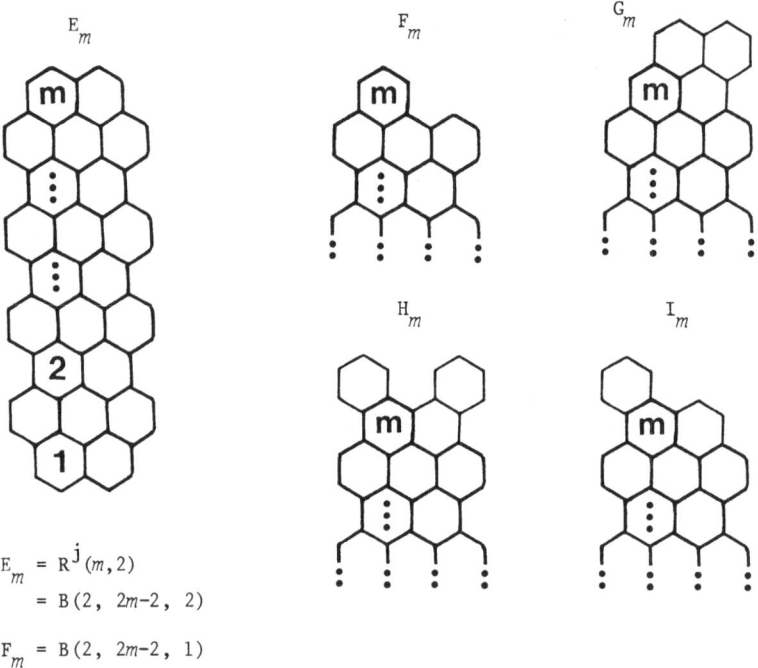

$$E_m = R^j(m,2)$$
$$\quad = B(2, \ 2m-2, \ 2)$$
$$F_m = B(2, \ 2m-2, \ 1)$$
$$G_m = B(2, \ 2m, \ 0)$$
$$H_m = B(2, \ 2m, \ -1)$$

Fig. 12.5. The class of oblate rectangles with $n=2$ (E_m) and four related classes.

the suitable initial conditions are accessible, which make it possible to deduce the explicit formulas. CHART III includes the explicit formula for $E_m = K\{R^j(m,2)\}$. Below we give only one more, which pertains to the class of mirror-symmetrical benzenoids, viz. H_m.

$$H_m = \frac{1}{2\sqrt{8}}\left[(4 + \sqrt{8})^{m+1} + (4 - \sqrt{8})^{m+1}\right] \tag{12.23}$$

The class J_m, consisting of the $n=2$ oblate rectangles with symmetrical modifications at both ends (see CHART VII), was considered by Zhang and Chen (1986a) among five additional related classes of less symmetry. Their explicit formula for $J_m = K\{J_m\}$ is entered (in a simplified form) in CHART VII. Also this quantity follows the second recurrence relation of Table 2; hence it is linearly coupled to the system of equations (22). Specifically:

$$J_m = 8E_m \tag{12.24}$$

CHART 12-VII. Two classes of dihedral benzenoids: oblate rectangles $R^j(m,2)$ modified at both ends

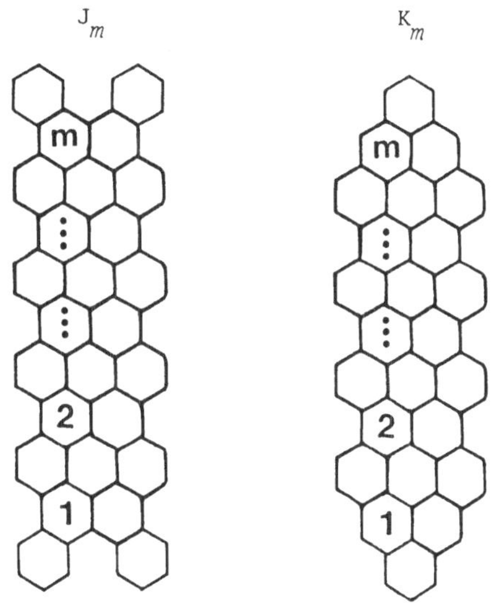

$$^aK\{J_m\} = \frac{1}{2}\left[(4 + \sqrt{8})^{m+1} + (4 - \sqrt{8})^{m+1}\right]$$

$$K\{K_m\} = \frac{1}{2}\left[(2 + \sqrt{2})^{m+1} + (2 - \sqrt{2})^{m+1}\right]$$

[a]Zhang FJ, Chen RS (1986). J Xinjiang Univ 3(3): 10

Finally in this paragraph we have studied the class K_m, which also is defined in CHART VII. It was found to obey a new recurrence relation, viz.

$$K_m = 4K_{m-1} - 2K_{m-2} \tag{12.25}$$

The explicit formula of CHART VII was deduced. We emphasize that K_m does not follow the same recurrence relation as J_m and the quantities of eqn. (22); hence it is not linearly coupled to them. Instead one has the non-linear, but simple connections:

$$J_m = 2^{m+1} K_m , \qquad E_m = 2^{m-2} K_m \tag{12.26}$$

12.5.4 Modifications of $R^j(m,3)$

The classes L_m, M_m, N_m, O_m and P_m defined in Fig. 6 have been treated by Chen (1986a) and by Cyvin SJ, Cyvin and Bergan (1986), Q_m only by Chen (1986a), while R_m is new. Four additional (unsymmetrical) related classes are included in the work of

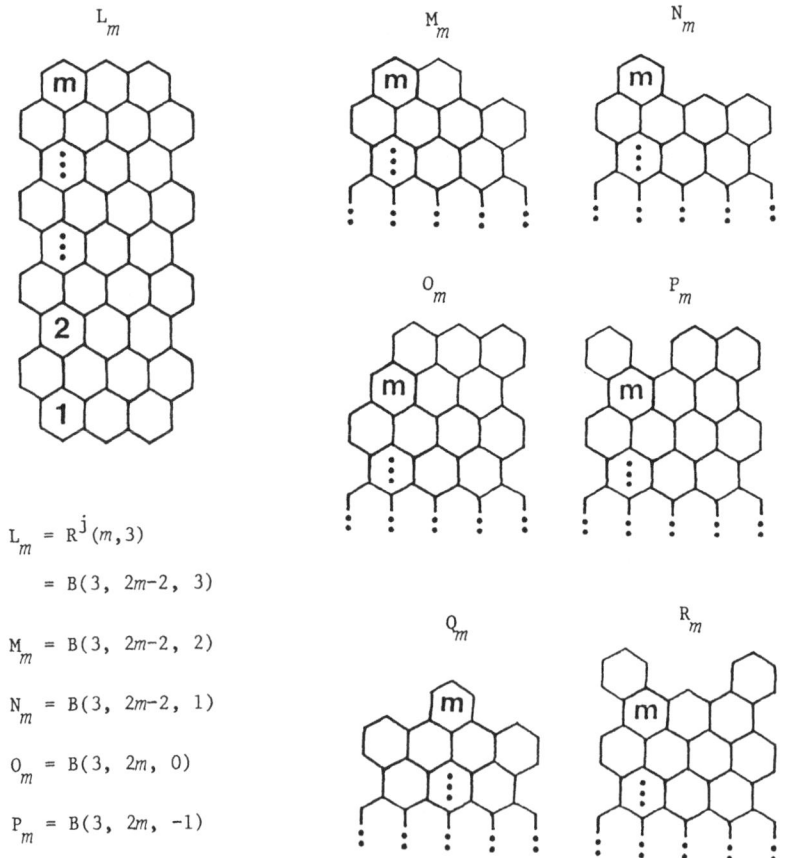

$$L_m = R^j(m,3)$$

$$\quad = B(3, 2m-2, 3)$$

$$M_m = B(3, 2m-2, 2)$$

$$N_m = B(3, 2m-2, 1)$$

$$O_m = B(3, 2m, 0)$$

$$P_m = B(3, 2m, -1)$$

Fig. 12.6. The class of oblate rectangles with $n=3$ (L_m) and six related classes.

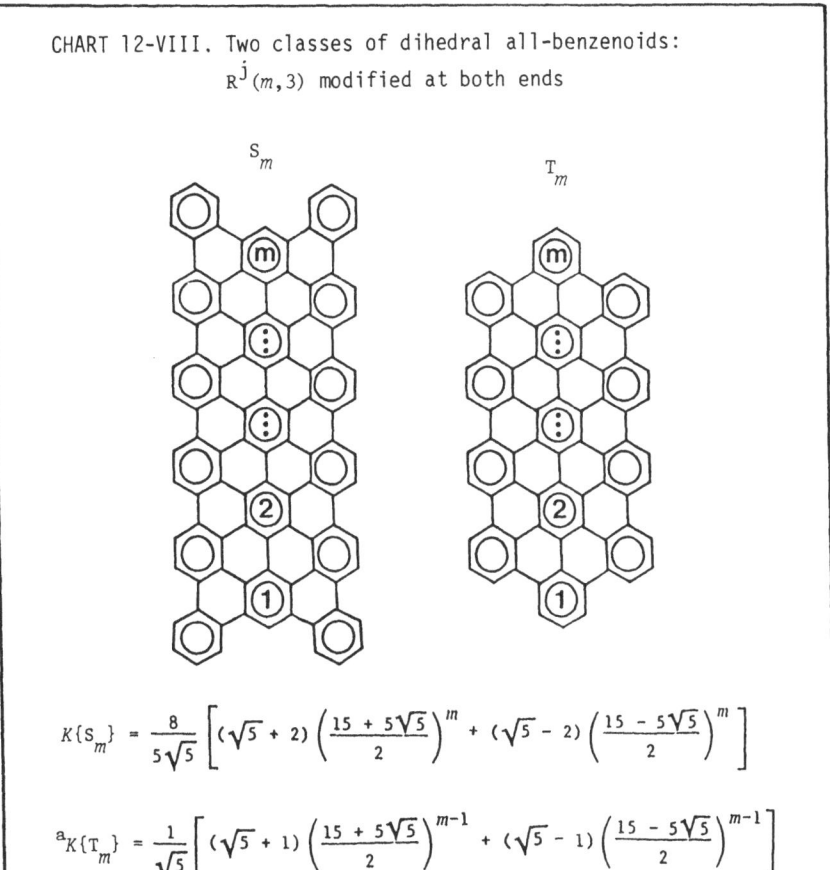

CHART 12-VIII. Two classes of dihedral all-benzenoids:
$R^j(m,3)$ modified at both ends

S_m

T_m

$$K\{S_m\} = \frac{8}{5\sqrt{5}}\left[(\sqrt{5}+2)\left(\frac{15+5\sqrt{5}}{2}\right)^m + (\sqrt{5}-2)\left(\frac{15-5\sqrt{5}}{2}\right)^m\right]$$

$$^aK\{T_m\} = \frac{1}{\sqrt{5}}\left[(\sqrt{5}+1)\left(\frac{15+5\sqrt{5}}{2}\right)^{m-1} + (\sqrt{5}-1)\left(\frac{15-5\sqrt{5}}{2}\right)^{m-1}\right]$$

[a]Zhang FJ, Chen RS (1986). J Xinjiang Univ 3(3): 10

Chen (1986a), who produced explicit K formulas for all the ten classes in question. All these classes obey the third recurrence relation of Table 2. Linear dependencies of K numbers:

$$M_m = L_m - \frac{5}{2}L_{m-1}, \qquad N_m = \frac{1}{2}L_m,$$

$$O_m = \frac{5}{2}L_m, \qquad P_m = 5L_m - \frac{25}{2}L_{m-1}, \qquad (12.27)$$

$$Q_m = L_m - 5L_{m-1}, \qquad R_m = 4L_m - 10L_{m-1}$$

The explicit formula for $L_m = K\{R^j(m,3)\}$ is given in CHART III. With regard to the additional explicit formulas we give only those for the classes of mirror-symmetrical benzenoids:

$$Q_m = \frac{2}{5\sqrt{5}}\left[\left(\frac{15 + 5\sqrt{5}}{2}\right)^m - \left(\frac{15 - 5\sqrt{5}}{2}\right)^m\right] \,,$$

(12.28)

$$R_m = \frac{4}{25\sqrt{5}}\left[\left(\frac{15 + 5\sqrt{5}}{2}\right)^{m+1} - \left(\frac{15 - 5\sqrt{5}}{2}\right)^{m+1}\right]$$

Two modifications of the $n=3$ oblate rectangle at both ends as shown in CHART VIII are of special interest inasmuch as they are all-benzenoids. One of them (T_m) has been treated by Zhang and Chen (1986a) among additional forty-four related, less symmetrical classes. The K numbers of both of these classes follow the third recurrence relation of Table 2. They are linearly coupled to (27) through:

$$S_m = 12L_m - 20L_{m-1} = \frac{4}{5}L_{m+1}, \qquad T_m = \frac{1}{2}L_m - \frac{5}{2}L_{m-1} + \frac{25}{2}L_{m-2}$$

(12.29)

The explicit formulas are given in CHART VIII.

12.5.5 Modifications of $R^j(m,4)$

Figure 7 defines eight benzenoid classes including the oblate rectangle with $n=4$, viz. U_m. Su (1986) solved the K enumeration problem for all these and ten additional, related classes. She gave the explicit formulas for all of them. Three of these formulas are given here (in a simplified form); see CHART III and Fig. 7. Equations for the linear dependencies are included (Fig. 7).

It is instructive to show the derivation of one of the explicit formulas (as a supplement to Paragraph 6.4.2). In the present case we are faced with a four-term recurrence relation (cf. Table 2), which leads to a cubic characteristic equation, viz.

$$a^3 - 27a^2 + 108a - 108 = 0$$

(12.30)

It has the solution

$$a_1 = 3, \qquad a_2 = 12 + 6\sqrt{3}, \qquad a_3 = 12 - 6\sqrt{3}$$

(12.31)

Consequently

$$U_m = 3^m A + B(12 + 6\sqrt{3})^m + C(12 - 6\sqrt{3})^m$$

(12.32)

and the same expression for all the classes of Fig. 7, only with differing values of the constants A, B and C.

Let us consider, as an example, the class $\mathcal{E}_m$. Non-degenerate benzenoids of this class occur for $m \geq 1$. For the K numbers of $\mathcal{E}_m$ one finds $\mathcal{E}_1 = 27$, $\mathcal{E}_2 = 621$ and $\mathcal{E}_3 = 13959$. It is possible to determine the constants in (32) with these numbers as initial conditions together with $m = 1$, 2 and 3, respectively. But the calculation is

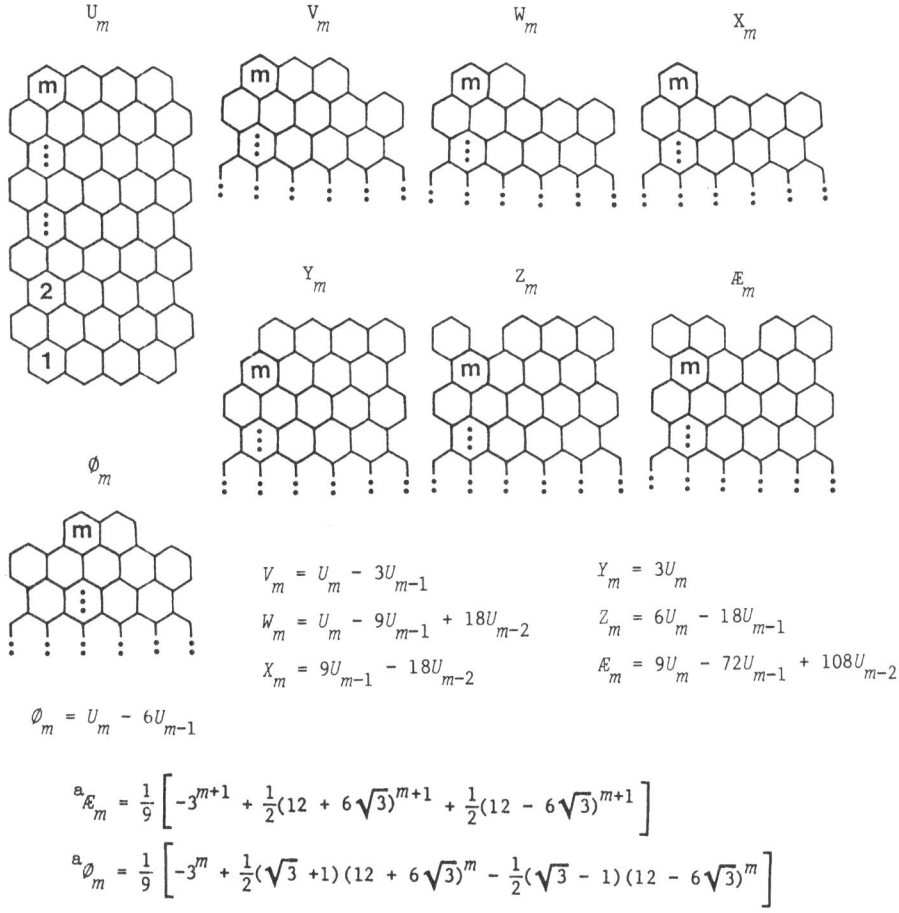

$$V_m = U_m - 3U_{m-1}$$

$$W_m = U_m - 9U_{m-1} + 18U_{m-2}$$

$$X_m = 9U_{m-1} - 18U_{m-2}$$

$$Y_m = 3U_m$$

$$Z_m = 6U_m - 18U_{m-1}$$

$$Æ_m = 9U_m - 72U_{m-1} + 108U_{m-2}$$

$$\emptyset_m = U_m - 6U_{m-1}$$

$$^a Æ_m = \frac{1}{9}\left[-3^{m+1} + \frac{1}{2}(12 + 6\sqrt{3})^{m+1} + \frac{1}{2}(12 - 6\sqrt{3})^{m+1} \right]$$

$$^a \emptyset_m = \frac{1}{9}\left[-3^m + \frac{1}{2}(\sqrt{3}+1)(12 + 6\sqrt{3})^m - \frac{1}{2}(\sqrt{3} - 1)(12 - 6\sqrt{3})^m \right]$$

Fig. 12.7. The class of oblate rectangles with $n=4$ (U_m) and seven related classes.
[a]Su LX (1986). Match 20: 229

facilitated on introducing nominal K values, which need not correspond to actual benzenoids. On extrapolating it was found $Æ_0 = 1$ and $Æ_{-1} = 0$ in consistence with the fourth recurrence relation of Table 2. Consequently it is expedient to assume

$$Æ_m = 3^{m+1}A' + B'(12 + 6\sqrt{3})^{m+1} + C'(12 - 6\sqrt{3})^{m+1} \qquad (12.33)$$

and use $m = -1$, 0 and 1. This gives a relatively simple set of linear equations, from which it is found:

$$A' = -\frac{1}{9}, \qquad B' = C' = \frac{1}{18} \qquad (12.34)$$

Inserting into (33) leads to the appropriate explicit formula of Fig. 7.

12.5.6 *Modifications of* $R^j(m,5)$

Chen, Cyvin and Cyvin (1987) gave the explicit solution of $K\{R^j(m,5)\}$. Furthermore, they expressed the K numbers of thirteen modified classes in terms of those of the mentioned rectangle, seven of them being the different auxiliary classes for $n=5$.

12.6 SOME GENERAL FORMULATIONS CONCERNING OBLATE RECTANGLES

12.6.1 *Summation K Formulas in Terms of Auxiliary Benzenoids*

From eqn. (14) it is obtained on inserting $l=n$:

$$K\{R^j(m,n)\} = \sum_{i=0}^{n} K\{B(n, 2m-2, -i)\} \tag{12.35}$$

A more general formula (Cyvin SJ, Cyvin and Bergan 1986) reads

$$K\{R^j(m,n)\} = \sum_{i=0}^{n} K\{B(n, 2p, -i)\} \cdot K\{B(n, 2q, -i)\}; \qquad p+q = m-1 \tag{12.36}$$

This formula is easily deduced by the method of chopping (see Section 10.6). Figure 8 shows an illustration of eqns. (35) and (36).

Another useful summation formula was derived by Chen, Cyvin and Cyvin (1987);

$$K\{B(n, 2m-2, 0)\} = \sum_{i=0}^{n} (i+1)K\{B(n, 2m-4, -i)\}; \qquad m > 1 \tag{12.37}$$

Here we finally give the summation formula of Chen (1986b):

$$K\{B(n, 2m-2s, -l)\} = (n-l+1) \sum_{i=0}^{l} (i+1)K\{B(n, 2m-2s-2, -i)\}$$

$$+ (l+1) \sum_{i=l+1}^{n} (n-i+1)K\{B(n, 2m-2s-2, -i)\}; \qquad 1 \le s \le m-2 \tag{12.38}$$

12.6.2 *Matrix Formulation*

The developments in this paragraph are due to Chen, Cyvin and Cyvin (1987); more details are found in the cited work. The exposition may be compared to Paragraphs 9.9.2, 9.9.3 and 9.9.4 concerning zigzag chains.

Notation. Let us introduce the notation

$$K\{B(n, 2m-2, t)\} = R_n^{(t)}(m) \tag{12.39}$$

It was already used (as alternative notation) in CHARTS IV-VI and Table 3. We also

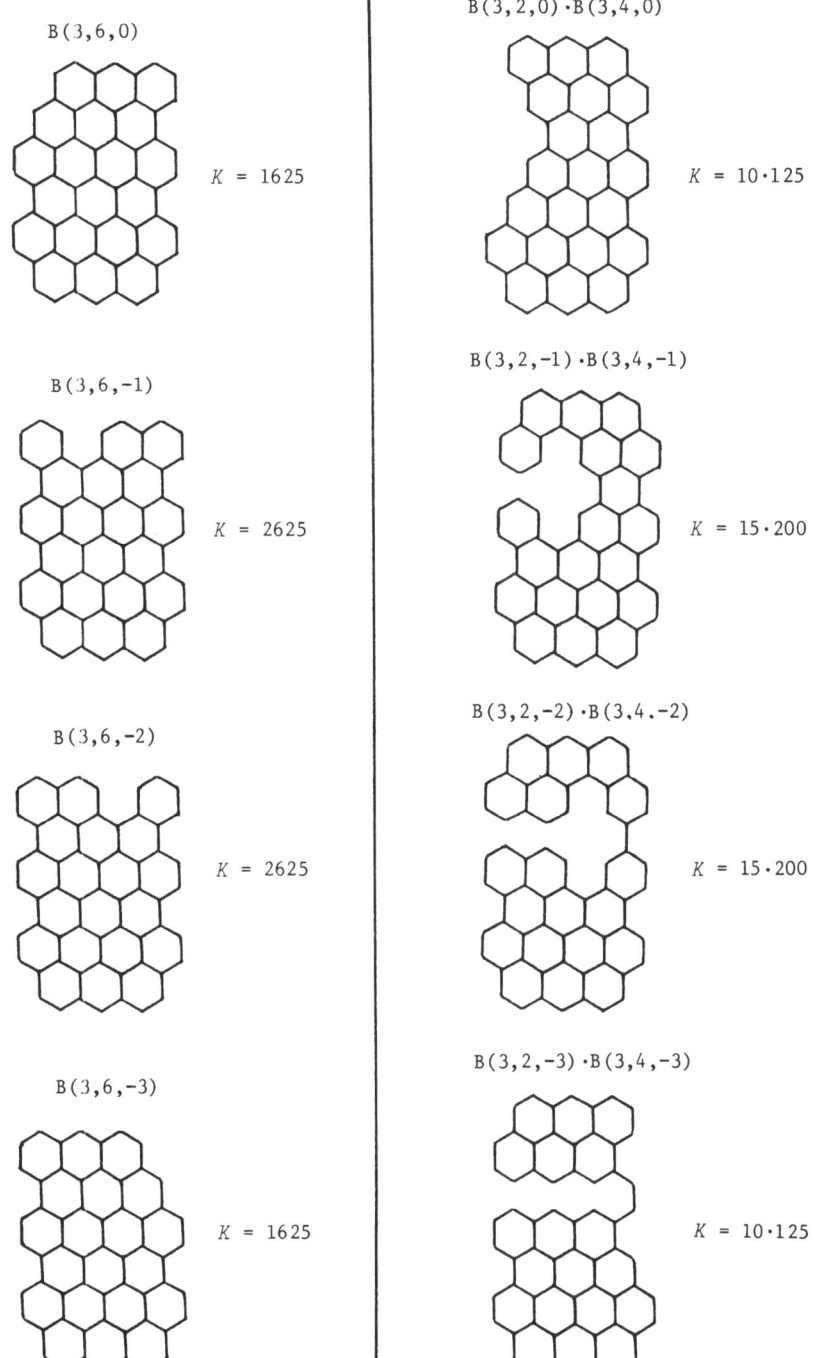

Fig. 12.8. Illustration of eqns. (35) and (36). The sum of K numbers for the four systems in each column is $K\{R\dot{J}(4,3)\} = 8500$.

recall the notation in the heading of Table 2, viz.

$$K\{R^j(m,n)\} = R_n(m) \tag{12.40}$$

In consistence with the formulas of Section 12.4 we have in the present notation:

$$R_n^{(n)}(m) = R_n(m) \tag{12.41}$$

and

$$R_n^{(l)}(1) = l+1, \qquad R_n^{(-l)}(1) = 1 \tag{12.42}$$

Remember that $0 \le l \le n$.

A Set of Linear Equations. The formula (35) now assumes the form

$$R_n(m) = \sum_{i=0}^{n} R_n^{(-i)}(m) \tag{12.43}$$

A more general formula reads

$$R_n(m+j) = \sum_{i=0}^{n} R_n^{(-i)}(j+1) R_n^{(-i)}(m) \tag{12.44}$$

where j is (primarily) an arbitrary positive integer or zero; in the latter case ($j=0$) eqn. (44) coincides with (43). (Later we shall allow for negative integer values of j.)

Let (44) be applied for $j = 0, 1, 2, \ldots, n$. Then a set of linear equations is obtained, which may be written as

$$\begin{bmatrix} R_n(m) \\ R_n(m+1) \\ R_n(m+2) \\ \cdot \\ \cdot \\ \cdot \\ R_n(m+n) \end{bmatrix} = M \begin{bmatrix} R_n^{(0)}(m) \\ R_n^{(-1)}(m) \\ R_n^{(-2)}(m) \\ \cdot \\ \cdot \\ \cdot \\ R_n^{(-n)}(m) \end{bmatrix} \tag{12.45}$$

Here M is a square $(n+1) \times (n+1)$ matrix with the general element equal to

$$(M)_{rs} = R_n^{(1-s)}(r) \tag{12.46}$$

The number of equations is drastically reduced by virtue of the symmetry properties (12) of the auxiliary benzenoids, or in the present notation

$$R_n^{(-l)} = R_n^{(l-n)} \tag{12.47}$$

The cases with even and odd n behave slightly differently. We will therefore exemplify both of these cases.

The Case of n=4. Here we are faced with the three unknowns $R_4^{(0)}$, $R_4^{(-1)}$ and $R_4^{(-2)}$, while $R_4^{(-3)} = R_4^{(-1)}$ and $R_4^{(-4)} = R_4^{(0)}$. The set of linear equations then reduces to three, viz.

$$
\begin{bmatrix} R_4(m) \\ R_4(m+1) \\ R_4(m+2) \end{bmatrix}
=
\begin{bmatrix}
2R_4^{(0)}(1) & 2R_4^{(-1)}(1) & R_4^{(-2)}(1) \\
2R_4^{(0)}(2) & 2R_4^{(-1)}(2) & R_4^{(-2)}(2) \\
2R_4^{(0)}(3) & 2R_4^{(-1)}(3) & R_4^{(-2)}(3)
\end{bmatrix}
\begin{bmatrix} R_4^{(0)}(m) \\ R_4^{(-1)}(m) \\ R_4^{(-2)}(m) \end{bmatrix}
\tag{12.48}
$$

The numerical values of the square matrix elements are available from eqns. (42), (15) and (16).

The Case of n=5. There are again three unknowns, viz. $R_5^{(0)}$, $R_5^{(-1)}$ and $R_5^{(-2)}$, while $R_5^{(-3)} = R_5^{(-2)}$, $R_5^{(-4)} = R_5^{(-1)}$, and $R_5^{(-5)} = R_5^{(0)}$. The set of linear equations may be written as

$$
\frac{1}{2}\begin{bmatrix} R_5(m) \\ R_5(m+1) \\ R_5(m+2) \end{bmatrix}
=
\begin{bmatrix}
R_5^{(0)}(1) & R_5^{(-1)}(1) & R_5^{(-2)}(1) \\
R_5^{(0)}(2) & R_5^{(-1)}(2) & R_5^{(-2)}(2) \\
R_5^{(0)}(3) & R_5^{(-1)}(3) & R_5^{(-2)}(3)
\end{bmatrix}
\begin{bmatrix} R_5^{(0)}(m) \\ R_5^{(-1)}(m) \\ R_5^{(-2)}(m) \end{bmatrix}
\tag{12.49}
$$

or with numerical values inserted:

$$
\frac{1}{2}\begin{bmatrix} R_5(m) \\ R_5(m+1) \\ R_5(m+2) \end{bmatrix}
=
\begin{bmatrix}
1 & 1 & 1 \\
21 & 35 & 42 \\
686 & 1225 & 1519
\end{bmatrix}
\begin{bmatrix} R_5^{(0)}(m) \\ R_5^{(-1)}(m) \\ R_5^{(-2)}(m) \end{bmatrix}
\tag{12.50}
$$

This case of $n=5$ will be elaborated somewhat further. We give therefore the linear equations obtained from (50);

$$R_5^{(0)}(m) = \frac{1}{98} R_5(m+2) - \frac{3}{7} R_5(m+1) + \frac{5}{2} R_5(m) \tag{12.51}$$

$$R_5^{(-1)}(m) = -\frac{3}{98} R_5(m+2) + \frac{17}{14} R_5(m+1) - \frac{9}{2} R_5(m) \tag{12.52}$$

$$R_5^{(-2)}(m) = \frac{1}{49} R_5(m+2) - \frac{11}{14} R_5(m+1) + \frac{5}{2} R_5(m) \tag{12.53}$$

Connection Between R_n and $R_n^{(0)}$. Eqn. (37) reads in the present notation:

$$R_n^{(0)}(m) = \sum_{i=0}^{n} (i+1) R_n^{(-i)}(m-1)$$

(12.54)

It was used to establish the following connection between R_n and $R_n^{(0)}$:

$$R_n(m) = \frac{2}{n+2} R_n^{(0)}(m+1)$$

(12.55)

Nominal Values of $R_n^{(-l)}$. Nominal values of $R_n^{(-l)}(m)$ should fit the system of equations, but are extrapolated to m values for which no benzenoid systems exist. The values in question for $m=0$, for instance, are obtained on inserting $j = -1$ in (44) and relating the obtained summation to (55). The result is

$$R_n^{(-l)}(0) = 0; \quad 0 < l < n, \qquad R_n^{(0)}(0) = R_n^{(-n)}(0) = \frac{1}{n+2}$$

(12.56)

Recurrence Relation. It is of interest to deduce the recurrence relation for R_n, i.e. the linear dependence between the quantities $R_n(m+j)$; see Table 2 for the results through $n=10$. The set of linear equations (see above) is independent when the $n+1$ equations (45) are reduced by virtue of the symmetry properties (47). Their number then becomes $\left[\frac{n+2}{2}\right]$, i.e. 1 for $n=1$, 2 for $n=2$ and $n=3$, 3 for $n=4$ and $n=5$, etc. The dependence between the R_n quantities is introduced by adding one equation more to the set. Consequently we can predict at once the number of terms in the recurrence relations; it is $\left[\frac{n+4}{2}\right]$ in consistence with Table 2.

An obvious way to derive the recurrence relation would be to add an equation by increasing j in $R_n(m+j)$ by one unit. The computation becomes substantially easier, however, when the matrix M is augmented by a row on top of it, i.e. $j = -1$. Here we take advantage of the nominal values (56).

For the case of $m=5$ eqn. (50) augmented in the described way reads:

$$\frac{1}{2} \begin{bmatrix} R_5(m-1) \\ R_5(m) \\ R_5(m+1) \\ R_5(m+2) \end{bmatrix} = \begin{bmatrix} \frac{1}{7} & 0 & 0 \\ 1 & 1 & 1 \\ 21 & 35 & 42 \\ 686 & 1225 & 1519 \end{bmatrix} \begin{bmatrix} R_5^{(0)}(m) \\ R_5^{(-1)}(m) \\ R_5^{(-2)}(m) \end{bmatrix}$$

(12.57)

The first row yields

$$R_5^{(0)}(m) = \frac{7}{2} R_5(m-1)$$

(12.58)

in consistence with eqn. (55). On equating (51) and (58) the recurrence relation for R_5 (see Table 2) is readily obtained.

Incorporation of the Quantities $R^{(l)}$. From eqn. (13) it is deduced

$$R_n^{(n-k)}(m) = R_n^{(n-k-1)}(m) + R_n^{(-k)}(m); \quad k = 0, 1, 2, \ldots, n-1 \qquad (12.59)$$

It is again expedient to employ the symmetry properties (47) in practical applications of (59). For $n=5$ in particular one has:

$$R_5(m) = R_5^{(4)}(m) + R_5^{(0)}(m), \qquad R_5^{(4)}(m) = R_5^{(3)}(m) + R_5^{(-1)}(m),$$

$$R_5^{(3)}(m) = R_5^{(2)}(m) + R_5^{(-2)}(m), \qquad R_5^{(2)}(m) = R_5^{(1)}(m) + R_5^{(-2)}(m), \qquad (12.60)$$

$$R_5^{(1)}(m) = R_5^{(0)}(m) + R_5^{(-1)}(m)$$

As nominal values for $m=0$ one attains at

$$R_n^{(l)}(0) = \frac{1}{n+2}; \quad 0 \le l < n, \qquad R_n(0) = \frac{2}{n+2} \quad (l=n) \qquad (12.61)$$

Relations for $R_5^{(t)}(m)$. It is important to realize that the quantities $R_n^{(t)}(m)$ are linearly dependent to the effect that they obey the same recurrence relation for a given n.

Without going into further details we finally give the expressions for the different $R_5^{(t)}(m)$ quantities as linear combinations of $R_5(m-j)$.

$$R_5^{(4)}(m) = R_5(m) - \frac{7}{2} R_5(m-1), \qquad R_5^{(3)}(m) = R_5(m) - \frac{21}{2} R_5(m-1) + \frac{49}{2} R_5(m-2),$$

$$R_5^{(2)}(m) = \frac{1}{2} R_5(m), \qquad R_5^{(1)}(m) = \frac{21}{2} R_5(m-1) - \frac{49}{2} R_5(m-2),$$

$$R_5^{(0)}(m) = \frac{7}{2} R_5(m-1), \qquad R_5^{(-1)}(m) = 7 R_5(m-1) - \frac{49}{2} R_5(m-2), \qquad (12.62)$$

$$R_5^{(-2)}(m) = \frac{1}{2} R_5(m) - \frac{21}{2} R_5(m-1) + \frac{49}{2} R_5(m-2)$$

All these quantities obey recurrence relations of the same form as the one of $R_5(m)$ in Table 2.

12.6.3 *Further Developments*

Further investigations by Cyvin SJ, Chen and Cyvin (1987), supplemented by still newer results of R.S. Chen and S.J. Cyvin, have led to several general formulations in connection with the recurrence relation for R_n (Table 2). Some of these results are reviewed here very briefly. The special classes of auxiliary benzenoids for $t = -n'$ play an important role in that work; here $n' = \frac{1}{2} n$ for even n and $n' = \frac{1}{2}(n-1)$ or $n' = \frac{1}{2}(n+1)$ for odd n.

Let the recurrence relation be written

$$R_n(m+1) = \sum_{j=0}^{n'} c_j R_n(m-j); \qquad n' = \left[\frac{n}{2}\right] \qquad (12.63)$$

Here the form (63) was chosen so that the j values also indicate the different quantities $R_n^{(-l)}$ (reduced by virtue of symmetry). They are in other words $R_n^{(0)}$, $R_n^{(-1)}$, $R_n^{(-2)}$, ..., $R_n^{(-n')}$.

It was observed that

$$c_0 = R_n^{(-n')}(2) \qquad (12.64)$$

Furthermore:

$$c_j > 0; \quad j = 0, 2, 4, ..., \qquad c_j < 0; \quad j = 1, 3, 5, ... \qquad (12.65)$$

and

$$c_{n'} = -c_{n'-1}; \qquad n = 2, 4, 6, ... \qquad (12.66)$$

A general method for successive derivation of the coefficients c_j by means of $R_n^{(-n')}(m)$ values was developed. Eqn. (64) for c_0 is followed by

$$c_1 = R_n^{(-n')}(3) - c_0 R_n^{(-n')}(2)$$

$$c_2 = R_n^{(-n')}(4) - c_0 R_n^{(-n')}(3) - c_1 R_n^{(-n')}(2)$$

etc. In general:

$$c_j = R_n^{(-n')}(j+2) - \sum_{k=0}^{j-1} c_k R_n^{(-n')}(j-k+1) \qquad (12.67)$$

By means of eqns. (15)-(17) formulas for $R_n^{(-n')}(m)$ were expanded as polynomials in n for $m = 2$, 3 and 4. The result is:

$$R_n^{(-n')}(2) = \begin{cases} \frac{1}{8}(n+2)^3; & n = 0, 2, 4, ... \\ \frac{1}{8}(n+1)(n+2)(n+3); & n = 1, 3, 5, ... \end{cases} \qquad (12.68)$$

$$R_n^{(-n')}(3) = \begin{cases} \dfrac{1}{384}(n+2)^4(5n^2 + 20n + 24); & n = 0, 2, 4, \ldots \\[3mm] \dfrac{1}{384}(n+1)(n+2)^2(n+3)(5n^2 + 20n + 23); & n = 1, 3, 5, \ldots \end{cases} \qquad (12.69)$$

$$R_n^{(-n')}(4) = \begin{cases} \dfrac{1}{46080}(n+2)^5(61n^4 + 488n^3 + 1564n^2 + 2352n + 1440); & n = 0,2,4, \ldots \\[3mm] \dfrac{1}{46080}(n+1)(n+2)^3(n+3)(61n^4 + 488n^3 + 1550n^2 + 2296n + 1365); \\[2mm] \hspace{6cm} n = 1,3,5, \ldots \end{cases}$$

$$(12.70)$$

Finally we give an explicit formula for the coefficients c_j. It reproduces, as special cases, all the coefficients of Table 2.

$$c_j = \begin{cases} \dfrac{(-1)^j (n+2)^{j+2}}{4(j+1)} \begin{pmatrix} \frac{n}{2} + j + 1 \\ 2j + 1 \end{pmatrix}; & n = 0, 2, 4, \ldots \\[5mm] (-1)^j (n+2)^{j+1} \begin{pmatrix} \frac{n+1}{2} + j + 1 \\ 2j + 2 \end{pmatrix}; & n = 1, 3, 5, \ldots \end{cases} \qquad (12.71)$$

CHAPTER 13

REGULAR SIX-TIER STRIPS AND RELATED SYSTEMS

13.1 INTRODUCTION

A regular t-tier strip is defined in Chapter 10. In the present chapter we give the K formulas (as polynomials in n) for all classes of 6-tier strips. Some of these formulas have been given more or less sporadically in different connections (Cyvin and Gutman 1986a; Cyvin SJ, Cyvin and Bergan 1986; Gutman and Cyvin 1987b; Cyvin 1987b; Cyvin BN and Cyvin 1987), but a systematic treatment has not been published before.

Also some K formulas for 7-tier strips are reported here, in addition to more general formulas for t-tier strips.

13.2 SIX-TIER STRIPS

13.2.1 *Classification*

The 6-tier strips are divided into three series, viz. I, II and III, which all contain skew strips. The top-bottom shifts are $\frac{1}{2}$, $\frac{3}{2}$ and $\frac{5}{2}$, respectively. The leading hexagons are (I) 0(3,4,n), (II) 0(2,5,n) and (III) 0(1,6,n).

Series I. The first series (I) consists of 55 classes. A survey is shown in Fig. 1. The notation therein is systematic (cf. Paragraph 10.3.4). Especially the concept of towers (H) could not be avoided; in fact it dominates the whole scheme.

Series II. The second series (II) consists of 15 classes, which resemble those of the 5-tier strips (see Chapter 10). The concept of towers may almost be avoided. Table 1 gives a survey of the classes and their terminology.

Series III. The last series (here III) consists as always of a unique class of parallelograms, here L(6,n).

13.2.2 *Dictionary of K Formulas for Regular 6-Tier Strips*

Table 2 comprises the K formulas as factored polynomials in n for all classes of 6-tier strips. Several of the formulas are also given in terms of binomial coefficients. Numerical values [in brackets] for h and K at $n=3$ are given.

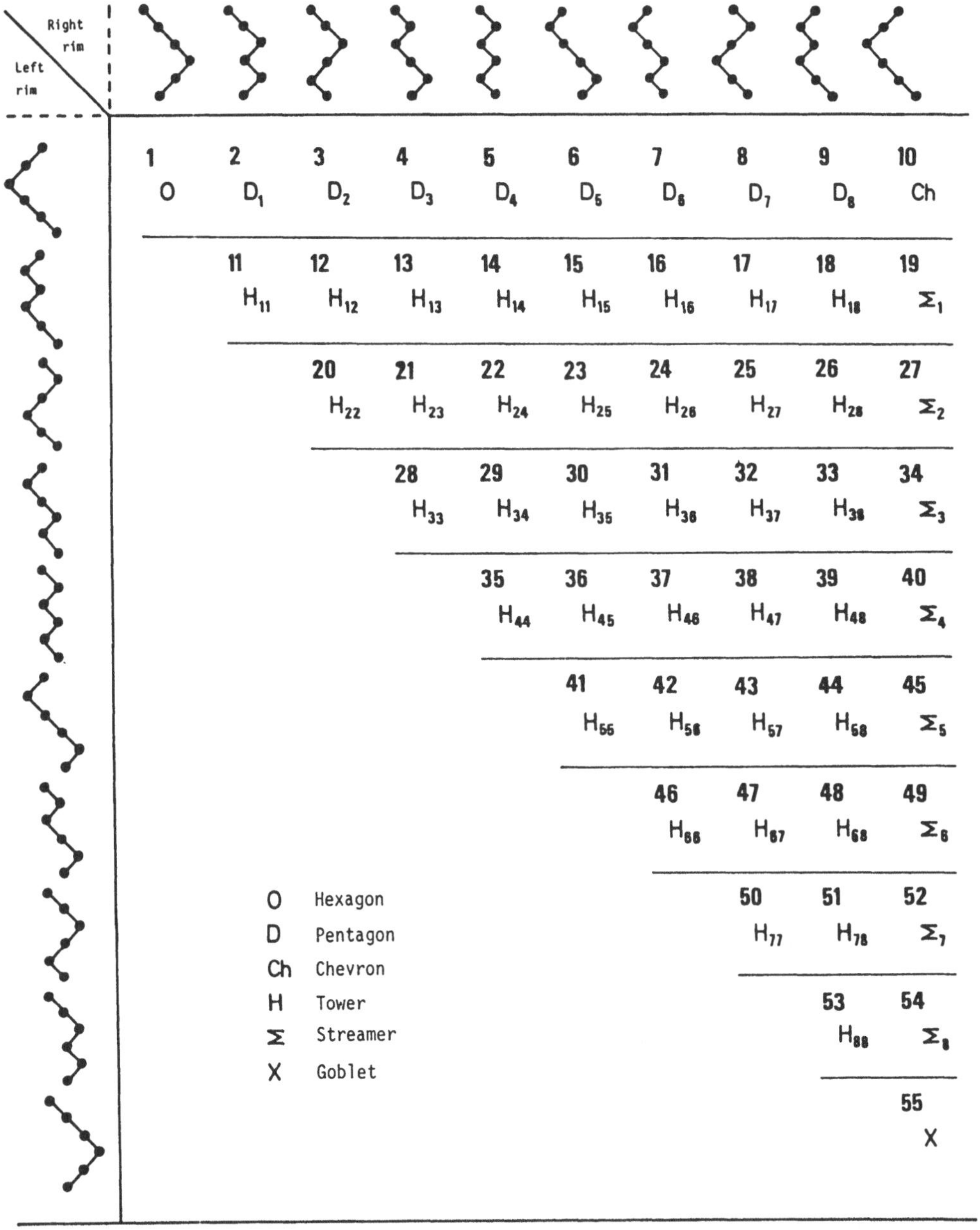

	Right rim Left rim	1	2	3	4	5	6	7	8	9	10
		O	D_1	D_2	D_3	D_4	D_5	D_6	D_7	D_8	Ch
			11	12	13	14	15	16	17	18	19
			H_{11}	H_{12}	H_{13}	H_{14}	H_{15}	H_{16}	H_{17}	H_{18}	Σ_1
				20	21	22	23	24	25	26	27
				H_{22}	H_{23}	H_{24}	H_{25}	H_{26}	H_{27}	H_{28}	Σ_2
					28	29	30	31	32	33	34
					H_{33}	H_{34}	H_{35}	H_{36}	H_{37}	H_{38}	Σ_3
						35	36	37	38	39	40
						H_{44}	H_{45}	H_{46}	H_{47}	H_{48}	Σ_4
							41	42	43	44	45
							H_{55}	H_{56}	H_{57}	H_{58}	Σ_5
								46	47	48	49
								H_{66}	H_{67}	H_{68}	Σ_6
									50	51	52
									H_{77}	H_{78}	Σ_7
										53	54
										H_{88}	Σ_8
											55
											X

O Hexagon
D Pentagon
Ch Chevron
H Tower
Σ Streamer
X Goblet

Fig. 13.1. Survey of the 55 classes of Series I of 6-tier strips: Indices: $(3,4,n)$.

Table 13.1. Terminology for the classes of Series II of 6-tier strips.

(Series) No.	Systematic notation	Practical notation and designation
(II)		
1.	$O(2,5,n)$	$O(2,5,n)$ Hexagon (centrosymmetrical)
2.	$D_1(2,5,n)$	$D^j(2,5,n)$ Oblate pentagon
3.	$D_2(2,5,n)$	$D(2,5,n)$ Intermediate pentagon
4.	$D_3(2,5,n)$	$D^i(2,5,n)$ Prolate pentagon
5.	$Ch(2,5,n)$	$Ch(2,5,n)$ Chevron (asymmetrical)
6.	$H_{11}(2,5,n)$	$H^j(2,5,n)$ Oblate tower
7.	$H_{12}(2,5,n)$	$H(2,5,n)$ Intermediate tower
8.	$H_{13}(2,5,n)$	$M_n(LLLAAL)$ Multiple chain; $H^i(2,5,n)$ Prolate tower
9.	$\Sigma_1(2,5,n)$	$\Sigma^j(2,5,n)$ Oblate streamer
10.	$H_{22}(2,5,n)$	$M_n(LLAALL)$ Multiple chain
11.	$H_{23}(2,5,n)$	$L(2,n)\cdot L(3,n)$ Essentially disconnected; see also Series I, No. 19
12.	$\Sigma_2(2,5,n)$	$\Sigma(2,5,n)$ Intermediate streamer; also $L(n)\cdot L(3,n)$ as Series I, No. 27
13.	$H_{33}(2,5,n)$	$L(2,n)\cdot L(2,n)$ Essentially disconnected; see also Series I, Nos. 34 and 50
14.	$\Sigma_3(2,5,n)$	$\Sigma^i(2,5,n)$ Prolate streamer; also $L(n)\cdot L(2,n)$ as Series I, Nos. 40, 45 and 51
15.	$X(2,5,n)$	$X(2,5,n)$ Goblet (centrosymmetrical); also $L(n)\cdot L(n)$ as Series I, Nos. 49 and 53

Table 13.2. Classes of regular 6-tier strips. Formulas for K.

---

1 $O(3,4,n)$ <u>Hexagon</u> (centrosymm.)

[24, 4116]

$$\frac{1}{75}\binom{n+4}{4}\binom{n+5}{4}\binom{n+6}{4}$$

$$= \frac{1}{1036800}(n+1)(n+2)^2(n+3)^3(n+4)^3(n+5)^2(n+6)$$

---

2 $D_1(3,4,n)$ <u>Pentagon</u>

[23, 3626]

$$\frac{1}{75}\binom{n+4}{4}\binom{n+5}{4}\left[\binom{n+6}{4} - \binom{n+3}{4}\right]$$

$$= \frac{1}{86400}(n+1)(n+2)^2(n+3)^3(n+4)^2(n+5)(n^2 + 6n + 10)$$

---

3 $D_2(3,4,n)$ <u>Pentagon</u>

[22, 2450]

$$\frac{1}{12}\binom{n+3}{2}\binom{n+4}{3}\binom{n+5}{5}$$

$$= \frac{1}{17280}(n+1)(n+2)^3(n+3)^3(n+4)^2(n+5)$$

---

4 $D_3(3,4,n)$ <u>Pentagon</u>

[22, 2646]

$$\frac{1}{43200}(n+1)(n+2)^2(n+3)^2(n+4)^2(n+5)(3n^2 + 13n + 15)$$

---

5 $D_4(3,4,n)$ <u>Pentagon</u>

[21, 1925]

$$\frac{1}{4320}(n+1)(n+2)^3(n+3)^3(n+4)(2n + 5)$$

---

6 $D_5(3,4,n)$ <u>Pentagon</u>

[21, 1470]

$$\frac{1}{8640}(n+1)(n+2)^2(n+3)^2(n+4)^2(n+5)(2n + 3)$$

---

7 $D_6(3,4,n)$ <u>Pentagon</u>

[20, 1120]

$$\frac{1}{2880}(n+1)(n+2)^2(n+3)^2(n+4)(5n^2 + 21n + 20)$$

---

8 $D_7(3,4,n)$ <u>Pentagon</u>

[20, 1015]

$$\frac{1}{720}(n+1)(n+2)^2(n+3)^2(n+4)(n^2 + 5n + 5)$$

---

9 $D_8(3,4,n)$ <u>Pentagon</u>

[19, 665]

$$\frac{1}{720}(n+1)(n+2)^2(n+3)(n+4)(5n^2 + 18n + 15)$$

---

10 $Ch(3,4,n)$ <u>Chevron</u> (unsymm.)

$M_n(LLLALL)$ Multiple chain

[18, 273]

$$\binom{n+2}{2}\binom{n+4}{4} - \binom{n+1}{1}\binom{n+4}{5} + \binom{n+4}{6}$$

$$= \frac{1}{360}(n+1)(n+2)(n+3)(n+4)(5n^2 + 19n + 15)$$

---

11 $H_{11}(3,4,n)$ <u>Tower</u>

[22, 3171]

$$\frac{1}{86400}(n+1)(n+2)^2(n+3)^2(n+4)(11n^4 + 110n^3 + 439n^2 + 820n + 600)$$

---

12 $H_{12}(3,4,n)$ <u>Tower</u>

[21, 2100]

$$\frac{1}{8640}(n+1)(n+2)^3(n+3)^2(n+4)(5n^2 + 23n + 30)$$

13

$H_{13}(3,4,n)$ Tower

[21, 2275]

$$\frac{1}{1440}(n+1)(n+2)^3(n+3)^2(n+4)(n^2+4n+5)$$

14

$H_{14}(3,4,n)$ Tower

[20, 1625]

$$\frac{1}{240}(n+1)(n+2)^4(n+3)(n^2+4n+5)$$

15

$H_{15}(3,4,n)$ Tower

[20, 1225]

$$\frac{1}{1440}(n+1)(n+2)^3(n+3)^2(n+4)(3n+5)$$

16

$H_{16}(3,4,n)$ Tower

[19, 915]

$$\frac{1}{720}(n+1)(n+2)^2(n+3)(2n+3)(5n^2+19n+20)$$

17

$H_{17}(3,4,n)$ Tower

[19, 825]

$$\frac{1}{360}(n+1)(n+2)^3(n+3)(2n+3)(2n+5)$$

18

$H_{18}(3,4,n)$ Tower
$M_n(LLAAAL)$ Multiple chain

[18, 523]

$$\binom{n+2}{1}\left[\frac{1}{2}\binom{n+2}{2}\binom{n+3}{3}+\binom{n+3}{5}\right]-\binom{n+4}{6}$$
$$=\frac{1}{720}(n+1)(n+2)(n+3)(35n^3+153n^2+232n+120)$$

19

$\Sigma_1(3,4,n)$ Streamer
$L(2,n)\cdot L(3,n)$ Essent. disconn.

[17, 200]

$$\frac{1}{12}(n+1)^2(n+2)^2(n+3)$$

20

$H_{22}(3,4,n)$ Tower

[20, 1330]

$$\frac{1}{2880}(n+1)(n+2)^2(n+3)^2(n+4)(7n^2+23n+20)$$

21

$H_{23}(3,4,n)$ Tower

[20, 1435]

$$\frac{1}{720}(n+1)(n+2)^2(n+3)^2(n+4)(2n^2+6n+5)$$

22

$H_{24}(3,4,n)$ Tower

[19, 985]

$$\frac{1}{720}(n+1)(n+2)^2(n+3)(11n^3+58n^2+101n+60)$$

23

$H_{25}(3,4,n)$ Tower
$L(n)\cdot O(2,3,n)$ Essent. disconn.

[19, 700]

$$\frac{1}{144}(n+1)^2(n+2)^2(n+3)^2(n+4)$$

24

$H_{26}(3,4,n)$ Tower
$L(n)\cdot D(2,3,n)$ Essent. disconn.

[18, 500]

$$\frac{1}{24}(n+1)^2(n+2)^3(n+3)$$

25

$H_{27}(3,4,n)$ Tower
$M_n(LLALAL)$ Multiple chain

[18, 457]

$$\frac{1}{360}(n+1)(n+2)(n+3)(13n^3+69n^2+113n+60)$$

26

$H_{28}(3,4,n)$ Tower
$L(n)\cdot Ch(2,3,n)$ Essent. disconn.

[17, 260]

$$\frac{1}{24}(n+1)^2(n+2)(n+3)(3n+4)$$

27 $\Sigma_2(3,4,n)$ Streamer

$L(n)\cdot L(3,n)$ Essent. disconn.

[16, 80]

$$\frac{1}{6}(n+1)^2(n+2)(n+3)$$

28 $H_{33}(3,4,n)$ Tower

$E(3,4,n)$ Étagère

[20, 1575]

$$\binom{n+3}{2}^2\binom{n+4}{4} - 2\binom{n+2}{1}\binom{n+3}{2}\binom{n+5}{5} + \binom{n+2}{1}^2\binom{n+6}{6}$$

$$= \frac{1}{1440}(n+1)(n+2)^3(n+3)(n+4)(5n^2 + 16n + 15)$$

29 $H_{34}(3,4,n)$ Tower

[19, 1075]

$$\frac{1}{720}(n+1)(n+2)^3(n+3)(13n^2 + 37n + 30)$$

30 $H_{35}(3,4,n)$ Tower

[19, 805]

$$\frac{1}{720}(n+1)(n+2)^2(n+3)(n+4)(7n^2 + 20n + 15)$$

31 $H_{36}(3,4,n)$ Tower

$M_n(LAALAL)$ Multiple chain

[18, 573]

$$\frac{1}{360}(n+1)(n+2)(n+3)(2n + 3)(10n^2 + 27n + 20)$$

32 $H_{37}(3,4,n)$.   Tower

$L(2,n)\cdot O(2,n)$ Essent. disconn.

[18, 500]

$$\frac{1}{24}(n+1)^2(n+2)^3(n+3)$$

33 $H_{38}(3,4,n)$ Tower

$L(2,n)\cdot Ch(2,n)$ Essent. disconn.

[17, 300]

$$\frac{1}{12}(n+1)^2(n+2)^2(2n + 3)$$

34 $\Sigma_3(3,4,n)$ Streamer

$L(2,n)\cdot L(2,n)$ Essent. disconn.

[16, 100]

$$\frac{1}{4}(n+1)^2(n+2)^2$$

35 $H_{44}(3,4,n)$ Tower

$Z(6,n)$ Zigzag chain

$M_n(LAAAAL)$ Multiple chain

[18, 707]

$$\binom{n+2}{3}^3 - 2\binom{n+2}{2}\binom{n+3}{4} + \binom{n+4}{6}$$

$$= \frac{1}{720}(n+1)(n+2)(61n^4 + 366n^3 + 845n^2 + 888n + 360)$$

36 $H_{45}(3,4,n)$ Tower

$L(n)\cdot D(2,3,n)$ Essent. disconn.

[18, 500]

$$\frac{1}{24}(n+1)^2(n+2)^3(n+3)$$

37 $H_{46}(3,4,n)$ Tower

$L(n)\cdot Z(4,n)$

Essent. disconn.

[17,340]

$$\frac{1}{24}(n+1)^2(n+2)(5n^2 + 15n + 12)$$

38 $H_{47}(3,4,n)$ Tower

$L(2,n)\cdot Ch(2,n)$

Essent. disconn.

[17,300]

$$\frac{1}{12}(n+1)^2(n+2)^2(2n + 3)$$

39 $H_{48}(3,4,n)$ Tower

$L(n)\cdot L(n)\cdot L(2,n)$

Essent.disconn.

[16, 160]

$$\frac{1}{2}(n+1)^3(n+2)$$

40 $\Sigma_4(3,4,n)$ Streamer

$L(n)\cdot L(2,n)$ Essent. disconn.

[15, 40]

$$\frac{1}{2}(n+1)^2(n+2)$$

41 $H_{55}(3,4,n)$ Tower

$M_n(LALLAL)$ Multiple chain

[18, 399]

$$\binom{n+1}{1}\left[\binom{n+1}{1}\binom{n+4}{4} - 2\binom{n+4}{5}\right] + \binom{n+4}{6}$$

$$= \frac{1}{720}(n+1)(n+2)(n+3)(n+4)(19n^2 + 47n + 30)$$

42	$H_{56}(3,4,n)$ Tower $L(n)\cdot Ch(2,3,n)$ Essent. <u>disconn.</u> [17, 260] $\frac{1}{24}(n+1)^2(n+2)(n+3)(3n+4)$	**43**	$H_{57}(3,4,n)$ Tower $L(n)\cdot O(2,n)$ Essent. <u>disconn.</u> [17, 200] $\frac{1}{12}(n+1)^2(n+2)^2(n+3)$

44	$H_{58}(3,4,n)$ Tower $L(n)\cdot Ch(2,n)$ Essent. disconn. [16, 120] $\frac{1}{6}(n+1)^2(n+2)(2n+3)$	**45**	$\Sigma_5(3,4,n)$ Streamer $L(n)\cdot L(2,n)$ Essent. disconn. [15, 40] $\frac{1}{2}(n+1)^2(n+2)$	**46**	$H_{66}(3,4,n)$ $L(n)\cdot L(2,n)\cdot L(n)$ Essent. disconn. [16, 160] $\frac{1}{2}(n+1)^3(n+2)$
47	$H_{67}(3,4,n)$ Tower $L(n)\cdot Ch(2,n)$ Essent. disconn. [16, 120] $\frac{1}{6}(n+1)^2(n+2)(2n+3)$	**48**	$H_{68}(3,4,n)$ Tower $L(n)\cdot L(n)\cdot L(n)$ Essent. disconn. [15, 64] $(n+1)^3$	**49**	$\Sigma_6(3,4,n)$ Streamer $L(n)\cdot L(n)$ Essent. disconn. [14, 16] $(n+1)^2$
50	$H_{77}(3,4,n)$ Tower $L(2,n)\cdot L(2,n)$ Essent. disconn. [16, 100] $\frac{1}{4}(n+1)^2(n+2)^2$	**51**	$H_{78}(3,4,n)$ Tower $L(n)\cdot L(2,n)$ Essent. disconn. [15, 40] $\frac{1}{2}(n+1)^2(n+2)$	**52**	$\Sigma_7(3,4,n)$ Streamer [14, 0] 0
53	$H_{88}(3,4,n)$ Tower $L(n)\cdot L(n)$ Essent. <u>disconn.</u> [14, 16] $(n+1)^2$	**54**	$\Sigma_8(3,4,n)$ Streamer [13, 0] 0	**55**	$X(3,4,n)$ Goblet (centro- symm.) [12, 0] 0

⓪⓪ **1**	$O(2,5,n)$ Hexagon (centrosymm.) [22, 1176] $\frac{1}{6}\binom{n+5}{5}\binom{n+6}{5}$ $= \frac{1}{86400}(n+1)(n+2)^2(n+3)^2(n+4)^2(n+5)^2(n+6)$	**2**	$D^j(2,5,n)$ <u>Oblate</u> <u>pentagon</u> [21, 980] $\frac{1}{6}\binom{n+5}{5}\left[\binom{n+4}{4}+\binom{n+5}{4}\right]$ $= \frac{1}{8640}(n+1)(n+2)^2(n+3)^3(n+4)^2(n+5)$
3	$D(2,5,n)$ <u>Intermediate</u> <u>pentagon</u> [20, 700] $\frac{1}{6}\binom{n+2}{1}\binom{n+3}{2}\binom{n+5}{5}$ $= \frac{1}{1440}(n+1)(n+2)^3(n+3)^2(n+4)(n+5)$	**4**	$D^i(2,5,n)$ <u>Prolate</u> <u>pentagon</u> [19, 420] $\binom{n+2}{2}\binom{n+5}{5} - \binom{n+2}{1}\binom{n+5}{6}$ $= \frac{1}{720}(n+1)(n+2)^2(n+3)(n+4)(n+5)(2n+3)$

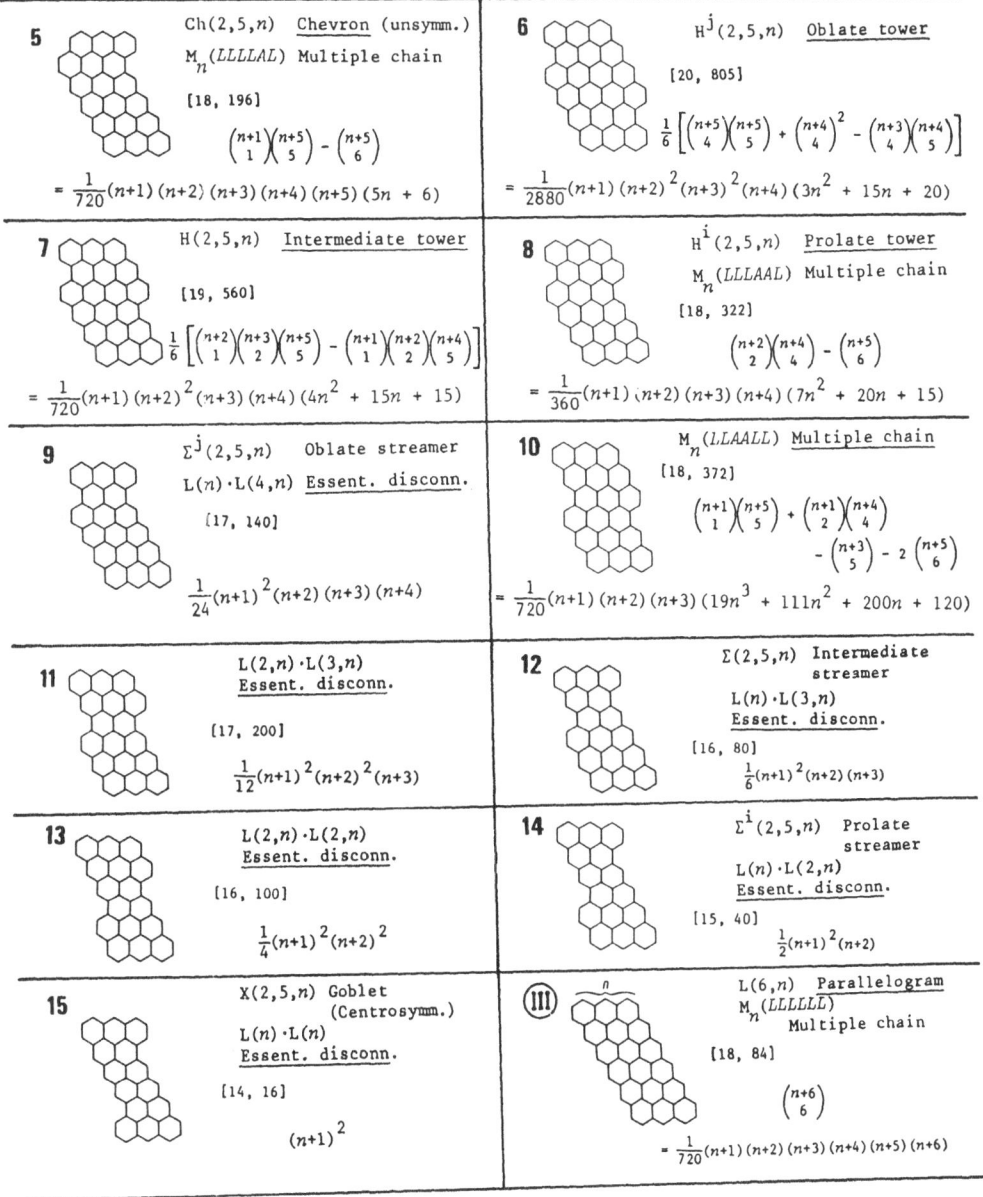

5 Ch(2,5,n) Chevron (unsymm.)
M_n(LLLLAL) Multiple chain
[18, 196]
$$\binom{n+1}{1}\binom{n+5}{5} - \binom{n+5}{6}$$
$$= \frac{1}{720}(n+1)(n+2)(n+3)(n+4)(n+5)(5n+6)$$

6 H^j(2,5,n) Oblate tower
[20, 805]
$$\frac{1}{6}\left[\binom{n+5}{4}\binom{n+5}{5} + \binom{n+4}{4}^2 - \binom{n+3}{4}\binom{n+4}{5}\right]$$
$$= \frac{1}{2880}(n+1)(n+2)^2(n+3)^2(n+4)(3n^2+15n+20)$$

7 H(2,5,n) Intermediate tower
[19, 560]
$$\frac{1}{6}\left[\binom{n+2}{1}\binom{n+3}{2}\binom{n+5}{5} - \binom{n+1}{1}\binom{n+2}{2}\binom{n+4}{5}\right]$$
$$= \frac{1}{720}(n+1)(n+2)^2(n+3)(n+4)(4n^2+15n+15)$$

8 H^i(2,5,n) Prolate tower
M_n(LLLAAL) Multiple chain
[18, 322]
$$\binom{n+2}{2}\binom{n+4}{4} - \binom{n+5}{5}$$
$$= \frac{1}{360}(n+1)(n+2)(n+3)(n+4)(7n^2+20n+15)$$

9 Σ^j(2,5,n) Oblate streamer
L(n)·L(4,n) Essent. disconn.
[17, 140]
$$\frac{1}{24}(n+1)^2(n+2)(n+3)(n+4)$$

10 M_n(LLAALL) Multiple chain
[18, 372]
$$\binom{n+1}{1}\binom{n+5}{5} + \binom{n+1}{2}\binom{n+4}{4}$$
$$- \binom{n+3}{5} - 2\binom{n+5}{6}$$
$$= \frac{1}{720}(n+1)(n+2)(n+3)(19n^3+111n^2+200n+120)$$

11 L(2,n)·L(3,n)
Essent. disconn.
[17, 200]
$$\frac{1}{12}(n+1)^2(n+2)^2(n+3)$$

12 Σ(2,5,n) Intermediate streamer
L(n)·L(3,n)
Essent. disconn.
[16, 80]
$$\frac{1}{6}(n+1)^2(n+2)(n+3)$$

13 L(2,n)·L(2,n)
Essent. disconn.
[16, 100]
$$\frac{1}{4}(n+1)^2(n+2)^2$$

14 Σ^i(2,5,n) Prolate streamer
L(n)·L(2,n)
Essent. disconn.
[15, 40]
$$\frac{1}{2}(n+1)^2(n+2)$$

15 X(2,5,n) Goblet
(Centrosymm.)
L(n)·L(n)
Essent. disconn.
[14, 16]
$$(n+1)^2$$

(III) L(6,n) Parallelogram
M_n(LLLLLL̄)
Multiple chain
[18, 84]
$$\binom{n+6}{6}$$
$$= \frac{1}{720}(n+1)(n+2)(n+3)(n+4)(n+5)(n+6)$$

13.3 SUPPLEMENT TO THE METHODS OF DERIVATION OF
K FORMULAS FOR t-TIER STRIPS

13.3.1 *Introduction*

Methods of derivation of K formulas for t-tier strips were treated in Section
10.6. In the present section we mainly give a supplement showing a further exploitation of the fragmentation method supported by algebraic computations. This supplement is actually to be classified under the method called chopping; cf. Paragraph
10.6.4 and especially Fig. 10.8. It is also referred to Figs. 11.4 and 12.8.

This advanced method of chopping was first applied to $R^j(4,n)$, the seven-tier
oblate rectangle (Cyvin SJ, Cyvin and Bergan 1986). Eqn. (12.36) with $p=2$ and $q=1$
gives

$$K\{R^j(4,n)\} = \sum_{i=0}^{n} K\{B(n,\ 4,\ -i)\}\cdot K\{B(n,\ 2,\ -i)\} \tag{13.1}$$

By means of the expression for $K\{B(n,\ 2,\ -l)\}$ given in CHART 12-V and certain manipulations, which are explained in the next paragraph, it was attained at

$$K\{R^j(4,n)\} = \binom{n+3}{2} K\{H_{14}(3,4,n)\} - (n+2)K\{H*(3,5,n)\} \tag{13.2}$$

Hence one has avoided to introduce an explicit expression for $K\{B(n,\ 4,\ -i)\}$. No
such expression was known to Cyvin SJ, Cyvin and Bergan (1986), but has since been
derived; cf. CHART 12-V. In eqn. (2) $H_{14}(3,4,n)$ is a six-tier tower; see Table 2
(I-14). If its K formula is assumed to be known, the K enumeration problem for
seven-tier oblate rectangles is reduced to the same problem for another class of
seven-tier strips, viz. H*(3,5,n); see Fig. 2. Therefore it may seem that nothing
has been gained by this approach. However, $R^j(4,n)$ is a straight strip ($sh = 0$)
while H*(3,5,n) is skew ($sh = 1$), and
in general it is (loosely speaking)
easier to derive the combinatorial K H*(3,5,n)
formulas for skew strips than for
straight strips.

The advanced method of chopping
may be exploited in the derivation of
several of the K formulas of Table 2
(six-tier strips). In the next paragraphs
we show two examples.

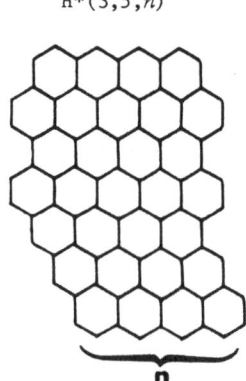

Fig. 13.2. A seven-tier tower.

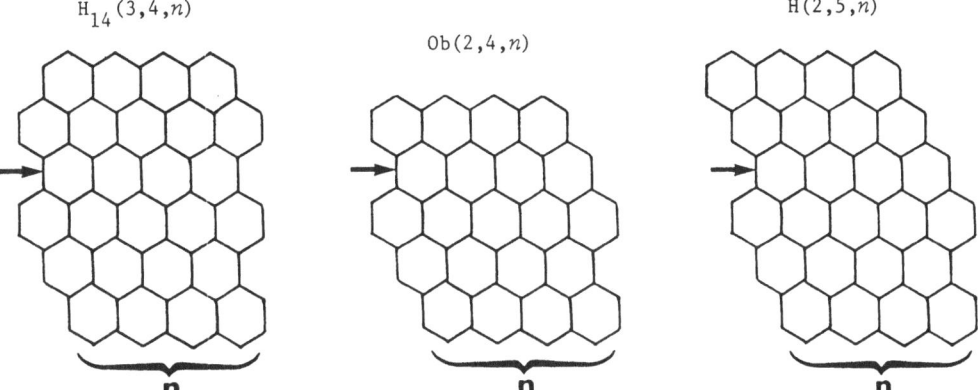

Fig. 13.3. The benzenoids appearing in eqn. (13.3).

13.3.2 *First Example*

Consider one of the very classes of eqn. (2), viz. $H_{14}(3,4,n)$. It is by the way a six-tier problate pentagon without apex, Da$(3,4,n)$; cf. Paragraph 11.2.3, especially Fig. 11.3.

In analogy with eqn. (1) one obtains the fundamental relation

$$K\{H_{14}(3,4,n)\} = \sum_{i=0}^{n} K\{B(n, 3, -i)\} \cdot K\{B(n, 2, -i)\} \qquad (13.3)$$

The result is obtained by the method of chopping of $H_{14}(3,4,n)$ along the row indicated by an arrow in Fig. 3. In eqn. (3) $B(n, 2, -i)$ pertains to the auxiliary class defined in more general terms in Fig. 12.2. For the sake of clarity we have reproduced the special case considered here in CHART I. The chart also includes the definition of $B(n, 3, -l)$.

On inserting the expression for $K\{B(n, 2, -i)\}$ from CHART 12-V or CHART I into (3) one obtains

$$K\{H_{14}(3,4,n)\} = \binom{n+2}{2} \sum_{i=0}^{n} (i+1)K\{B(n,3,-i)\} - (n+2) \sum_{i=0}^{n} \binom{i+1}{2} K\{B(n,3,-i)\} \qquad (13.4)$$

By making use of the identities $\binom{i+1}{2} = \binom{i+2}{2} - (i+1)$ and $\binom{n+2}{2} + (n+2) = \binom{n+3}{2}$ it was arrived at

$$K\{H_{14}(3,4,n)\} = \binom{n+3}{2} \sum_{i=0}^{n} (i+1)K\{B(n,3,-i)\} - (n+2) \sum_{i=0}^{n} \binom{i+2}{2} K\{B(n,3,-i)\}$$

$$= \binom{n+3}{2} \sum_{i=0}^{n} K\{B(n,3,-i)\} \cdot K\{L(i)\} - (n+2) \sum_{i=0}^{n} K\{B(n,3,-i)\} \cdot K\{L(2,i)\} \qquad (13.5)$$

Here also the well-known K formulas for the single linear chain (CHART 6-I) and the parallelogram (CHART 8-IV) were utilized.

The idea of the above manipulations was to avoid the inserting of an explicit expression for $K\{B(n, 3, -i)\}$ into the fundamental relation (although such an expression actually is known; cf. CHART I).

The term in question is eliminated on identifying the two last summations of (5) with the K formulas for certain classes so that the equation is rendered into the form

$$K\{H_{14}(3,4,n)\} = \binom{n+3}{2} K\{Ob(2,4,n)\} - (n+2)K\{H(2,5,n)\} \qquad (13.6)$$

The identification of the mentioned summations becomes clear when $Ob(2,4,n)$ and $H(2,5,n)$ are subjected to a chopping as indicated by the pertinent arrows in Fig. 3. Here $Ob(2,4,n)$ is a five-tier hexagon without two corners; see Table 10.5(II-5). $H(2,5,n)$ is a six-tier strip like $H_{14}(3,4,n)$. However, we have arrived at a class with a larger top-bottom shift, viz. $sh = \frac{3}{2}$ versus $\frac{1}{2}$.

In the following we only give a brief summary of one of the ways how the K formula for $H(2,5,n)$ may be determined. At the same time we demonstrate some of the very many inter-relations between the K numbers of different t-tier strips.

By the method of stripping (Paragraph 10.6.3) with multiple fragmentation it is obtained

$$K\{H(2,5,n)\} - K\{H(2, 5, n-1)\}$$

$$= K\{H^i(2,5,n)\} + K\{Ob(2, 4, n-1)\} + K\{D(2, 3, n-1)\}; \qquad n \geq 1 \qquad (13.7)$$

With regard to the benzenoids of the right-hand side of this equation we have already encountered the five-tier hexagon without two corners, $Ob(2,4,n)$. Neither the four-tier pentagon, viz. $D(2,3,n)$, which is found in Table 10.4(I-2), should represent any serious problem. The crucial point here is to determine the K formula for $H^i(2,5,n)$; cf. Table 2(II-8). It is often expedient to start to entangle the inter-relations between classes of t-tier strips, if possible, from a multiple chain. Here $H^i(2,5,n)$ represents just a class like that since $H^i(2,5,n) = M_n(LLLAAL)$. The combinatorial formula given in Table 2(II-8) for this class is obtainable by stripping with multiple fragmentation according to a scheme like the one of Fig. 10.7. Hence the right-hand side of (7) may now in principle be expanded as a polynomial in n. The last step from eqn. (7) to the explicit formula of $K\{H(2,5,n)\}$ is executed by a summation, viz.

$$K\{H(2,5,n)\} = \sum_{i=0}^{n} K\{H^i(2,5,i)\} + \sum_{i=1}^{n} K\{Ob(2,4,i-1)\} + \sum_{i=1}^{n} K\{D(2,3,i-1)\} \qquad (13.8)$$

Each of the summations in (8) may in fact be identified with the K formula for a certain benzenoid class; one has:

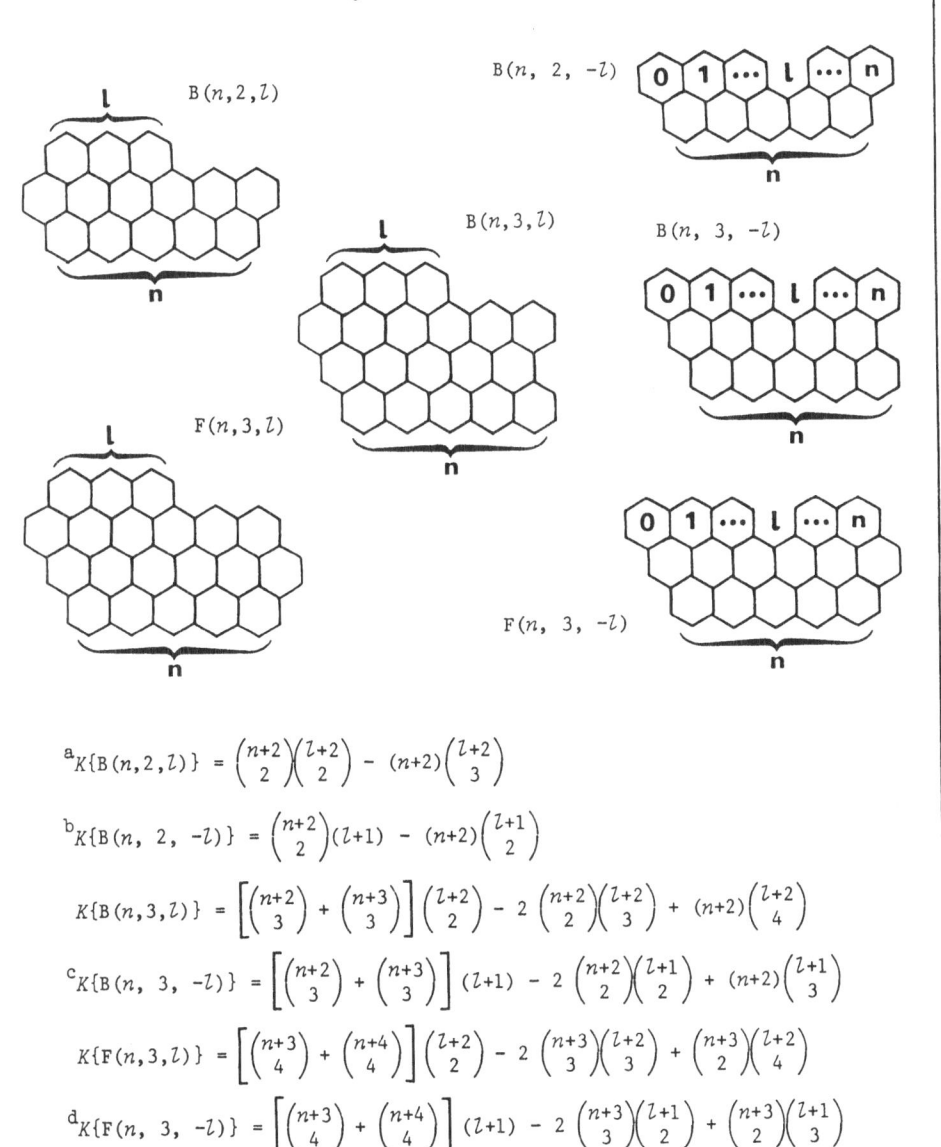

CHART 13-I. Some auxiliary benzenoids

$$^{a}K\{B(n,2,l)\} = \binom{n+2}{2}\binom{l+2}{2} - (n+2)\binom{l+2}{3}$$

$$^{b}K\{B(n,\ 2,\ -l)\} = \binom{n+2}{2}(l+1) - (n+2)\binom{l+1}{2}$$

$$K\{B(n,3,l)\} = \left[\binom{n+2}{3} + \binom{n+3}{3}\right]\binom{l+2}{2} - 2\binom{n+2}{2}\binom{l+2}{3} + (n+2)\binom{l+2}{4}$$

$$^{c}K\{B(n,\ 3,\ -l)\} = \left[\binom{n+2}{3} + \binom{n+3}{3}\right](l+1) - 2\binom{n+2}{2}\binom{l+1}{2} + (n+2)\binom{l+1}{3}$$

$$K\{F(n,3,l)\} = \left[\binom{n+3}{4} + \binom{n+4}{4}\right]\binom{l+2}{2} - 2\binom{n+3}{3}\binom{l+2}{3} + \binom{n+3}{2}\binom{l+2}{4}$$

$$^{d}K\{F(n,\ 3,\ -l)\} = \left[\binom{n+3}{4} + \binom{n+4}{4}\right](l+1) - 2\binom{n+3}{3}\binom{l+1}{2} + \binom{n+3}{2}\binom{l+1}{3}$$

[a] See CHART 12-IV

[b] See CHART 12-V

[c] Cyvin SJ, Cyvin BN (1986). Match 21: 295

[d] Cyvin BN, Cyvin SJ (1987). Match 22: 157

$$K\{H(2,5,n)\} = K\{D^i(2,5,n)\} + K\{D^j(2,4,n-1)\} + K\{0(2,3,n-1)\}; \qquad n \geq 1 \qquad (13.9)$$

13.3.3 Second Example

Consider the benzenoid class of $H_{13}(3,4,n)$; see Table 2(I-13). It obeys the fundamental relation

$$K\{H_{13}(3,4,n)\} = \sum_{i=0}^{n} K\{F(n, 3, -i)\} \cdot K\{B(n, 2, -i)\} \qquad (13.10)$$

where $F(n, 3, -i)$ represents another of the auxiliary classes defined in CHART I. Following the procedure of the preceding paragraph it was arrived at

$$K\{H_{13}(3,4,n)\} = \binom{n+3}{2} K\{D^j(2,4,n)\} - (n+2)K\{D(2,5,n)\} \qquad (13.11)$$

Here $D^j(2,4,n)$ is the five-tier oblate pentagon; see Table 10.5(II-2). $D(2,5,n)$ is the six-tier strip found in Table 2(II-3). When the K formula for $H(2,5,n)$ has been determined as explained briefly in the preceding paragraph, the $K\{D(2,5,n)\}$ formula may be obtained by the method of stripping with single fragmentation;

$$K\{D(2,5,n)\} - K\{D(2, 5, n-1)\} = K\{H(2,5,n)\}; \qquad n \geq 1 \qquad (13.12)$$

$$K\{D(2,5,n)\} = \sum_{i=0}^{n} K\{H(2,5,i)\} \qquad (13.13)$$

13.4 AUXILIARY BENZENOID CLASSES

The auxiliary classes defined in CHART I are very similar to the incomplete oblate rectangles and their associated classes; cf. Section 12.4. It is assumed $0 \leq l \leq n$. The classes $B(n,2,l)$ and $B(n, 2, -l)$ are in fact the special cases of $B(n, 2m-2, l)$ and $B(n, 2m-2, -l)$, respectively, for $m=2$. We have therefore:

$$B(n,2,n) = R^j(2,n) = 0(2,n), \qquad B(n,2,0) = L(2,n)$$

Similarly for the other classes:

$$B(n,3,n) = D(2,3,n), \qquad B(n,3,0) = Ch(2,n)$$

$$F(n,3,n) = 0(2,3,n), \qquad F(n,3,0) = 0(2,n)$$

For $n=0$, which implies $l=0$, all the benzenoids considered degenerate to the trivial case of no hexagons $(K=1)$.

The fundamental relation (12.13) for the special case of $m=2$ (also $m' = 2$) reads

$$K\{B(n, 2, -l)\} = K\{B(n,2,l)\} - K\{B(n, 2, l-1)\}; \qquad l \geq 1 \qquad (13.14)$$

Similarly:

$$K\{B(n, 3, -l)\} = K\{B(n,3,l)\} - K\{B(n, 3, l-1)\}; \qquad l \geq 1 \qquad (13.15)$$

$$K\{F(n, 3, -l)\} = K\{F(n,3,l)\} - K\{F(n, 3, l-1)\}; \qquad l \geq 1 \qquad (13.16)$$

The following inter-relations exist between the K numbers for some of the considered classes.

$$K\{F(n,3,l)\} = K\{B(n,3,l)\} + K\{F(n-1, 3,l)\}; \qquad n \geq 1 \qquad (13.17)$$

$$K\{F(n, 3, -l)\} = K\{B(n, 3, -l)\} + K\{F(n-1, 3, -l)\}; \qquad n \geq 1 \qquad (13.18)$$

13.5 TWO-PARAMETER K FORMULAS FOR SOME MULTIPLE CHAINS

It is referred to the fragmentation scheme of Fig. 10.7 for $Z(4,n) = M_n(LAAL)$. It was mentioned in the preceding section (Paragraph 13.3.2) that a similar scheme is applicable to $M_n(LLLAAL)$ in order to determine the K formula of the corresponding class. The procedure is in fact applicable to the more general case of $M_n(L^{r-1}AAL)$, which represents a class with two parameters (r, n). One obtains $(r > 1)$:

$$K\{M_n(L^{r-1}AAL)\} = K\{M_{n-1}(L^{r-1}AAL)\} + K\{L(2,n)\} \cdot K\{L(r-1,n)\} + K\{Ch(2,r,n-1)\};$$
$$n \geq 1 \quad (13.19)$$

The relation may be used to derive the two-parameter K formula for the class of multiple chains $M_n(L^{r-1}AAL)$. The result is entered into CHART II together with other two-parameter formulas.

13.6 GENERALIZED AUXILIARY CLASS

Consider the three-parameter benzenoid class defined in Fig. 4 $(0 \leq l \leq n)$. It is referred to as a multiple $(r\text{-tuple})$-incomplete strip. Primarily it is assumed $r > 0$. It is clear that $B(n,2,l)$ - cf. Section 13.4 and CHART I - is the special case of $r=1$:

$$B_1(n,2,l) = B(n,2,l)$$

For $l=0$ one has by definition

$$B_r(n,2,0) = L(2,n)$$

irrespective of r. The strip is "completed" on inserting $l=n$. For the first four r values one has:

CHART 13-II. Multiple chains and related benzenoids
including étagère

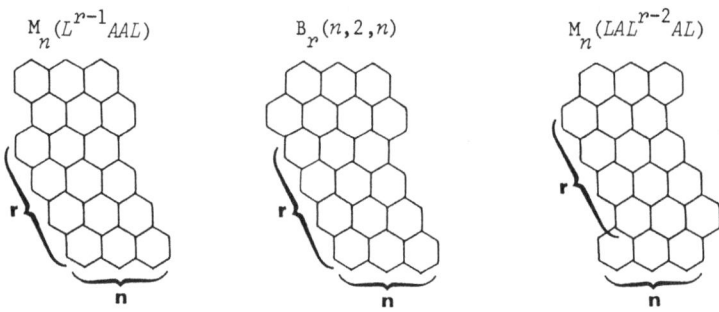

$M_n(L^{r-1}AAL)$

$B_r(n,2,n)$

$M_n(LAL^{r-2}AL)$

$$^aK\{M_n(L^{r-1}AAL)\} = \binom{n+2}{2}\binom{n+r}{n} - \binom{n+r+1}{n-1}$$

$$^aK\{B_r(n,2,n)\} = \binom{n+2}{2}\binom{n+r+1}{n} - (n+2)\binom{n+r+1}{n-1}$$

$$^aK\{M_n(LAL^{r-2}AL)\} = (n+2)^2\binom{n+r}{n} - 2(n+2)\binom{n+r+1}{n}\binom{n+r+2}{n}$$

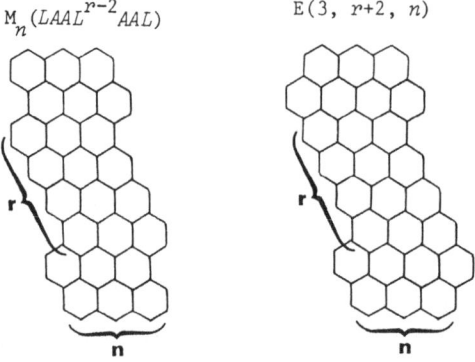

$M_n(LAAL^{r-2}AAL)$

$E(3, r+2, n)$

$$^aK\{M_n(LAAL^{r-2}AAL)\} = \binom{n+2}{2}^2\binom{n+r}{n} - 2\binom{n+2}{2}\binom{n+r+1}{n-1}\binom{n+r+1}{n-2}$$

$$^aK\{E(3, r+2, n)\} = \binom{n+3}{2}^2\binom{n+r+2}{n} - 2(n+2)\binom{n+3}{2}\binom{n+r+3}{n} + (n+2)^2\binom{n+r+4}{n}$$

[a]Cyvin SJ, Cyvin BN (1987). Monatsh Chem 118: 337

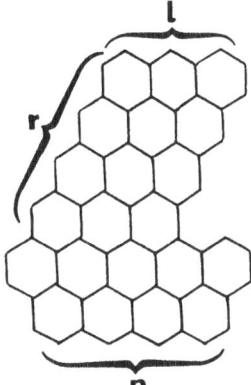

Fig. 13.4. Definition of the
auxiliary class $B_r(n,2,l)$.

$$B_1(n,2,n) = 0(2,n), \qquad B_2(n,2,n) = D(2,3,n)$$

$$B_3(n,2,n) = D^i(2,4,n), \qquad B_4(n,2,n) = D^i(2,5,n)$$

A fundamental relation for the multiple-incomplete strips reads

$$K\{B_r(n,2,l)\} = K\{B_r(n,\ 2,\ l-1)\} + K\{B_{r-1}(n,2,l)\}; \qquad l \geq 1 \qquad (13.20)$$

as is easily derived (by the method of fragmentation) for $r > 1$. In order to be valid
also for $r=1$ one has to define

$$B_0(n,2,l) = B(n,\ 2,\ -l)$$

Then the relation (20) for $r=1$ is consistent with (14).

From (20) one obtains the summation formula

$$K\{B_r(n,2,l)\} = \sum_{i=0}^{l} K\{B_{r-1}(n,2,i)\}; \qquad r \geq 1 \qquad (13.21)$$

and finally the explicit form (Cyvin SJ and Cyvin 1987b):

$$K\{B_r(n,2,l)\} = \binom{n+2}{2}\binom{l+r+1}{r+1} - (n+2)\binom{l+r+1}{r+2} \qquad (13.22)$$

The two first formulas of CHART I are seen to be the special cases of (22) for $r=1$
and $r=0$, respectively.

13.7 ÉTAGÈRE

The last formula of CHART II was derived (Cyvin SJ and Cyvin 1987b) by means of the generalized auxiliary classes starting from

$$K\{E(3, r+2, n)\} = \sum_{i=0}^{n} K\{B_{r-1}(n,2,i)\} \cdot K\{B_0(n,2,i)\}; \qquad r \geq 1 \qquad (13.23)$$

Here the method of chopping was employed.

The benzenoid $E(3,m,n)$ is referred to as an étagère. It is a regular $(m+2)$-tier strip, where it is assumed $m > 1$. With the parameter $m = r+2$ the combinatorial K formula for étagères reads:

$$K\{E(3,m,n)\} = \binom{n+3}{2}^2 \binom{m+n}{n} - 2(n+2)\binom{n+3}{2}\binom{m+n+1}{n} + (n+2)^2\binom{m+n+2}{n} \qquad (13.24)$$

For $m=2$ the étagère degenerates to the 4-tier pentagon: $E(3,2,n) = D(3,2,n)$. For $m=3$: $E(3,3,n) = R^j(3,n)$, the 5-tier oblate rectangle. For $m=4$: $E(3,4,n) = H_{33}(3,4,n)$; cf. Table 2(I-28).

An interesting special case of étagères occurs for $n=1$:

$$K\{E(3,m,1)\} = 9(m-1) \qquad (13.25)$$

These systems represent two-sided annelation of *pyrenes* to a linear chain of $m-4$ hexagons $(m \geq 4)$. On inserting $m = a+4$ it is seen that (25) coincides with $K\{P_3\}$ of CHART 7-III.

13.8 SOME SEVEN-TIER STRIPS: A SUMMING UP

The classes of regular 7-tier strips have not been treated systematically. However, many of these classes have been encountered in different connections. A survey of the available material with respect to the combinatorial K formulas is given in the following.

The classes of 7-tier strips are divided into four series with the following leading hexagons and top-bottom shifts.

(I): $0(4,4,n)$, $sh = 0$;   (II) $0(3,5,n)$, $sh = 1$;
(III): $0(2,6,n)$, $sh = 2$;   (IV) $0(1,7,n)$, $sh = 3$.

In Chapter 8 a general K formula for hexagons, $O(k,m,n)$, is given (CHART 8-I). Also an explicit form for $O(4,4,n) = O(4,n)$, a dihedral hexagon, is given (Paragraph 8.2.3). From eqn. (ii) of CHART 8-II one obtains for the mirror-symmetrical 7-tier chevron:

$$K\{Ch(4,n)\} = \frac{1}{2520}\,(n+1)\,(n+2)\,(n+3)\,(n+4)\,(2n+5)\,(5n^2+25n+21) \qquad (13.26)$$

The general K formula for parallelograms (CHART 8-IV) is also relevant since $O(1,7,n) = L(7,n)$. The formula for $K\{Z(7,n)\}$ pertaining to the 7-tier zigzag chain, is found in CHART 9-II. General K formulas for prolate pentagons are given in CHART 11-I, and especially the explicit form for $K\{D^i(4,n)\}$ in Paragraph 11.2.2. For the other mirror-symmetrical 7-tier pentagons, see CHART 11-IV for $D^j(4,n)$ - the oblate pentagon - and CHART 11-VI for the additional four systems of this kind. The K formulas for all mirror-symmetrical 7-tier pentagons without apex are also available (Section 11.2). For the 7-tier prolate rectangle one has

$$K\{R^i(4,n)\} = (n+1)^4 \qquad (13.27)$$

in accord with CHART 12-I, while the formula of $K\{R^j(4,n)\}$ for the oblate rectangle

Table 13.3. Supplementary classes of regular 7-tier strips. Formulas for K.

$H*(4,4,n)$ [16, 632] $\frac{1}{5040}(n+1)(n+2)^3(n+3)(13n^4 + 104n^3 + 311n^2 + 412n + 210)$	$H*(3,5,n)$ [16, 448] $\frac{1}{20160}(n+1)(n+2)^3(n+3)(n+4)(25n^3 + 142n^2 + 295n + 210)$
$M_n(LLAAAAL)$ [14, 275] $\frac{1}{5040}(n+1)(n+2)(n+3)(155n^4 + 911n^3 + 2062n^2 + 2122n + 840)$	$H*(2,6,n)$ [15, 193] $\frac{1}{20160}(n+1)(n+2)^2(n+3)(n+4)(31n^3 + 236n^2 + 545n + 420)$
$M_n(LLLAALL)$ [14, 142] $\frac{1}{5040}(n+1)(n+2)(n+3)(n+4)(34n^3 + 199n^2 + 355n + 210)$	

is found in CHART 12-II. The classes of multiple chains and related benzenoids including étagère, as defined in CHART II, all have members among 7-tier strips. Of course there exist also a great number of 7-tier strips of essentially disconnected benzenoids (in addition to the prolate rectangles) as well as non-Kekuléans; see, e.g. Fig. 11.6 for mirror-symmetrical streamers. With regard to goblets (Paragraph 11.4.3) one has:

$$K\{X(4,n)\} = K\{X(3,5,n)\} = 0, \qquad K\{X(2,6,n)\} = (n+1)^2 \tag{13.28}$$

Finally we give a supplementary (highly incomplete) material for 7-tier strips in Table 3. The bracketed figures give h and K values at $n=2$. The material includes the K formula for the benzenoid of Fig. 2. Three of the K formulas have been published before (Cyvin SJ, Cyvin and Bergan 1986).

14.1 INTRODUCTION

Application of the John-Sachs theorem (cf. *Theorem 9* of Chapter 3 and Paragraph 4.10.2) in order to produce determinant combinatorial formulas for the number
of Kekulé structures (K) is among the methods outside the main lines of this book.
Nevertheless, the method is very useful, and an extensive report on different determinant formulas, which have been derived, is warranted on this place.

The mentioned theorem (John and Sachs 1985a; John and Rempel 1985) was employed to deduce a new method of enumeration of Kekulé structures (Gutman and Cyvin
1987a). Shortly thereafter a number of applications have seen the light (Gutman, Su
and Cyvin 1987; Cyvin BN and Cyvin 1987; Cyvin SJ, Cyvin, Brunvoll, Chen and Su
1987; Cyvin SJ, Cyvin, Brunvoll and Gutman 1987). The present chapter contains,
apart from quotations, a number of original contributions to the determinant formulas.

14.2 HEXAGON (Section 8.2)

The following determinant formula holds for the hexagon-shaped benzenoids,
$O(k,m,n)$.

$$
K\{O(k,m,n)\} =
\begin{vmatrix}
\binom{n+k}{n} & \binom{n+k}{n-1} & \binom{n+k}{n-2} & \cdots & \binom{n+k}{n-m+1} \\
\binom{n+k}{n+1} & \binom{n+k}{n} & \binom{n+k}{n-1} & \cdots & \binom{n+k}{n-m+2} \\
\binom{n+k}{n+2} & \binom{n+k}{n+1} & \binom{n+k}{n} & \cdots & \binom{n+k}{n-m+3} \\
\cdot & \cdot & \cdot & \cdots & \cdot \\
\binom{n+k}{n+m-1} & \binom{n+k}{n+m-2} & \binom{n+k}{n+m-3} & \cdots & \binom{n+k}{n}
\end{vmatrix}
\tag{14.1}
$$

In this formulation, (1) is an $m \times m$ determinant. The centrosymmetry of $O(k,m,n)$ is reflected in the determinant by its symmetry around the secondary diagonal (from top-right to bottom-left). On equating the determinant (1) with the K formula (i) of CHART 8-I one attains at an interesting identity. It has recently been proved mathematically (Bodroža et al. 1988), thus establishing for the first time a rigorous derivation of the mentioned K formula. As a simple example we have the following application to the special hexagon depicted in CHART 8-I (*circumanthanthrene*; $k=4$, $m=2$, $n=3$):

$$K\{0(4,2,3)\} = \frac{\begin{vmatrix} \binom{7}{3} & \binom{7}{2} \\ \binom{7}{4} & \binom{7}{3} \end{vmatrix}}{\binom{7}{3}\binom{8}{3}} \cdot \binom{3}{3}\binom{4}{3} = 490 \tag{14.2}$$

The subsequent formulas in this chapter may be used similarly to produce a number of mathematical identities of similar kinds.

It may be of interest to study the determinant formulas for dihedral hexagons in particular. The following form, where the dimension of the determinant is $n\times n$, is obtained from (1) only after permuting the parameters.

$$K\{0(m,n)\} = \begin{vmatrix} \binom{2m}{m} & \binom{2m}{m+1} & \binom{2m}{m+2} & \cdots & \binom{2m}{m+n-1} \\ \binom{2m}{m+1} & \binom{2m}{m} & \binom{2m}{m+1} & \cdots & \binom{2m}{m+n-2} \\ \binom{2m}{m+2} & \binom{2m}{m+1} & \binom{2m}{m} & \cdots & \binom{2m}{m+n-3} \\ \cdot & \cdot & \cdot & \cdots & \cdot \\ \binom{2m}{m+n-1} & \binom{2m}{m+n-2} & \binom{2m}{m+n-3} & \cdots & \binom{2m}{m} \end{vmatrix} \tag{14.3}$$

As a consequence of the dihedral symmetry of $0(m,n)$ this determinant is symmetrical both around the main diagonal and the secondary diagonal.

In supplement to Section 8.2.3 we give here some K formulas for $0(m,n)$ with fixed (small) values of n. They emerge directly from eqn. (3) in the following way. For $n=2$:

$$K\{0(m,2)\} = \begin{vmatrix} \binom{2m}{m} & \binom{2m}{m+1} \\ \binom{2m}{m+1} & \binom{2m}{m} \end{vmatrix} = \binom{2m}{m}^2 - \binom{2m}{m+1}^2 = \binom{2m+1}{m}\left[\binom{2m}{m} - \binom{2m}{m+1}\right] =$$

$$= \frac{2m+1}{(m+1)^2}\binom{2m}{m}^2 \tag{14.4}$$

Similarly for $n=3$ and $n=4$:

$$K\{0(m,3)\} = \frac{4(2m+1)^2}{(m+1)^3(m+2)^2}\binom{2m}{m}^3 \qquad (14.5)$$

$$K\{0(m,4)\} = \frac{48(2m+1)^3(2m+3)}{(m+1)^4(m+2)^4(m+3)^2}\binom{2m}{m}^4 \qquad (14.6)$$

The formulas (4)-(6) are consistent with eqn. (8.7).

Each element (W_{ij}) of the determinant is identified with the K number of a subgraph of the original benzenoid. Usually the subgraphs represent benzenoids themselves, but may also degenerate to systems with acyclic chains of edges or the null graph.

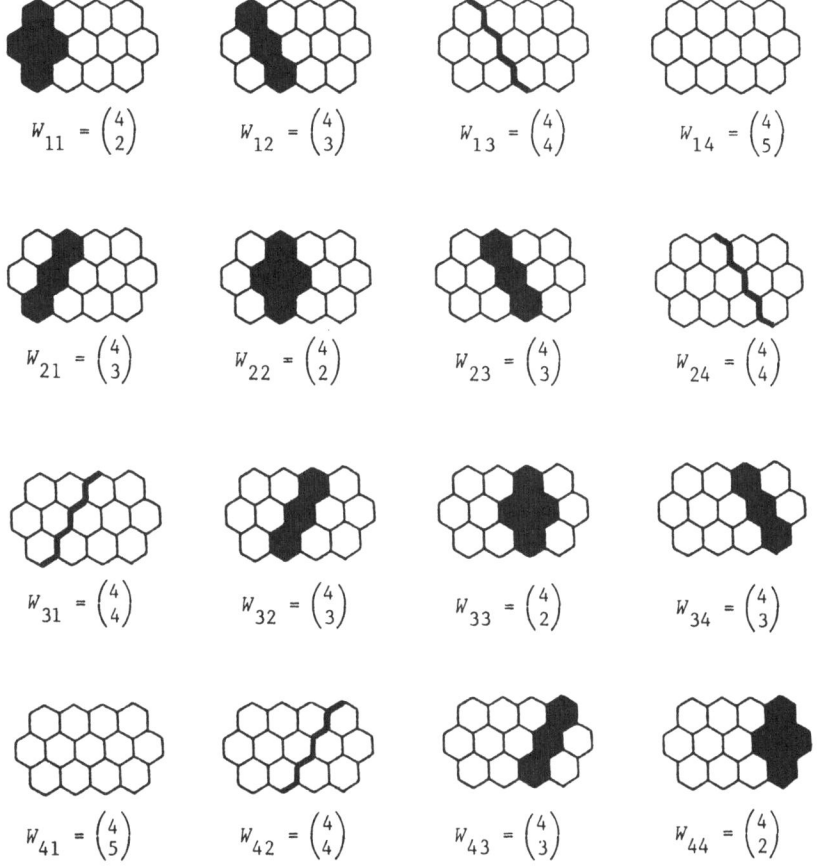

Fig. 14.1. Subgraphs and the corresponding determinant elements pertaining to *circumanthracene*, O(2,4).

In the present application to hexagons any subgraph of the considered type is represented by either (a) a parallelogram, occasionally degenerated to a single linear chain, (b) acyclic chain with $K=1$, or (c) the empty graph.

In Fig. 1 the subgraphs pertaining to *circumanthracene*, 0(2,4), are depicted as black silhouettes or heavy lines on the background of the original benzenoid. The indicated elements W_{ij} are consistent with eqn. (3).

The auxiliary drawings besides a determinant as in eqn. (1) and throughout this chapter are oriented in the appropriate way for the derivation of the determinant. In particular the (same) number of peaks and valleys on the drawing indicates the dimension of the determinant.

14.3 CHEVRON (Section 8.3)

The following $m \times m$ determinant was derived.

$$K\{Ch(k,m,n)\} = \begin{vmatrix} \binom{n+k}{n} & \binom{n+k}{n-1} & \binom{n+k}{n-2} & \cdots & \binom{n+k}{n-m+1} \\ \hline \binom{n+1}{n+1} & \binom{n+1}{n} & \binom{n+1}{n-1} & \cdots & \binom{n+1}{n-m+2} \\ \binom{n+1}{n+2} & \binom{n+1}{n+1} & \binom{n+1}{n} & \cdots & \binom{n+1}{n-m+3} \\ \cdot & \cdot & \cdot & \cdots & \cdot \\ \binom{n+1}{n+m-1} & \binom{n+1}{n+m-2} & \binom{n+1}{n+m-3} & \cdots & \binom{n+1}{n} \end{vmatrix} \qquad (14.7)$$

Here the first row is special, and therefore separated by a dashed line.

14.4 RIBBON (Section 8.4)

The following $m \times m$ determinant was derived for the ribbon.

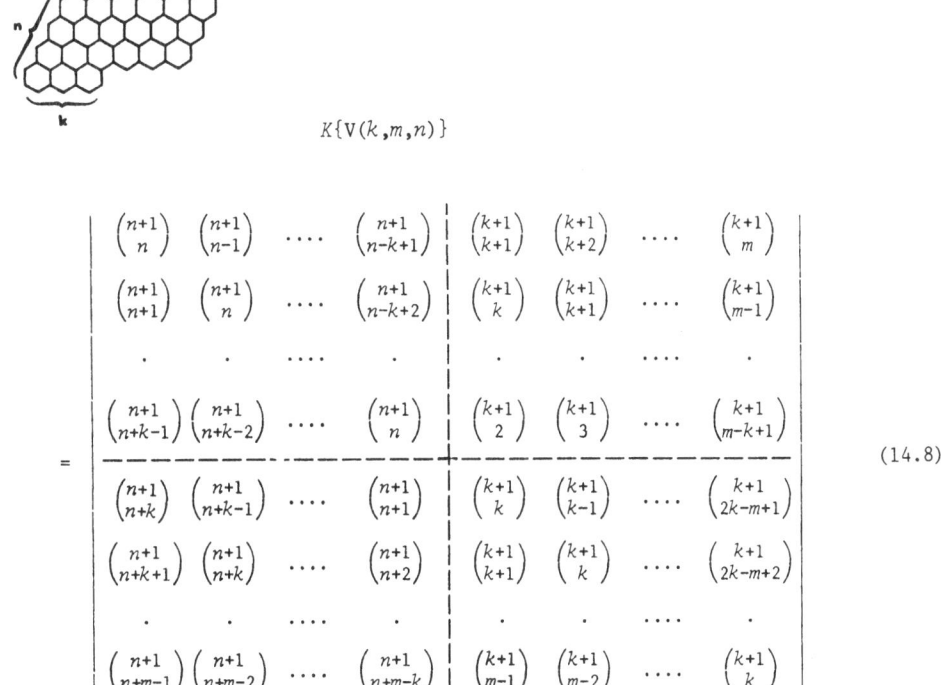

$$K\{V(k,m,n)\}$$

The determinant was put in a form which clearly shows the regularities within each quadrant. The dimensions of the quadrants are indicated. The form may be simplified substantially. For instance, all the elements in the lower-left quadrant vanish except one of them (in the upper-right corner), which is unity. We will not go into further details of the properties of the determinant (8) because it will not be applied in the following in its general form; the case where the ribbon degenerates to a parallelogram is mentioned in the next section.

14.5 PARALLELOGRAM (Section 8.5)

14.5.1 *Parallelogram* L(*m,n*)

The parallelogram, $L(m,n)$, appears as the degenerate case of a hexagon, $O(1,m,n)$ or a chevron, $Ch(1,m,n)$; here $k=1$ was chosen in both cases. Correspondingly the determinant for the parallelogram is obtained on inserting $k=1$ in eqn. (1) or (7). The parallelogram is also a degenerate ribbon for $k=m$, viz. $V(m,m,n)$, when $m \leq n$ has been assumed. In this case the determinant (8) reduces to one quadrant, viz. the upper-left, which has the dimension $k \times k$. All these cases are consistent with the result:

$$K\{L(m,n)\} = \begin{vmatrix} \binom{n+1}{n} & \binom{n+1}{n-1} & \binom{n+1}{n-2} & \cdots & \binom{n+1}{n-m+1} \\ \binom{n+1}{n+1} & \binom{n+1}{n} & \binom{n+1}{n-1} & \cdots & \binom{n+1}{n-m+2} \\ \binom{n+1}{n+2} & \binom{n+1}{n+1} & \binom{n+1}{n} & \cdots & \binom{n+1}{n-m+3} \\ \cdot & \cdot & \cdot & \cdots & \cdot \\ \binom{n+1}{n+m-1} & \binom{n+1}{n+m-2} & \binom{n+1}{n+m-3} & \cdots & \binom{n+1}{n} \end{vmatrix} \qquad (14.9)$$

This is an $m \times m$ determinant, and symmetrical around the secondary diagonal. The elements right below the main diagonal form a diagonal of unities. Furthermore, all elements below the diagonal of unities vanish. The determinant may in fact be rendered into the following form.

$$\begin{vmatrix} (n+1) & \binom{n+1}{2} & \binom{n+1}{3} & \cdots & \binom{n+1}{m} \\ 1 & (n+1) & \binom{n+1}{2} & \cdots & \binom{n+1}{m-1} \\ 0 & 1 & (n+1) & \cdots & \binom{n+1}{m-2} \\ \cdot & \cdot & \cdot & \cdots & \cdot \\ 0 & 0 & 0 & \cdots & (n+1) \end{vmatrix} = \binom{n+m}{m} \qquad (14.10)$$

Three zeros in the last row presuppose that the determinant has at least five rows $(m \geq 5)$. By using the combinatorial K formula for parallelograms (cf. CHART 8-IV) we have formulated eqn. (10) as a pure mathematical identity.

The determinant for $K\{L(m,n)\}$ is not very useful because the equivalent result in terms of one binomial coefficient is so much simpler. However, the determinant may easily be modified so that it applies to the parallelogram augmented with a row, as is shown in the next paragraph.

14.5.2 Parallelogram Augmented with a Row, L(n,m,l)

For the benzenoid $L(n,m,l)$, cf. CHART 8-V, it was found:

$$K\{L(n,m,l)\} = \begin{vmatrix} (l+1) & \binom{l+1}{2} & \binom{l+1}{3} & \cdots & \binom{l+1}{m+1} \\ 1 & (n+1) & \binom{n+1}{2} & \cdots & \binom{n+1}{n-1} \\ 0 & 1 & (n+1) & \cdots & \binom{n+1}{m-2} \\ \cdot & \cdot & \cdot & \cdots & \cdot \\ 0 & 0 & 0 & \cdots & (n+1) \end{vmatrix} \qquad (14.11)$$

This $(m+1) \times (m+1)$ determinant is obtained from the one of eqn. (10) by augmenting the latter with one row and one column, as is indicated by the dashed lines in (11).

It is instructive to show some applications of (11) for fixed values of m. We find for $m=1$:

$$K\{L(n,1,l)\} = \begin{vmatrix} (l+1) & \binom{l+1}{2} \\ 1 & (n+1) \end{vmatrix} = (n+1)(l+1) - \binom{l+1}{2} \qquad (14.12)$$

which is to be compared with eqn. (8.53). For $m=2$:

$$K\{L(n,2,l)\} = \begin{vmatrix} (l+1) & \binom{l+1}{2} & \binom{l+1}{3} \\ 1 & (n+1) & \binom{n+1}{2} \\ 0 & 1 & (n+1) \end{vmatrix} = \binom{n+2}{2}(l+1) - (n+1)\binom{l+1}{2} + \binom{l+1}{3} \qquad (14.13)$$

Here we have made use of the identity

$$\begin{vmatrix} (n+1) & \binom{n+1}{2} \\ 1 & (n+1) \end{vmatrix} = (n+1)^2 - \binom{n+1}{2} = \binom{n+2}{2} \qquad (14.14)$$

which is the special case of (10) for $m=2$. Finally we also give the application of (11) for $m=3$:

$$K\{L(n,3,l)\} = \binom{n+3}{3}(l+1) - \binom{n+2}{2}\binom{l+1}{2} + (n+1)\binom{l+1}{3} - \binom{l+1}{4} \qquad (14.15)$$

Eqs. (12), (13) and (15) are the special cases of eqn. (ii) of CHART 8-V for $m = 1$, 2 and 3, respectively.

14.6 ZIGZAG CHAINS (Chapters 4 and 9)

14.6.1 *Single Zigzag Chain* (CHART 6-II)

The determinant form of $Z_1(m) = K\{Z(m,1)\} = K\{A(m)\}$ is actually a determinant representation of the Fibonacci numbers. Examples for even and odd m ($m=8$ and $m=9$, respectively):

$$Z_1(8) = \begin{vmatrix} 3 & 1 & 0 & 0 \\ 1 & 3 & 1 & 0 \\ 0 & 1 & 3 & 1 \\ 0 & 0 & 1 & 3 \end{vmatrix} = \begin{vmatrix} 2 & 1 & 0 & 0 & 0 \\ 1 & 3 & 1 & 0 & 0 \\ 0 & 1 & 3 & 1 & 0 \\ 0 & 0 & 1 & 3 & 1 \\ 0 & 0 & 0 & 1 & 2 \end{vmatrix} = 55 \qquad (14.16)$$

$$Z_1(9) = \begin{vmatrix} 2 & 1 & 0 & 0 & 0 \\ 1 & 3 & 1 & 0 & 0 \\ 0 & 1 & 3 & 1 & 0 \\ 0 & 0 & 1 & 3 & 1 \\ 0 & 0 & 0 & 1 & 3 \end{vmatrix} = 89 \qquad (14.17)$$

In these determinants an arbitrary element W_{ij} equals either 3 (K for *naphthalene*), 2 (K for *benzene*), 1 (K for an acyclic chain), or it vanishes. In general the non-vanishing elements are situated on the main diagonal and its two neighbouring diagonals, as is indicated by the dotted lines in (16) and (17). In these examples the Fibonacci numbers $F_9 = 55$ and $F_{10} = 89$ are reproduced.

It is interesting that a purely mathematical property of Fibonacci numbers was derived "chemically" through the studies of Kekulé structures (Cyvin SJ, Cyvin, Brunvoll and Gutman 1987).

14.6.2 Double Zigzag Chain

Numerical Determinants. Here we show the extension of the determinant forms to $Z_2(m) = K\{Z(m,2)\} = K\{A(2,m)\}$. On going from the single to the double zigzag chain one nonvanishing diagonal is added above and to the right as shown in the below examples (for $m=8$ and $m=9$).

$$Z_2(8) = \begin{vmatrix} 6 & 5 & 1 & 0 \\ 1 & 6 & 5 & 1 \\ 0 & 1 & 6 & 5 \\ 0 & 0 & 1 & 6 \end{vmatrix} = \begin{vmatrix} 3 & 4 & 1 & 0 & 0 \\ 1 & 6 & 5 & 1 & 0 \\ 0 & 1 & 6 & 5 & 1 \\ 0 & 0 & 1 & 6 & 4 \\ 0 & 0 & 0 & 1 & 3 \end{vmatrix} = 793 \qquad (14.18)$$

$$Z_2(9) = \begin{vmatrix} 3 & 4 & 1 & 0 & 0 \\ 1 & 6 & 5 & 1 & 0 \\ 0 & 1 & 6 & 5 & 1 \\ 0 & 0 & 1 & 6 & 5 \\ 0 & 0 & 0 & 1 & 6 \end{vmatrix} = 1782 \qquad (14.19)$$

Again a very limited number of values of the W_{ij} elements occur. The nonvanishing elements greater than unity are either 6 (K for *pyrene*), 5 (K for *naphthacene*), 4 (K for *anthracene*) or 3 (K for *naphthalene*).

Nonlinearly Coupled Recurrence Relations. A treatment of double zigzag chains according to an orientation where each benzenoid has two peaks and two valleys (Cyvin SJ, Cyvin, Brunvoll and Gutman 1987) led to the definition of six new benzenoid classes and the derivation of nonlinearly coupled recurrence relations.

The six classes are illustrated in Fig. 2. The two fundamental classes $t_m(p)$

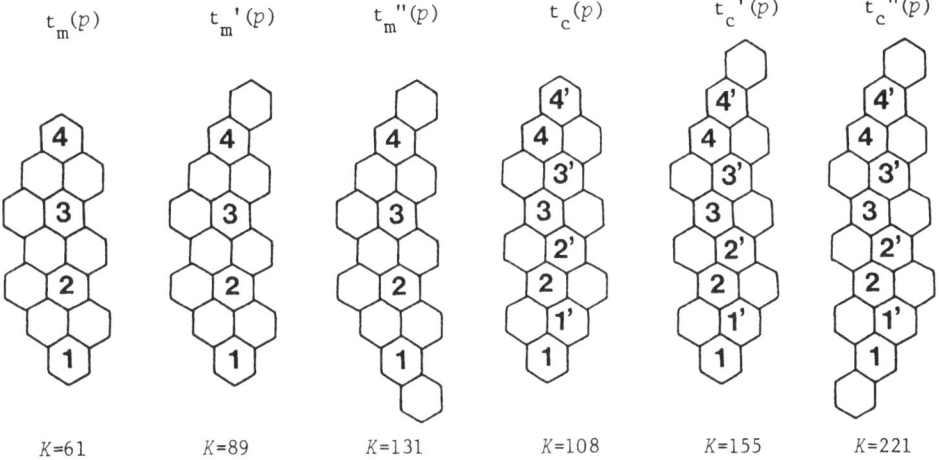

$t_m(p)$ $t_m'(p)$ $t_m''(p)$ $t_c(p)$ $t_c'(p)$ $t_c''(p)$

$K=61$ $K=89$ $K=131$ $K=108$ $K=155$ $K=221$

Fig. 14.2. Six benzenoid classes. In all the depicted examples $p=4$. The K numbers for this parameter value are given.

and $t_c(p)$ consist of mirror-symmetrical (m) and centrosymmetrical (c) benzenoids, respectively. In the former case (m) a double zigzag chain, $Z(m,2)$, where m is an odd number, is augmented by two hexagons condensed at the two ends. In the latter case (c) $Z(m,2)$ with m even is augmented in the same way. The other classes are obtained from $t_m(p)$ and $t_c(p)$ by fusing one hexagon to one or both hexagons at the ends.

The six classes (Fig. 2) were studied by the methods of linearly coupled recurrence relations. We apply the notation $t_m(p) = K\{t_m(p)\}$, $t_m'(p) = K\{t_m'(p)\}$, ..., etc. The four classes t_m, t_m', t_c and t_c' appear to constitute a self-consistent system; K numbers for them are linearly dependent, and it was found in terms of t_m':

$$t_m(p) = t_m'(p) - t_m'(p-1) \tag{14.20}$$

$$t_c(p) = 3t_m'(p) - 6t_m'(p-1) + t_m'(p-2) \tag{14.21}$$

$$t_c'(p) = 5t_m'(p) - 11t_m'(p-1) + 2t_m'(p-2) \tag{14.22}$$

while

$$t_m'(p) = \sum_{i=0}^{p} t_m(i) \tag{14.23}$$

The two remaining classes were coupled to this system with the result:

$$t_m''(p) = 5t_m'(p-1) - t_m'(p-2) \tag{14.24}$$

$$t_c''(p) = 9t_m'(p) - 22t_m'(p-1) + 4t_m'(p-2)$$

(14.25)

All the six classes obey recurrence relations of the same form, for which it was found:

$$t(p) = 5t(p-1) - 6t(p-2) + t(p-3)$$

(14.26)

All numerical values of K are obtainable successively from (26) along with the following initial values (which include nominal values and values for trivial and degenerate cases): $t_m(-1) = 2$, $t_m'(-1) = t_c'(-1) = t_c''(-1) = 0$, $t_m''(-1) = t_c(-1) = 1$; $t_m(0) = t_m'(0) = t_c(0) = t_c'(0) = t_c''(0) = 1$, $t_m''(0) = 2$; $t_m(1) = 2$, $t_m'(1) = t_c(1) = 3$, $t_m''(1) = t_c''(1) = 5$, $t_c'(1) = 4$.

In this treatment of double zigzag chains, $Z(m,2)$, the even and odd m values must be considered separately. Figure 3 illustrates the construction of the determinant for an even m, viz. $2p$. The analysis yields

$$Z_2(2p) = \begin{vmatrix} t_m'(p) & t_c''(p-1) \\ t_c(p) & t_m'(p) \end{vmatrix}$$

(14.27)

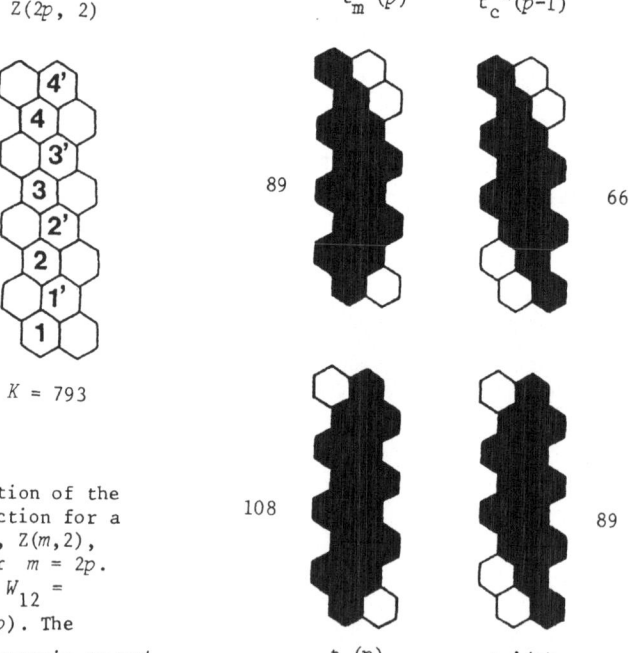

$Z(2p, 2)$

$K = 793$

Fig. 14.3. Illustration of the determinant construction for a double zigzag chain, $Z(m,2)$, with an even integer $m = 2p$. $W_{11} = W_{22} = t_m'(p)$, $W_{12} = t_c''(p-1)$, $W_{21} = t_c(p)$. The numerical K values pertain to $p=4$.

$t_m(p+1)$ $t_c'(p)$

Z(2p+1, 2)

K = 1782

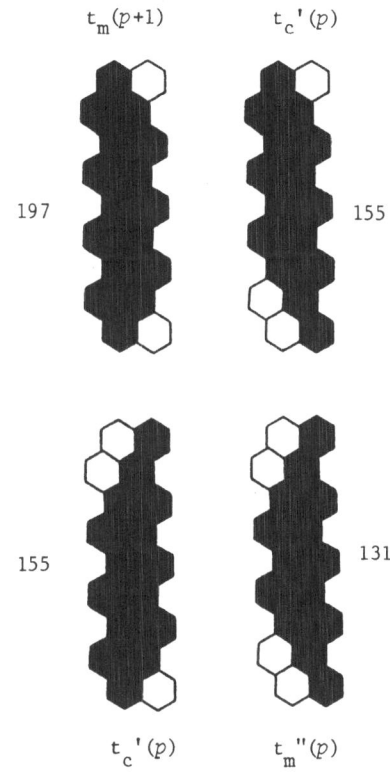

197 155

155 131

$t_c'(p)$ $t_m''(p)$

Fig. 14.4. Illustration of the determinant construction for a double zigzag chain, $Z(m,2)$, with an odd integer $m = 2p+1$. $W_{11} = t_m(p+1)$, $W_{12} = W_{21} = t_c'(p)$, $W_{22} = t_m''(p)$. The numerical K values pertain to $p=4$.

With the aid of (21), (25) and (26) it was attained at

$$Z_2(2p) = \tau_p{}^2 - (3\tau_p - 6\tau_{p-1} + \tau_{p-2})(4\tau_p - 11\tau_{p-1} + 2\tau_{p-2})$$

$$= -11\tau_p{}^2 + 57\tau_p\tau_{p-1} - 66\tau_{p-1}{}^2 - 10\tau_p\tau_{p-2} + 23\tau_{p-1}\tau_{p-2} - 2\tau_{p-2}{}^2 \qquad (14.28)$$

where the following abbreviation is employed.

$$\tau_p = t_m'(p) \qquad (14.29)$$

The corresponding analysis for $Z(m,2)$, where m is odd, viz. $2p+1$, leads to

$$Z_2(2p+1) = \begin{vmatrix} t_m(p+1) & t_c'(p) \\ t_c'(p) & t_m''(p) \end{vmatrix} \qquad (14.30)$$

cf. Fig. 4. With the aid of (20), (22), (24) and (26) it was attained at the following nonlinear dependency on the quantity (29).

$$Z_2(2p+1) = (4\tau_p - 6\tau_{p-1} + \tau_{p-2})(5\tau_{p-1} - \tau_{p-2}) - (5\tau_p - 11\tau_{p-1} + 2\tau_{p-2})^2$$

$$= -25\tau_p^2 + 130\tau_p\tau_{p-1} - 151\tau_{p-1}^2 - 24\tau_p\tau_{p-2} + 55\tau_{p-1}\tau_{p-2} - 5\tau_{p-2}^2 \qquad (14.31)$$

This equation is equivalent to:

$$Z_2(2p-1) = (\tau_p - \tau_{p-1})(-\tau_p + 5\tau_{p-1} - \tau_{p-2}) - (2\tau_p - 5\tau_{p-1} + \tau_{p-2})^2$$

$$= -5\tau_p^2 + 26\tau_p\tau_{p-1} - 30\tau_{p-1}^2 - 5\tau_p\tau_{p-2} + 11\tau_{p-1}\tau_{p-2} - \tau_{p-2}^2 \qquad (14.32)$$

14.6.3 *Multiple Zigzag Chain*

Considering the multiple zigzag chain $Z(m,n)$, where m is an even number, viz. $2p$, the following $p \times p$ determinant is found for the number of Kekulé structures.

$$Z_n(2p) = \begin{vmatrix} \binom{n+2}{n} & \binom{n+3}{n-1} & \binom{n+4}{n-2} & \cdots & \binom{n+p+1}{n-p+1} \\ \binom{n+1}{n+1} & \binom{n+2}{n} & \binom{n+3}{n-1} & \cdots & \binom{n+p}{n-p+2} \\ \binom{n}{n+2} & \binom{n+1}{n+1} & \binom{n+2}{n} & \cdots & \binom{n+p-1}{n-p+3} \\ \cdot & \cdot & \cdot & \cdots & \cdot \\ \binom{n-p+3}{n+p-1} & \binom{n-p+4}{n+p-2} & \binom{n-p+5}{n+p-3} & \cdots & \binom{n+2}{n} \end{vmatrix} \qquad (14.33)$$

The determinant is symmetrical around the secondary diagonal. Like the determinant (9) for a parallelogram, the above determinant (33) has a diagonal of unities and vanishing elements below it. It may be rendered into the form:

$$Z_n(2p) = \begin{vmatrix} \binom{n+2}{2} & \binom{n+3}{4} & \binom{n+4}{6} & \cdots & \binom{n+p+1}{2p} \\ 1 & \binom{n+2}{2} & \binom{n+3}{4} & \cdots & \binom{n+p}{2p-2} \\ 0 & 1 & \binom{n+2}{2} & \cdots & \binom{n+p-1}{2p-4} \\ \cdot & \cdot & \cdot & \cdots & \cdot \\ 0 & 0 & 0 & \cdots & \binom{n+2}{2} \end{vmatrix} \qquad (14.34)$$

The forms (33) and (34) are not adequate for $p=0$, in which case $Z_n(0) = 1$ as a trivial case. The 4×4 determinants of eqns. (16) and (18) are special cases of (34) for $p=4$. We recognize the unities and zeros in the lower-left part. There are also diagonals of zeros in the upper-right part if p is large enough in relation to n. In fact the addition of one nonvanishing diagonal (cf. Paragraph 14.6.2) continues for every unit n is increased.

The determinant for an odd number m, viz. $2p+1$, is obtained most easily from (33) or (34) by augmenting this $p×p$ determinant with a 0-th row and 0-th column:

$$Z_n(2p+1) = \begin{vmatrix} (n+1) & \binom{n+2}{3} & \binom{n+3}{5} & \cdots & \binom{n+p+1}{2p+1} \\ 1 & \binom{n+2}{2} & \binom{n+3}{4} & \cdots & \binom{n+p+1}{2p} \\ 0 & 1 & \binom{n+2}{2} & \cdots & \binom{n+p}{2p-2} \\ \cdot & \cdot & \cdot & \cdots & \cdot \\ 0 & 0 & 0 & \cdots & \binom{n+2}{2} \end{vmatrix} \qquad (14.35)$$

This is a $(p+1)×(p+1)$ determinant. For $p=0$ eqn. (35) reduces to $Z_n(1) = n+1$.

An application of (34) and (35) leads to the expressions for $Z_n(m) = K\{A(n,m)\}$ derived in a different way in Section 9.6 (also denoted A_n therein). The corresponding polynomial representations for m up to 10 are found in CHART 9-II.

Here we also give the general form of the determinant for $Z_n(2p)$ which corresponds to the second version (5×5 determinants) of eqns. (16) and (18).

$$Z_n(2p) = \begin{vmatrix} (n+1) & \binom{n+2}{3} & \binom{n+3}{5} & \cdots & \binom{n+p}{2p-1} & \binom{n+p}{2p} \\ 1 & \binom{n+2}{2} & \binom{n+3}{4} & \cdots & \binom{n+p}{2p-2} & \binom{n+p}{2p-1} \\ 0 & 1 & \binom{n+2}{2} & \cdots & \binom{n+p-1}{2p-4} & \binom{n+p-1}{2p-3} \\ \cdot & \cdot & \cdot & \cdots & \cdot & \cdot \\ 0 & 0 & 0 & \cdots & \binom{n+2}{2} & \binom{n+2}{3} \\ 0 & 0 & 0 & \cdots & 1 & (n+1) \end{vmatrix} \qquad (14.36)$$

This is a $(p+1)\times(p+1)$ determinant. The inner portion of $(p-1)\times(p-1)$ elements corresponds to $Z_n(2p-2)$ according to (34). The symmetry around the secondary diagonal is an aid in setting up this determinant.

Here we give only one example, namely $Z_n(4)$ according to the versions (34) and (36):

$$Z_n(4) = K\{A(n,4)\} = \begin{vmatrix} \binom{n+2}{4} & \binom{n+3}{4} \\ 1 & \binom{n+2}{2} \end{vmatrix} = \binom{n+2}{2}^2 - \binom{n+3}{4}$$

$$= \begin{vmatrix} (n+1) & \binom{n+2}{3} & \binom{n+2}{4} \\ 1 & \binom{n+2}{2} & \binom{n+2}{3} \\ 0 & 1 & (n+1) \end{vmatrix} = (n+1)\left[(n+1)\binom{n+2}{2} - 2\binom{n+2}{3}\right] + \binom{n+2}{4} \qquad (14.37)$$

The results are identical with eqns. (9.16) and (9.17).

14.6.4 *Multiple Zigzag Chain Augmented with a Row*

Here we will show the determinants for the auxiliary benzenoid classes introduced in Section 9.3.

For $Z_n^{(l)}(m) = K\{A(n,m,l)\}$ with an even m it is found:

A(n, 2p, l)

$$Z_n^{(l)}(2p) = \begin{vmatrix} (l+1) & \binom{l+2}{3} & \binom{l+3}{5} & \cdots & \binom{l+p+1}{2p+1} \\ 1 & \binom{n+2}{2} & \binom{n+3}{4} & \cdots & \binom{n+p+1}{2p} \\ 0 & 1 & \binom{n+2}{2} & \cdots & \binom{n+p}{2p-2} \\ \cdot & \cdot & \cdot & \cdots & \cdot \\ 0 & 0 & 0 & \cdots & \binom{n+2}{2} \end{vmatrix} \qquad (14.38)$$

This $(p+1)\times(p+1)$ determinant is seen to be identical with (35) except for the first row, where n is substituted by l. Therefore one obtains $Z_n^{(n)}(2p) = Z_n(2p+1)$. It is also easily seen that $Z_n^{(0)}(2p) = Z_n(2p)$, because the determinant (38) reduces to the framed portion in this case. These relations are consistent with the known properties of the auxiliary benzenoid classes; cf. especially eqn. (9.32).

In order to set up the determinant for $Z_n^{(l)}(m)$ where m is odd, viz. $2p+1$, we start with the determinant for $Z_n(2p)$ in the version (34) and augment it with a frame in the following way.

m = 2p+1   A(n, 2p+1, l)

$$Z_n^{(l)}(2p+1) = \begin{vmatrix} (l+1) & \begin{pmatrix} l+2 \\ 3 \end{pmatrix} & \begin{pmatrix} l+3 \\ 5 \end{pmatrix} & \cdots & \begin{pmatrix} l+p+1 \\ 2p+1 \end{pmatrix} & \begin{pmatrix} l+p+1 \\ 2p+2 \end{pmatrix} \\ 1 & \begin{pmatrix} n+2 \\ 2 \end{pmatrix} & \begin{pmatrix} n+3 \\ 4 \end{pmatrix} & \cdots & \begin{pmatrix} n+p+1 \\ 2p \end{pmatrix} & \begin{pmatrix} n+p+1 \\ 2p+1 \end{pmatrix} \\ 0 & 1 & \begin{pmatrix} n+2 \\ 2 \end{pmatrix} & \cdots & \begin{pmatrix} n+p \\ 2p-2 \end{pmatrix} & \begin{pmatrix} n+p-1 \\ 2p-1 \end{pmatrix} \\ \cdot & \cdot & \cdot & \cdots & \cdot & \cdot \\ 0 & 0 & 0 & \cdots & \begin{pmatrix} n+2 \\ 2 \end{pmatrix} & \begin{pmatrix} n+2 \\ 3 \end{pmatrix} \\ 0 & 0 & 0 & \cdots & 1 & (n+1) \end{vmatrix} \qquad (14.39)$$

In this $(p+2)\times(p+2)$ determinant the inner portion equals $Z_n(2p)$ as in (34). If p is substituted by $p-1$ in (39) the resulting $(p+1)\times(p+1)$ determinant for $Z_n(2p-1)$ comes out as identical to (36) except for the first row, where n is replaced by l. One has clearly $Z_n^{(n)}(2p-1) = Z_n(2p)$ or $Z_n^{(n)}(2p+1) = Z_n(2p+2)$. Furthermore, $l=0$ inserted in (38) leads to a determinant equivalent to (35); hence $Z_n^{(0)}(2p+1) = Z_n(2p+1)$. We have again verified some known properties of the auxiliary classes.

The applications of eqns. (38) and (39) lead exactly to the expressions of CHART 9-I. It should be sufficient to exemplify this feature by the two determinants for $p=1$:

$$Z_n^{(l)}(2) = K\{A(n,2,l)\} = \begin{vmatrix} (l+1) & \begin{pmatrix} l+2 \\ 3 \end{pmatrix} \\ 1 & \begin{pmatrix} n+2 \\ 2 \end{pmatrix} \end{vmatrix} \qquad (14.40)$$

$$Z_n^{(l)}(3) = K\{A(n,3,l)\} = \begin{vmatrix} (l+1) & \begin{pmatrix} l+2 \\ 3 \end{pmatrix} & \begin{pmatrix} l+2 \\ 4 \end{pmatrix} \\ 1 & \begin{pmatrix} n+2 \\ 2 \end{pmatrix} & \begin{pmatrix} n+2 \\ 3 \end{pmatrix} \\ 0 & 1 & (n+1) \end{vmatrix} \qquad (14.41)$$

14.7.1 *Prolate Pentagon*

Determinant formula for the K numbers of a prolate pentagon, $D^i(m,n)$:

$$K\{D^i(m,n)\} = \begin{vmatrix} \binom{n+1}{n} & \binom{n+2}{n-1} & \binom{n+3}{n-2} & \cdots & \binom{n+m}{n-m+1} \\[2mm] \binom{n+1}{n+1} & \binom{n+2}{n} & \binom{n+3}{n-1} & \cdots & \binom{n+m}{n-m+2} \\[2mm] \binom{n+1}{n+2} & \binom{n+2}{n+1} & \binom{n+3}{n} & \cdots & \binom{n+m}{n-m+3} \\[2mm] \cdot & \cdot & \cdot & \cdots & \cdot \\[2mm] \binom{n+1}{n+m-1} & \binom{n+2}{n+m-2} & \binom{n+3}{n+m-3} & \cdots & \binom{n+m}{n} \end{vmatrix} \qquad (14.42)$$

On equating this $m{\times}m$ determinant with one of the equations in CHART 11-I one attains again at interesting mathematical identities. Example for $D^i(3,2)$:

$$\begin{vmatrix} \binom{3}{2} & \binom{4}{1} & \binom{5}{0} \\[2mm] \binom{3}{3} & \binom{4}{2} & \binom{5}{1} \\[2mm] \binom{3}{4} & \binom{4}{3} & \binom{5}{2} \end{vmatrix} = \frac{\binom{8}{3}\binom{10}{3}}{\binom{4}{3}\binom{6}{3}} = 84 \qquad (14.43)$$

In Fig. 5 the subgraphs pertaining to $D^i(4,3)$ and the corresponding W_{ij} elements are displayed in the style of Fig. 1.

The determinant (42) may be put into the form

263

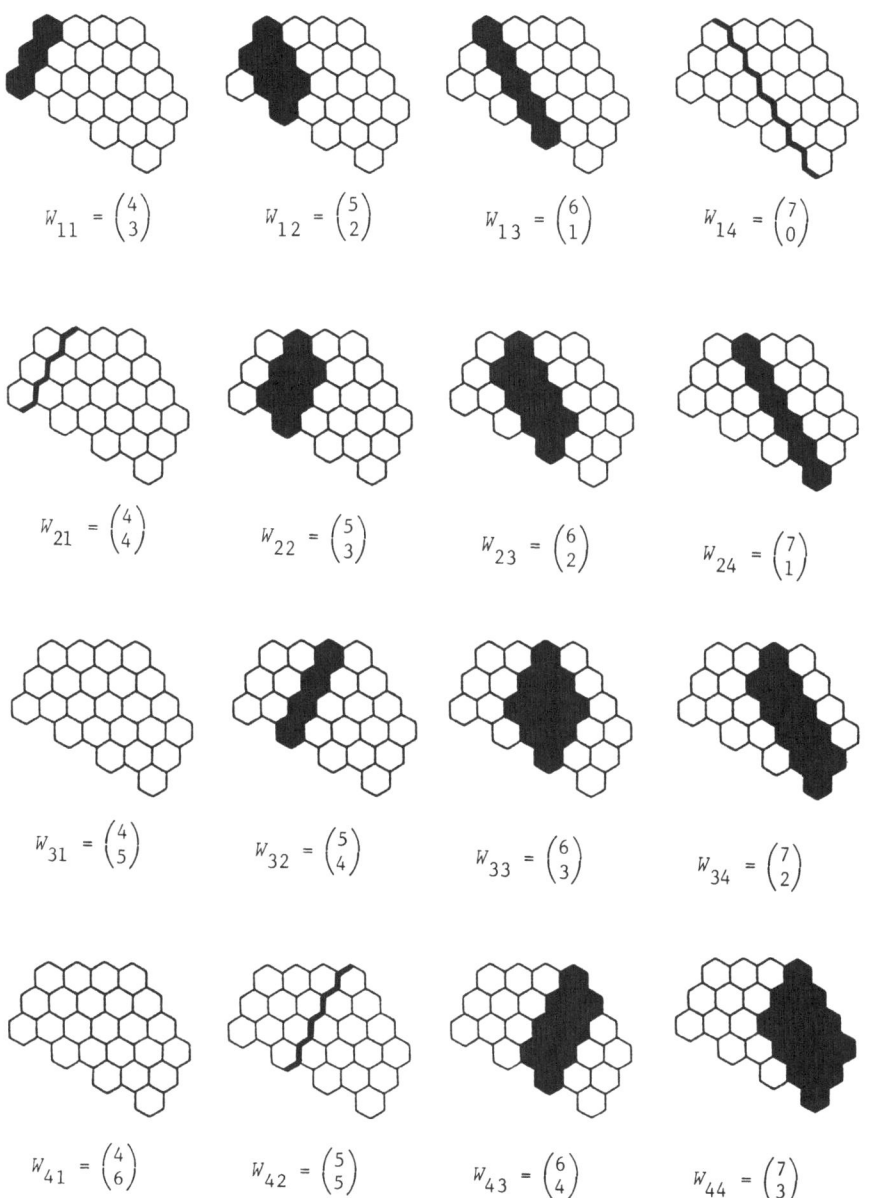

Fig. 14.5. Subgraphs and the corresponding determinant elements pertaining to $D^i(4,3)$.

$$K\{D^{i}(m,n)\} = \begin{vmatrix} (n+1) & \binom{n+2}{3} & \binom{n+3}{5} & \cdots & \binom{n+m}{2m-1} \\ 1 & \binom{n+2}{2} & \binom{n+3}{4} & \cdots & \binom{n+m}{2m-2} \\ 0 & (n+2) & \binom{n+3}{3} & \cdots & \binom{n+m}{2m-3} \\ \cdot & \cdot & \cdot & \cdots & \cdot \\ 0 & 0 & 0 & \cdots & \binom{n+m}{m} \end{vmatrix} \qquad (14.44)$$

Here the three zeros in the last row presuppose that the determinant has at least seven rows.

The application of eqn. (44) gives formulas for $K\{D^{i}(m,n)\}$ with fixed values of m in terms of binomial coefficients. They represent alternative forms of the equivalent formulas of Paragraph 11.2.2. Below we give the expanded determinants for m up to 4.

$$K\{D^{i}(1,n)\} = n+1, \qquad K\{D^{i}(2,n)\} = (n+1)\binom{n+2}{2} - \binom{n+2}{3},$$

$$K\{D^{i}(3,n)\} = \binom{n+3}{3}\left[(n+1)\binom{n+2}{2} - \binom{n+2}{3}\right] - (n+2)\left[(n+1)\binom{n+3}{4} - \binom{n+3}{5}\right],$$

$$K\{D^{i}(4,n)\} = \left[(n+1)\binom{n+2}{2} - \binom{n+2}{3}\right]\left[\binom{n+3}{3}\binom{n+4}{4} - \binom{n+3}{2}\binom{n+4}{5}\right]$$

$$- \left[(n+1)\binom{n+3}{4} - \binom{n+3}{5}\right]\left[(n+2)\binom{n+4}{4} - \binom{n+4}{5}\right] + \left[(n+2)\binom{n+3}{2} - \binom{n+3}{3}\right]\left[(n+1)\binom{n+4}{6} - \binom{n+4}{7}\right]$$

14.7.2 *Problate Pentagon*

An $m \times m$ determinant for the K numbers of the problate pentagon $D(m, m+1, n)$ is obtained from (42) or (44) by (a) deleting the first row and column and (b) adding a new m-th row and m-th column without changing the regularities:

$$K\{D(m, m+1, n)\} = \begin{vmatrix} \binom{n+2}{2} & \binom{n+3}{4} & \binom{n+4}{6} & \cdots\cdots & \binom{n+m+1}{2m} \\ (n+2) & \binom{n+3}{3} & \binom{n+4}{5} & \cdots\cdots & \binom{n+m+1}{2m-1} \\ 1 & \binom{n+3}{2} & \binom{n+4}{4} & \cdots\cdots & \binom{n+m+1}{2m-2} \\ \cdot & \cdot & \cdot & \cdots\cdots & \cdot \\ 0 & 0 & 0 & \cdots\cdots & \binom{n+m}{m+1} \end{vmatrix} \qquad (14.45)$$

An application of this equation gives binomial-coefficient expressions, which are equivalent to the polynomials of CHART 11-III.

It is useful to have an alternative version of the $K\{D(m, m+1, n)\}$ formula. The following $(m+1) \times (m+1)$ determinant is obtained by adding an $(m+1)$-th row and $(m+1)$-th column to the determinant (42) or (44). The $(m+1)$-th column is special, and therefore separated by a dashed line.

$$K\{D(m, m+1, n)\} = \begin{vmatrix} (n+1) & \binom{n+2}{3} & \cdots\cdots & \binom{n+m}{2m-1} & \bigm| & \binom{n+m}{2m} \\ 1 & \binom{n+2}{2} & \cdots\cdots & \binom{n+m}{2m-2} & \bigm| & \binom{n+m}{2m-1} \\ 0 & (n+2) & \cdots\cdots & \binom{n+m}{2m-3} & \bigm| & \binom{n+m}{2m-2} \\ \cdot & \cdot & \cdots\cdots & \cdot & \bigm| & \cdot \\ 0 & 0 & \cdots\cdots & \binom{n+m}{m-1} & \bigm| & \binom{n+m}{m} \end{vmatrix} \qquad (14.46)$$

14.7.3 *Oblate Pentagon*

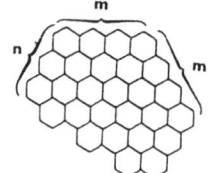

The following $m \times m$ determinant was derived.

$$K\{D^j(m,n)\} = \begin{vmatrix} \binom{n+2}{n} & \binom{n+3}{n-1} & \binom{n+4}{n-2} & \cdots & \binom{n+m}{n-m+2} & \bigg| & \binom{n+m}{n-m+1} \\ \binom{n+2}{n+1} & \binom{n+3}{n} & \binom{n+4}{n-1} & \cdots & \binom{n+m}{n-m+3} & \bigg| & \binom{n+m}{n-m+2} \\ \binom{n+2}{n+2} & \binom{n+3}{n+1} & \binom{n+4}{n} & \cdots & \binom{n+m}{n-m+4} & \bigg| & \binom{n+m}{n-m+3} \\ \cdot & \cdot & \cdot & \cdots & \cdot & \bigg| & \cdot \\ \binom{n+2}{n+m-1} & \binom{n+3}{n+m-2} & \binom{n+4}{n+m-3} & \cdots & \binom{n+m}{n+1} & \bigg| & \binom{n+m}{n} \end{vmatrix} \qquad (14.47)$$

It is equivalent to:

$$K\{D^j(m,n)\} = \begin{vmatrix} \binom{n+2}{2} & \binom{n+3}{4} & \binom{n+4}{6} & \cdots & \binom{n+m}{2m-2} & \bigg| & \binom{n+m}{2m-1} \\ (n+2) & \binom{n+3}{3} & \binom{n+4}{5} & \cdots & \binom{n+m}{2m-3} & \bigg| & \binom{n+m}{2m-2} \\ 1 & \binom{n+3}{2} & \binom{n+4}{4} & \cdots & \binom{n+m}{2m-4} & \bigg| & \binom{n+m}{2m-3} \\ \cdot & \cdot & \cdot & \cdots & \cdot & \bigg| & \cdot \\ 0 & 0 & 0 & \cdots & \binom{n+m}{m-1} & \bigg| & \binom{n+m}{m} \end{vmatrix} \qquad (14.48)$$

The occurrence of three zeros in the last row presupposes that the determinant has at least eight rows.

For the four first m values one has the following expressions, which are equivalent with the polynomial forms of CHART 11-IV.

$$K\{D^j(1,n)\} = |n+1|, \qquad K\{D^j(2,n)\} = \begin{vmatrix} \binom{n+2}{2} & \binom{n+2}{3} \\ (n+2) & \binom{n+2}{2} \end{vmatrix},$$

$$K\{D^j(3,n)\} = \begin{vmatrix} \binom{n+2}{2} & \binom{n+3}{4} & \binom{n+3}{5} \\ (n+2) & \binom{n+3}{3} & \binom{n+3}{4} \\ 1 & \binom{n+3}{2} & \binom{n+3}{3} \end{vmatrix} , \qquad K\{D^j(4,n)\} = \begin{vmatrix} \binom{n+2}{2} & \binom{n+3}{4} & \binom{n+4}{6} & \binom{n+4}{7} \\ (n+2) & \binom{n+3}{3} & \binom{n+4}{5} & \binom{n+4}{6} \\ 1 & \binom{n+3}{2} & \binom{n+4}{4} & \binom{n+4}{5} \\ 0 & (n+3) & \binom{n+4}{3} & \binom{n+4}{4} \end{vmatrix}$$

14.7.4 *Problate or Oblate Pentagon Augmented with a Row*

The auxiliary benzenoid classes of this title are defined in Paragraph 11.2.6.

The problate pentagon augmented with a row, $C(n, 2m-2, l)$, is an incomplete prolate pentagon. An $m \times m$ determinant for $K\{C(n, 2m-2, l)\}$ is obtained from (44) on substituting n by l in the first row.

The oblate pentagon augmented with a row, $C(n, 2m-1, l)$, is an incomplete problate pentagon. An $(m+1) \times (m+1)$ determinant for $K\{C(n, 2m-1, l)\}$ is obtained on substituting n by l in (46), the second version of the determinant formula for a problate pentagon.

When the new determinants described above are properly expanded for special values of m, one arrives at the formulas of CHART 11-VII.

14.8 OBLATE RECTANGLE (Section 12.3)

14.8.1 *Oblate Rectangle* $R^j(m,n)$

Cyvin SJ, Cyvin, Brunvoll, Chen and Su (1987) reported some details of the derivation of the determinant K formula for oblate rectangles. They started from a determinant for prolate rectangles, which is a more trivial case since these benzenoids are essentially disconnected (cf. Section 12.2). The result is:

$$K\{R^j(m,n)\} = \begin{vmatrix} \binom{n+2}{n} & \binom{n+3}{n-1} & \binom{n+4}{n-2} & \cdots & \binom{n+m}{n-m+2} & \binom{n+m}{n-m+1} \\ \binom{n+2}{n+1} & \binom{n+3}{n} & \binom{n+4}{n-1} & \cdots & \binom{n+m}{n-m+3} & \binom{n+m}{n-m+2} \\ \binom{n+1}{n+2} & \binom{n+2}{n+1} & \binom{n+3}{n} & \cdots & \binom{n+m-1}{n-m+4} & \binom{n+m-1}{n-m+3} \\ \cdot & \cdot & \cdot & \cdots & \cdot & \cdot \\ \binom{n-m+4}{n+m-1} & \binom{n-m+5}{n+m-2} & \binom{n-m+6}{n+m-3} & \cdots & \binom{n+2}{n+1} & \binom{n+2}{n} \end{vmatrix} \quad (14.49)$$

This $m \times m$ determinant is symmetrical around the secondary diagonal. It can be rendered into the form

$$K\{R^j(m,n)\} = \begin{vmatrix} \binom{n+2}{2} & \binom{n+3}{4} & \binom{n+4}{6} & \cdots & \binom{n+m}{2m-2} & \binom{n+m}{2m-1} \\ (n+2) & \binom{n+3}{3} & \binom{n+4}{5} & \cdots & \binom{n+m}{2m-3} & \binom{n+m}{2m-2} \\ 0 & (n+2) & \binom{n+3}{3} & \cdots & \binom{n+m-1}{2m-5} & \binom{n+m-1}{2m-4} \\ \cdot & \cdot & \cdot & \cdots & \cdot & \cdot \\ 0 & 0 & 0 & \cdots & (n+2) & \binom{n+2}{2} \end{vmatrix} \quad (14.50)$$

In the following we give the application of (50) for the four first values of m. Formulas in terms of binomial coefficients are produced as alternatives to the polynomial forms of CHART 12-II.

$$K\{R^j(1,n)\} = |n+1| = n+1, \qquad\qquad K\{R^j(2,n)\} = \begin{vmatrix} \binom{n+2}{2} & \binom{n+2}{3} \\ (n+2) & \binom{n+2}{2} \end{vmatrix} = \binom{n+2}{2}^2 - (n+2)\binom{n+2}{3},$$

$$K\{R^j(3,n)\} = \begin{vmatrix} \binom{n+2}{2} & \binom{n+3}{4} & \binom{n+3}{5} \\ (n+2) & \binom{n+3}{3} & \binom{n+3}{4} \\ 0 & (n+2) & \binom{n+2}{2} \end{vmatrix} = \binom{n+2}{2}\left[\binom{n+2}{2}\binom{n+3}{3} - 2(n+2)\binom{n+3}{4}\right] + (n+2)^2\binom{n+3}{5},$$

$$K\{R^j(4,n)\} = \begin{vmatrix} \binom{n+2}{2} & \binom{n+3}{4} & \binom{n+4}{6} & \binom{n+4}{7} \\ (n+2) & \binom{n+3}{3} & \binom{n+4}{5} & \binom{n+4}{6} \\ 0 & (n+2) & \binom{n+3}{3} & \binom{n+3}{4} \\ 0 & 0 & (n+2) & \binom{n+2}{2} \end{vmatrix} = \left[\binom{n+2}{2}\binom{n+3}{3} - (n+2)\binom{n+3}{4}\right]^2 + (n+2)^2\left[2\binom{n+2}{2}\binom{n+4}{6} - (n+2)\binom{n+4}{7}\right] - (n+2)\binom{n+2}{2}^2\binom{n+4}{5}$$

14.8.2 *Incomplete Oblate Rectangles and Their Associated Class,* B$(n, 2m-2, l)$ *and* B$(n, 2m-2, -l)$

The title auxiliary benzenoid classes are defined in Section 12.4.

For $R_n^{(l)}(m) = K\{B(n, 2m-2, l)\}$ one obtains an $m \times m$ determinant on substituting n by l in (49) or (50);

$$R_n^{(l)}(m) = \begin{vmatrix} \binom{l+2}{2} & \binom{l+3}{4} & \binom{l+4}{6} & \cdots & \binom{l+m}{2m-2} & \binom{l+m}{2m-1} \\ (n+2) & \binom{n+3}{3} & \binom{n+4}{5} & \cdots & \binom{n+m}{2m-3} & \binom{n+m}{2m-2} \\ 0 & (n+2) & \binom{n+3}{3} & \cdots & \binom{n+m-1}{2m-5} & \binom{n+m-1}{2m-4} \\ \cdot & \cdot & \cdot & \cdots & \cdot & \cdot \\ 0 & 0 & 0 & \cdots & (n+2) & \binom{n+2}{2} \end{vmatrix} \qquad (14.51)$$

For $R_n^{(-l)}(m) = K\{B(n, 2m-2, -l)\}$ one obtains from eqn. (12.13) and the determinant (51):

$$R_n^{(-l)}(m) = \begin{vmatrix} (l+1) & \binom{l+2}{3} & \binom{l+3}{5} & \cdots & \binom{l+m-1}{2m-3} & \binom{l+m-1}{2m-2} \\ \hline (n+2) & \binom{n+3}{3} & \binom{n+4}{5} & \cdots & \binom{n+m}{2m-3} & \binom{n+m}{2m-2} \\ 0 & (n+2) & \binom{n+3}{3} & \cdots & \binom{n+m-1}{2m-5} & \binom{n+m-1}{2m-4} \\ \cdot & \cdot & \cdot & \cdots & \cdot & \cdot \\ 0 & 0 & 0 & \cdots & (n+2) & \binom{n+2}{2} \end{vmatrix} \qquad (14.52)$$

When (51) and (52) are properly expanded for the different (fixed) values of m, one obtains the expressions of CHART 12-IV and CHART 12-V, respectively.

14.8.3 *Further Developments*

The studies of determinants for oblate rectangles and their auxiliary classes were pursued further (Cyvin SJ, Cyvin, Brunvoll, Chen and Su 1987; Cyvin SJ, Cyvin and Chen 1987).

It was found that the binomial-coefficient formulas of CHART 12-IV and CHART 12-V may be generalized in the following way for $m \geq 2$.

$$R_n^{(l)}(m) = (-1)^{m-1}(n+2)^{m-1}\binom{l+m}{l-m+1} + \sum_{i=0}^{m-2}(-1)^i(n+2)^i R_n^{(0)}(m-i)\binom{l+i+2}{l-i} \qquad (14.53)$$

$$R_n^{(-l)}(m) = (-1)^{m-1}(n+2)^{m-1}\binom{l+m-1}{l-m+1} + \sum_{i=0}^{m-2}(-1)^i(n+2)^i R_n^{(0)}(m-i)\binom{l+i+1}{l-i} \qquad (14.54)$$

The formulas (53) and (54) are also valid for the degenerate case of $m=1$ if the summations are omitted in that case. Below we show how the formulas of CHART 12-IV and CHART 12-V actually are simplified by means of (53) and (54). One finds for $m = 2$, 3, 4, 5:

$$R_n^{(l)}(2) = L\binom{l+2}{2} - (n+2)\binom{l+2}{3},$$

$$R_n^{(l)}(3) = D\binom{l+2}{2} - (n+2)L\binom{l+3}{4} + (n+2)^2\binom{l+3}{5},$$

$$R_n^{(l)}(4) = H\binom{l+2}{2} - (n+2)D\binom{l+3}{4} + (n+2)^2L\binom{l+4}{6} - (n+2)^3\binom{l+4}{7},$$

$$R_n^{(l)}(5) = G\binom{l+2}{2} - (n+2)H\binom{l+3}{4} + (n+2)^2D\binom{l+4}{6} - (n+2)^3L\binom{l+5}{8} + (n+2)^4\binom{l+5}{9}$$

and

$$R_n^{(-l)}(2) = L\ (l+1) - (n+2)\binom{l+1}{2},$$

$$R_n^{(-l)}(3) = D\ (l+1) - (n+2)L\ \binom{l+2}{3} + (n+2)^2\binom{l+2}{4},$$

$$R_n^{(-l)}(4) = H\ (l+1) - (n+2)D\ \binom{l+2}{3} + (n+2)^2 L\ \binom{l+3}{5} - (n+2)^3\binom{l+3}{6},$$

$$R_n^{(-l)}(5) = G\ (l+1) - (n+2)H\ \binom{l+2}{3} + (n+2)^2 D\ \binom{l+3}{5} - (n+2)^3 L\ \binom{l+4}{7} + (n+2)^4\binom{l+4}{8}$$

Here

$$L = K\{L(2,n)\} = R_n^{(0)}(2),$$

$$D = K\{D(2,3,n)\} = R_n^{(0)}(3),$$

$$H = K\{H_{14}(3,4,n)\} = R_n^{(0)}(4),$$

$$G = K\{B(n,8,0)\} = R_n^{(0)}(5)$$

It is observed that all terms in the summations of (53) and (54) vanish for $i > l$. This situation occurs if m is large enough ($m > l+2$). The external term vanishes for $m > l+1$. These features lead to a considerable simplification in practical applications of the formulas.

A convenient formula for the number of Kekulé structures of oblate rectangles, $K\{R^j(m,n)\} = R_n(m)$, is obtained by inserting $l=n$ into (53):

$$R_n(m) = (-1)^{m-1}(n+2)^{m-1}\binom{n+m}{n-m+1} + \sum_{i=0}^{m-2}(-1)^i(n+2)^i R_n^{(0)}(m-i)\binom{n+i+2}{n-i} \tag{14.55}$$

This equation was used as the basis for deducing $R_n(8)$, viz. the last formula of CHART 12-II. But first it was necessary to deduce an algorithm to find $R_n^{(0)}(m)$ for arbitrarily large m values.

The expansion of (51) for $l=0$ gives for $m \geq 3$:

$$R_n^{(0)}(m) = (-1)^m(n+2)^{m-2}\binom{n+m}{n-m+2} + \sum_{i=0}^{m-3}(-1)^i(n+2)^i R_n^{(0)}(m-i-1)\binom{n+i+3}{n-i} \tag{14.56}$$

This is a recurrence relation for $R_n^{(0)}(m)$. It gives successively the pertinent formulas for increasing m values in the following way, where the first formula (for $m=2$) is an initial condition.

$$R_n^{(0)}(2) = L = \binom{n+2}{2},$$

$$R_n^{(0)}(3) = D = L\ \binom{n+3}{3} - (n+2)\binom{n+3}{4},$$

$$R_n^{(0)}(4) = H = D\ \binom{n+3}{3} - (n+2)L\ \binom{n+4}{5} + (n+2)^2\binom{n+4}{6},$$

$$R_n^{(0)}(5) = G = H\ \binom{n+3}{3} - (n+2)D\ \binom{n+4}{5} + (n+2)^2 L\ \binom{n+5}{7} - (n+2)^3\binom{n+5}{8}$$

The polynomial forms of $R_n^{(0)}(m)$ are found (for m up to 8) in CHART 12-VI.

Finally we give the deduced pattern for a successive derivation of the formulas of $K\{R^j(m,n)\} = R_n(m)$.

$$R_n(2) = L \binom{n+2}{2} - (n+2)\binom{n+2}{3} \,,$$

$$R_n(3) = D \binom{n+2}{2} - (n+2)L \binom{n+3}{4} + (n+2)^2\binom{n+3}{5} \,,$$

$$R_n(4) = H \binom{n+2}{2} - (n+2)D \binom{n+3}{4} + (n+2)^2 L \binom{n+4}{6} - (n+2)^3\binom{n+4}{7} \,,$$

$$R_n(5) = G \binom{n+2}{2} - (n+2)H \binom{n+3}{4} + (n+2)^2 D \binom{n+4}{6} - (n+2)^3 L \binom{n+5}{8} + (n+2)^4\binom{n+5}{9}$$

CHAPTER 15
ALGORITHM: A GENERALIZATION

15.1 INTRODUCTION

Most (but not all) of the algorithms treated under the relevant sections of the preceding text are special cases of the most general algorithm presented here.

The most general algorithm applies to benzenoids which may be built up row by row (conventionally starting from the bottom) and hexagon by hexagon within each row. During every step of this building-up process all the systems should be Kekuléan. However, allowance will be made for disconnected benzenoids. We may refer to this class of systems as "constructable".

15.2 GENERAL PRINCIPLES

1. The algorithm starts with one external unity.

2. All hexagons are filled by numerals, where each numeral indicates the increment in the K number (number of Kekulé structures) by adding the particular hexagon.

3. Consequently, during each step the sum of numerals until the one of the last added hexagon gives the K number of the system which has been built up.

4. The total K number (for the whole benzenoid) is the grand total of the algorithm numerals.

5. All the algorithmic rules (see below) prescribe that the numerals for each row are filled into the hexagons in the opposite direction of the building-up process. The rows themselves are treated in the same sequence as they are built up.

15.3 MULTIPLE CHAINS

The algorithms for single chains (Section 4.6; Paragraph 6.2.2), parallelograms (Paragraph 4.8.2), chevrons (Paragraph 8.3.3) and multiple zigzag chains (Paragraph 9.8.1) are straightforwardly generalized to a multiple chain with an arbitrary LA-sequence. The examples of Fig. 1 are supposed to be self-explanatory. It should just be remembered that the hexagon right after a kink (single chains) or the row right after a change in direction of the building-up process (multiple chains) should be treated in a special way. The cited section and paragraphs may be consulted for further details.

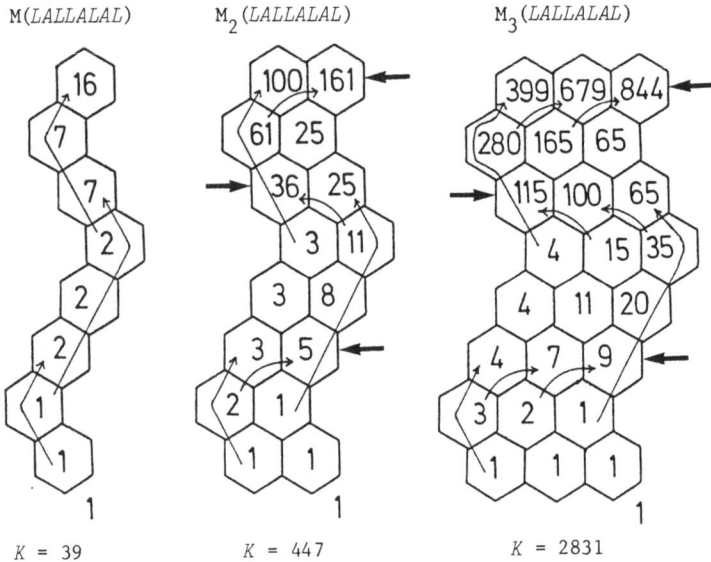

M(*LALLALAL*) M₂(*LALLALAL*) M₃(*LALLALAL*)

K = 39   *K* = 447   *K* = 2831

Fig. 15.1. Illustration of the algorithm for a single, double and triple chain, all with the same *LA*-sequence. In the multiple chains, the changes of direction in the building-up process are indicated by thick arrows.

15.4 MULTIPLE CHAINS WITH TRUNCATED ROWS

We have also seen the adaptation of the algorithm to truncated rows. More pre-cisely it is understood that any row should either have the same length or be shor-ter than the preceding row. It is referred to (a) parallelograms with truncated rows: ribbon (Paragraph 8.4.4), generalized ribbon (Paragraph 8.4.5), other trunca-ted parallelograms including parallelogram with an incomplete row, prolate and ob-late triangles (Paragraph 8.5.5); (b) multiple zigzag chain with truncated rows (Section 9.8).

A straightforward generalization is applicable to strips of truncated rows where the directions of the building-up are arbitrary. In Fig. 2 the benzenoid has rows of the same lengths, viz.
6, 5, 3, 3, 1, as in the
example of Fig. 9.6.

K = 412

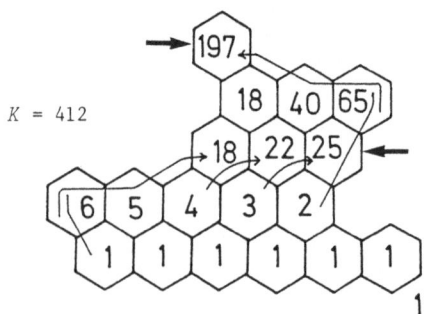

Fig. 15.2. Illustration of the algorithm for a truncated multiple chain with the *LA*-sequence *LALAL*. Changes of direction in the building-up process are indicated by thick arrows.

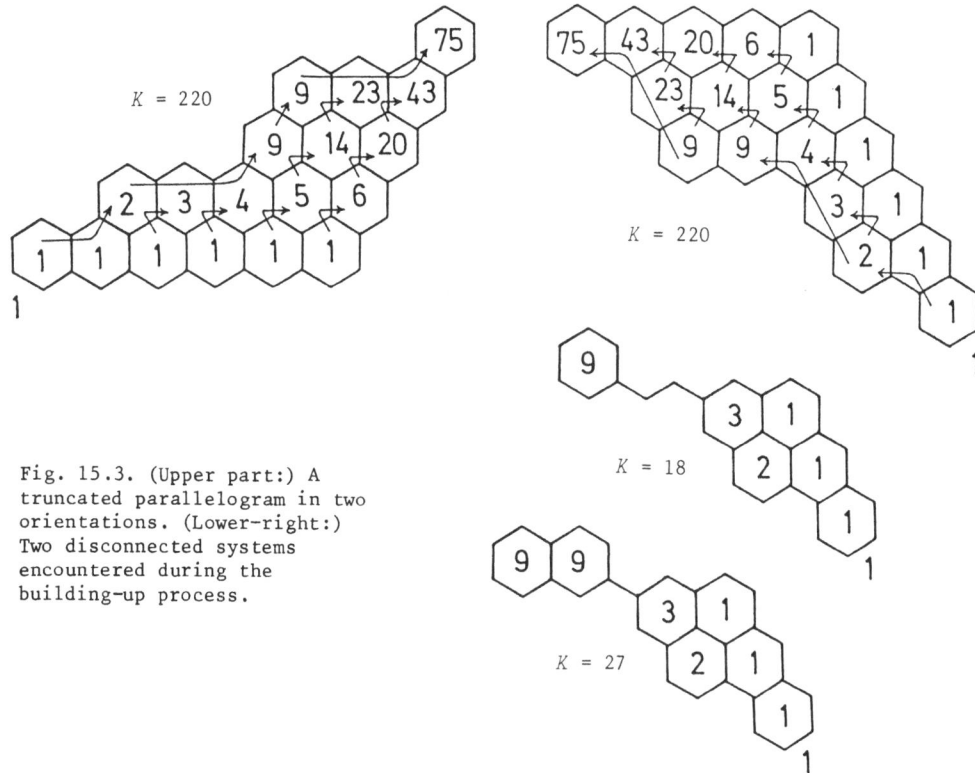

Fig. 15.3. (Upper part:) A truncated parallelogram in two orientations. (Lower-right:) Two disconnected systems encountered during the building-up process.

15.5 PARALLELOGRAM WITH TRUNCATED AND AUGMENTED ROWS

15.5.1 *Introduction*

Consider a parallelogram with truncated rows (or a truncated parallelogram; cf. Paragraph 8.5.5 and Fig. 9.6). In Fig. 3 the benzenoid from the left-hand part of Fig. 9.6 is depicted in two new orientations. The upper-left part of Fig. 3 empha-sizes that the algorithmic rules also should be mastered in the mirror-image fashion. Here the rows are built up from right to left. In the upper-right orientation the rows are (as usual) supposed to be built up from the bottom, while each row is built up from left to right. Hence the benzenoid may be characterized as having augmented rows. The algorithm numerals, although being the same as those of the preceding case, give different information because of the new way of building up the (same) benze-noid. The lower parts of Fig. 3 show two steps of the building-up process; the sys-tems are disconnected in the chosen examples. It is seen that the numbers of Kekulé structures obtained by adding algorithm numerals are consistent with the result ob-tained by multiplying the K numbers for the two disconnected units.

15.5.2 *Generalization*

It is clear that a benzenoid with both truncated and augmented rows may be constructed when the terms are used in the same sense as above. In general we refer to such systems as parallelograms with truncated and augmented rows. It is assumed that all the rows are built up in the same direction. Figure 4 shows an example where the algorithm numerals are filled in.

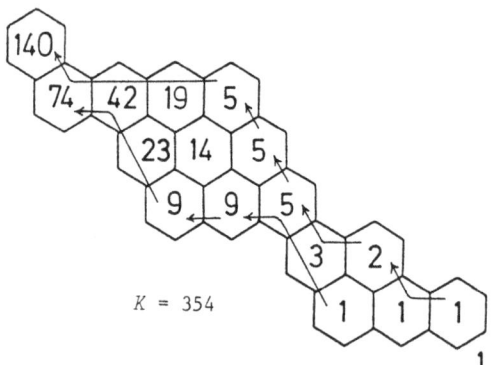

Fig. 15.4. A parallelogram with truncated and augmented rows. The rows have been built up from left to right.

$K = 354$

15.5.3 *Algorithmic Rules*

Here we summarize four rules by which the algorithm numerals of a parallelogram with truncated and augmented rows may be determined (Cyvin and Gutman 1986b). The benzenoid used as example (Fig. 5) is the same as in Fig. 4. It has six rows and six columns.

Let the numeral in the r-th row and s-th column be denoted by X_{rs}. The particular number $X_{53} = 19$ may be determined by all the four rules as indicated on Fig. 5(a)-(d).

(a) Addition of two numerals:

$$X_{rs} = X_{r-1\ s} + X_{r\ s+1}; \qquad r > 1 \qquad (15.1)$$

(b) Summation along a row:

$$X_{rs} = \sum_{j=s}^{j_{max}} X_{r-1\ j}; \qquad r > 1 \qquad (15.2)$$

Here j_{max} is the largest column index in the $(r-1)$-th row. It may happen that $j_{max} = s$. Then eqn. (2) reduces to only one term, and the algorithm numeral is simply repeated; $X_{rs} = X_{r-1\ s}$.

(c) Summation along a column:

$$X_{rs} = \sum_{i=i_{min}}^{r} X_{i\ s+1} \tag{15.3}$$

Here i_{min} is the lowest row index in the $(s+1)$-th column. If $i_{min} = r$, then eqn. (3) reduces to one term, and the algorithm numeral is repeated; $X_{rs} = X_{r\ s+1}$.

(d) Method of subunits: Mark pertinent parts of the rows and columns through the hexagon with X_{rs} as indicated in Fig. 5 (hatched hexagons). Suppose that the rows of the benzenoid are built up (as usual) from the bottom and each row from the left. Then X_{rs} is equal to the K number of the benzenoid unit below and to the right of the marked hexagons. This number is obtainable by addition of algorithm numerals as

$$X_{rs} = 1 + \sum_{i=i_{min}}^{r-1} \sum_{j=s+1}^{j_{max}} X_{ij}; \qquad r > 1 \tag{15.4}$$

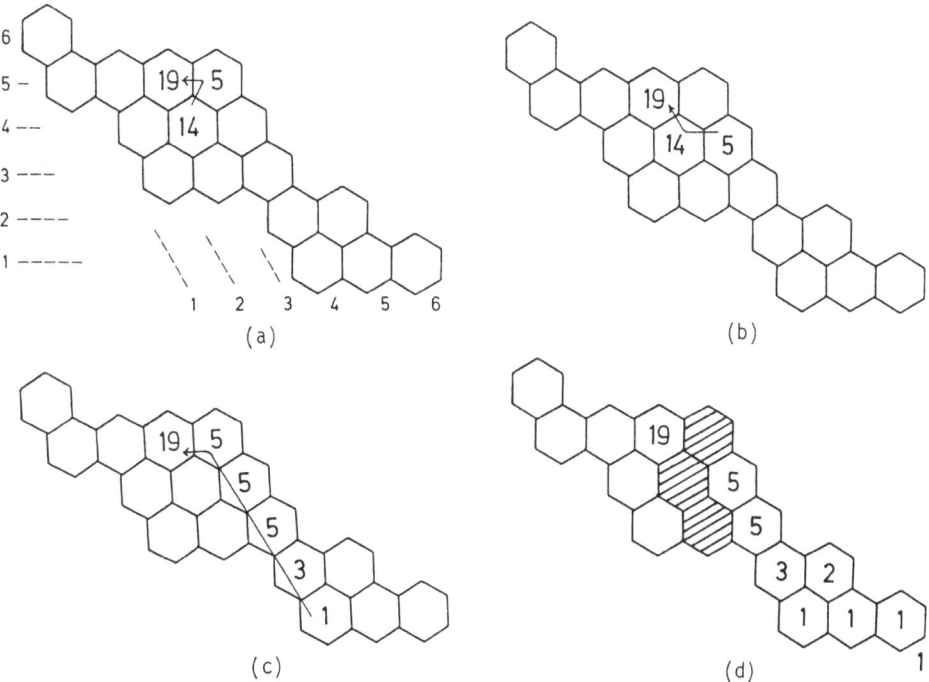

Fig. 15.5. Four ways how to determine an algorithm numeral in a parallelogram with truncated and augmented rows; see also Fig. 15.4. Row and column indices are shown in (a).

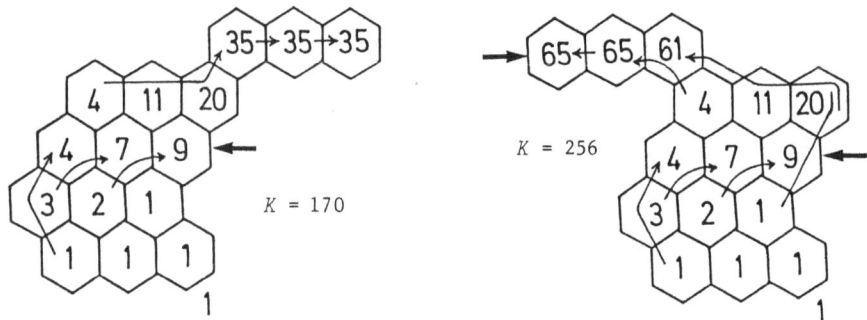

Fig. 15.6. Exemplification of the algorithm: In the left-hand drawing the top row has not changed direction from the preceding one; in the right-hand drawing it has; cf. the thick arrows.

from the adding of two numerals in a parallelogram; cf. Fig. 5(a). If this parallelogram-rule (without change of direction) is called a *phenalene*-type summation, then the present rule (after change in direction) may properly be termed *phenanthrene*-type summation.

If the row in question (index r) is augmented by more than one hexagon one finds that the hexagon in the $(r-1)$-th row is absent when the *phenalene*-type summation is attempted. In that case the pertinent numeral of the r-th row is simply repeated.

Figure 6 will help to explain the above rules.

Finally we give two more examples in Fig. 7. The benzenoids therein have the same lengths of the rows as in the example of Fig. 4, and they are also truncated and augmented in the same way. Only the directions in the building-up of the rows are varied; in the right-hand part of Fig. 7 the benzenoid is a multiple zigzag chain with truncated and augmented rows, while the left-hand part shows an intermediate case between a parallelogram and multiple zigzag chain with truncated and augmented rows.

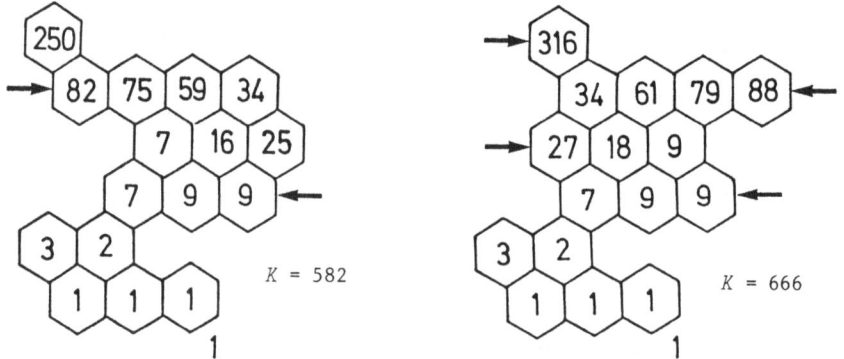

Fig. 15.7. Two illustrations of the algorithm for constructable benzenoids. Changes in direction of the building up are indicated by arrows.

The first rule (a), which is the simplest one, is not applicable to (α) the last hexagon of a row (i.e. with the largest column index) or (β) the first hexagon of a column (i.e. with the smallest row index). In the former case (α) rule (b) is applicable. In the latter case (β) rule (c) is applicable. Whenever (a) is applicable it also is the case with both (b) and (c). The three first rules, (a)-(c) are sufficient for deducing the numerals, while (d) is a useful supplement and generally applicable. For the first row the subunit pertaining to (d) degenerates to no hexagons, as represented by the external unity.

15.6 CONSTRUCTABLE BENZENOIDS

The above algorithm (Section 15.5) may be adapted to multiple zigzag chains with truncated and augmented rows. More generally it may be extended to arbitrary multiple chains with truncated and augmented rows, or in other words all constructable benzenoids. That is the most general class of benzenoids treated here.

The indexing of columns when the direction of building-up changes from one row to another is not so straightforward as in the case of parallelograms (Fig. 5). Therefore we refrain from a mathematical description of the algorithmic rules as in (1)-(4), but resort to verbal descriptions.

All we need to specify (in addition to the preceding rules) are the rules for determining algorithm numerals in the first row after a change in direction. In the following the terms first and last hexagon of a row are related to the building-up process.

Assume that the first row after a change in direction is not truncated. Let its index be r. The numeral of the last hexagon in this row (which should be filled in first) is determined by a summation along a column as explained by Fig. 1 (and in some of the preceding chapters). If the $(r-1)$-th row is augmented this summation reduces to one number.

If the first row after a change in direction is truncated the numeral of its last hexagon is determined by a slight generalization of the (somewhat complicated) procedure which was explained for truncated multiple zigzag chains (Paragraph 9.8.2). It is referred to the right-hand part of Fig. 6. (i) Let again the row in question have the index r, and let its last hexagon be Y. (ii) Concentrate upon the first hexagon, say X, in the $(r-1)$-th row, and let its algorithm numeral be X. (iii) Add all numerals along the column to which X belongs through the preceding row as far as possible without deviating from a linear-chain conformation. This summation will reduce to one number (X) if the $(r-1)$-th row is augmented. (iv) Take X once more, and add numerals along the $(r-1)$-th row for all hexagons (if any) until the nearest one to Y, but not in contact with it.

The remaining numerals of the first row after a change in direction are obtained by adding two numerals as shown, e.g. in Fig. 1, a rule which is different

CHAPTER 16

PERICONDENSED ALL-BENZENOIDS AND RELATED SYSTEMS

16.1 INTRODUCTORY REMARKS

For a definition of all-benzenoids it is referred to Paragraph 2.2.8. Cata-condensed all-benzenoids are treated in Section 6.5. Two classes of (generally pericondensed) all-benzenoids are found in CHART 12-VIII.

An algorithmic formula for K, the number of Kekulé structures, generally applicable to all-benzenoids is due to Polansky and Gutman (1980). It is described briefly in Paragraph 4.10.7.

Several references to previous works are given in the subsequent sections in connection with special classes of all-benzenoid systems.

16.2 ALL-BENZENOID CLASSES INCLUDING MODIFICATIONS

16.2.1 *Introduction*

Classes of homologs may be constructed among the pericondensed (as well as catacondensed) all-benzenoid systems. We shall treat some classes of repeated units, where *pyrene* is recognized as a characteristic subunit, a class among "ladders", viz. $\mathbb{K}(m)$, and finally a more "compact" class, $S(n)$.

We shall speak about linearly and angularly annelated *pyrenes* in the same spirit as the term "annelated" was used in CHAPTER 7. Several units (P_1, P_2, ...) may be annelated in one system and not only two as in the two-sided annelation to a single chain. The annelated systems in the present chapter are restricted to P = *pyrene* units fused to *benzenes*. The system P:L(1):P:L(1): .... represents a linear (resp. angular) annelation if L(1) designates a hexagon of the mode L_2 (resp. A_2).

The system of linearly annelated *pyrenes* L(1):P:L(1):P: .... :L(1) is a celebrated example of a class of all-benzenoids sometimes referred to as "*pyrenes* on a string" (Randić 1980; Polansky and Gutman 1980; Ohkami and Hosoya 1983; Cyvin SJ, Cyvin and Gutman 1986; 1987); cf. Fig. 1. Another class, which has attached much interest (Randić 1980; Polansky and Gutman 1980; Ohkami and Hosoya 1983) is here denoted $S(n)$. These two classes may cause confusion because *dibenzo*[e,l]*pyrene* is a member in both. We have therefore drawn the first few members of both classes in Fig. 1. Notice also that catacondensed and disconnected systems may occur as degenerate cases in the classes which we refer to as (generally) pericondensed.

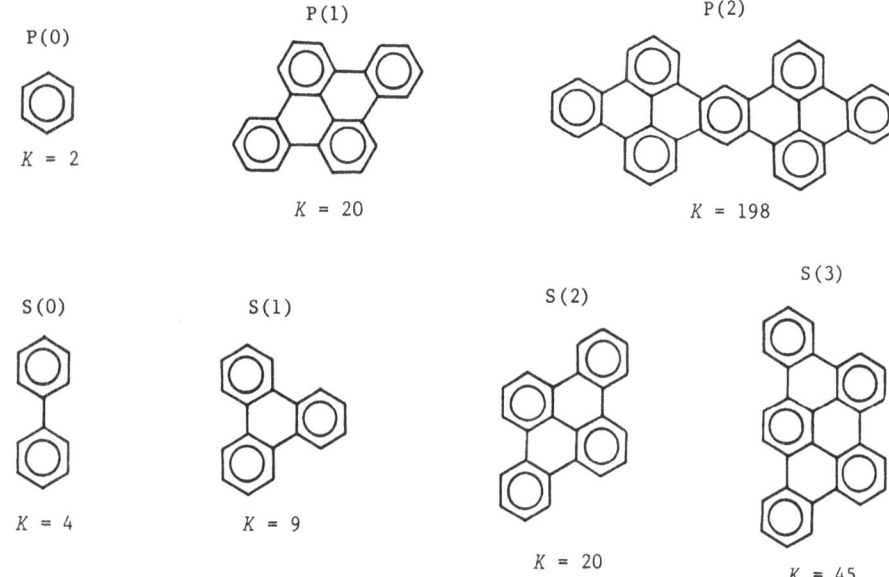

Fig. 16.1. Two classes of all-benzenoids: "*pyrenes* on a string", P(n), and the more "compact" S(n). Notice that P(1) = S(2).

16.2.2 *Annelated Pyrenes*

Linear Annelation. The class P(n), *pyrenes* on a string, is defined by Fig. 1 and CHART I. Notice that n designates the number of *pyrene* units. A recurrence formula for the K numbers is implied in the work of Ohkami and Hosoya (1983);

$$K\{P(n)\} = 10K\{P(n-1)\} - K\{P(n-2)\}; \qquad n \geq 2 \qquad (16.1)$$

Figure 1 contains the initial conditions, which were used together with (1) to establish the explicit formula given in CHART I. Numerical K values for n up to 10 are given in Table 1.

We may remove the terminal hexagons from P(n) to produce P'(n) and P"(n) in the same way as was done for the catacondensed all-benzenoid classes of CHART 6-V. These classes (cf. Fig. 2) are not all-benzenoid. It seems, however, to be of more interest to remove two hexagons from one or both sides of P(n) as shown in CHART I. Then the resulting classes, viz. $P_1(n)$ and $P_2(n)$, again consist of all-benzenoids. Notice that $P_2(2)$ = 10(2), a catacondensed benzenoid (cf. CHART 6-V). Another (not all-benzenoid) variant, viz. $P_1'(n)$, is included in Fig. 2. All the six classes under consideration (CHART I and Fig. 2) fit into a system of linearly coupled recurrence relations as far as the K numbers are concerned. Hence for every one of these classes a recurrence relation of the form (1) is sound. Below we give the K numbers of the five new classes as linear combinations of $K\{P(n-i)\}$; $i = 0, 1$.

$$K\{P'(n)\} = \frac{1}{2} [K\{P(n)\} + K\{P(n-1)\}]; \qquad n \geq 1 \qquad (16.2)$$

$$K\{P''(n)\} = 3K\{P(n-1)\}; \qquad n \geq 1 \qquad (16.3)$$

$$K\{P_1(n)\} = \frac{1}{2} [K\{P(n)\} - K\{P(n-1)\}]; \qquad n \geq 1 \qquad (16.4)$$

$$K\{P_2(n)\} = 2K\{P(n-1)\}; \qquad n \geq 1 \qquad (16.5)$$

CHART 16-I. All-benzenoids: linearly annelated *pyrenes*

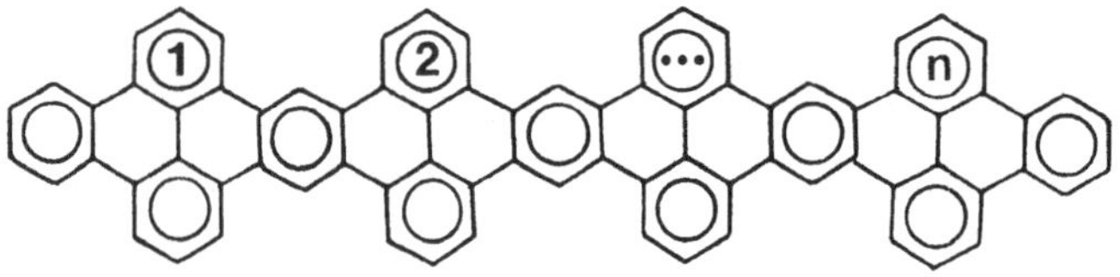

$P(n)$

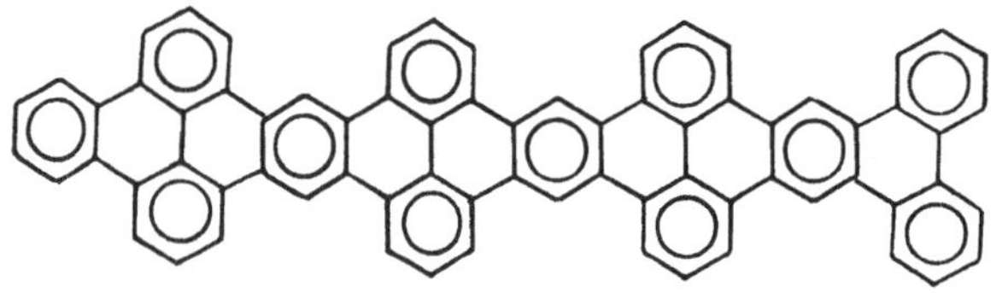

$P_1(n)$

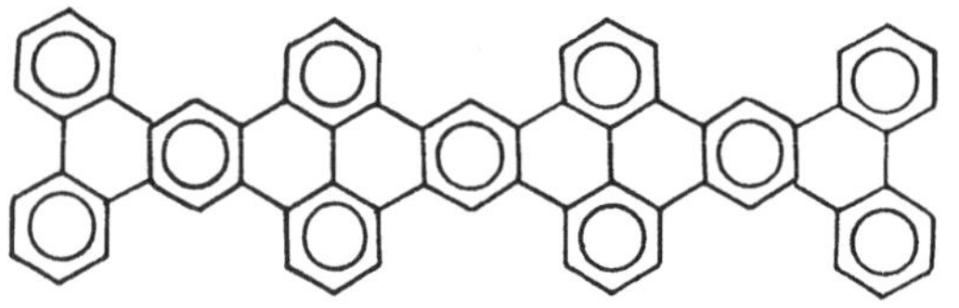

$P_2(n)$

$$^{a}K\{P(n)\} = \frac{1}{2\sqrt{6}} \left[(5 + 2\sqrt{6})^{n+1} - (5 - 2\sqrt{6})^{n+1} \right]$$

$$K\{P_1(n)\} = \frac{1}{2\sqrt{6}} \left[(\sqrt{6} + 2)(5 + 2\sqrt{6})^{n} + (\sqrt{6} - 2)(5 - 2\sqrt{6})^{n} \right]$$

$$K\{P_2(n)\} = \frac{1}{\sqrt{6}} \left[(5 + 2\sqrt{6})^{n} - (5 - 2\sqrt{6})^{n} \right]$$

[a]Cyvin SJ, Cyvin BN, Gutman I (1986). Coll Sci Papers Fac Sci Kragujevac 7: 5

Table 16.1. Numerical values of K for the benzenoids $P(n)$, $P_1(n)$, $P_2(n)$ and $Q(n)$, $Q_1(n)$, $Q_2(n)$.

n	$K\{P(n)\}$	$K\{P_1(n)\}$	$K\{P_2(n)\}$	$K\{Q(n)\}$	$K\{Q_1(n)\}$	$K\{Q_2(n)\}$
0	2	1	0	2	1	1
1	20	9	4	20	9	4
2	198	89	40	202	91	41
3	1960	881	396	2040	919	414
4	19402	8721	3920	20602	9281	4181
5	192060	86329	38804	208060	93729	42224
6	1901198	854569	384120	2101202	946571	426421
7	18819920	8459361	3802396	21220080	9559439	4306434
8	186298002	83739041	37639840	214302002	96540961	43490761
9	1844160100	828931049	372596004	2164240100	974969049	439214044
10	18255302998	8205571449	3688320200	21856703002	9846231451	4435631201

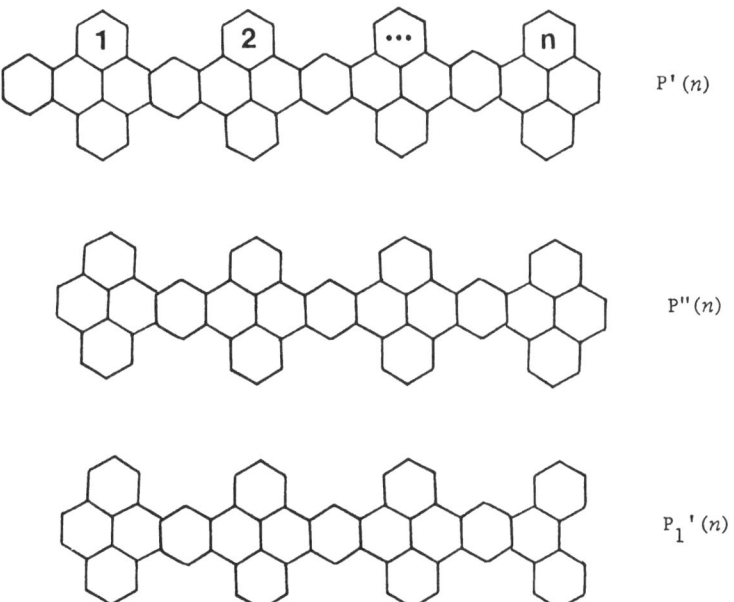

Fig. 16.2. Definition of three benzenoid classes related to $P(n)$.

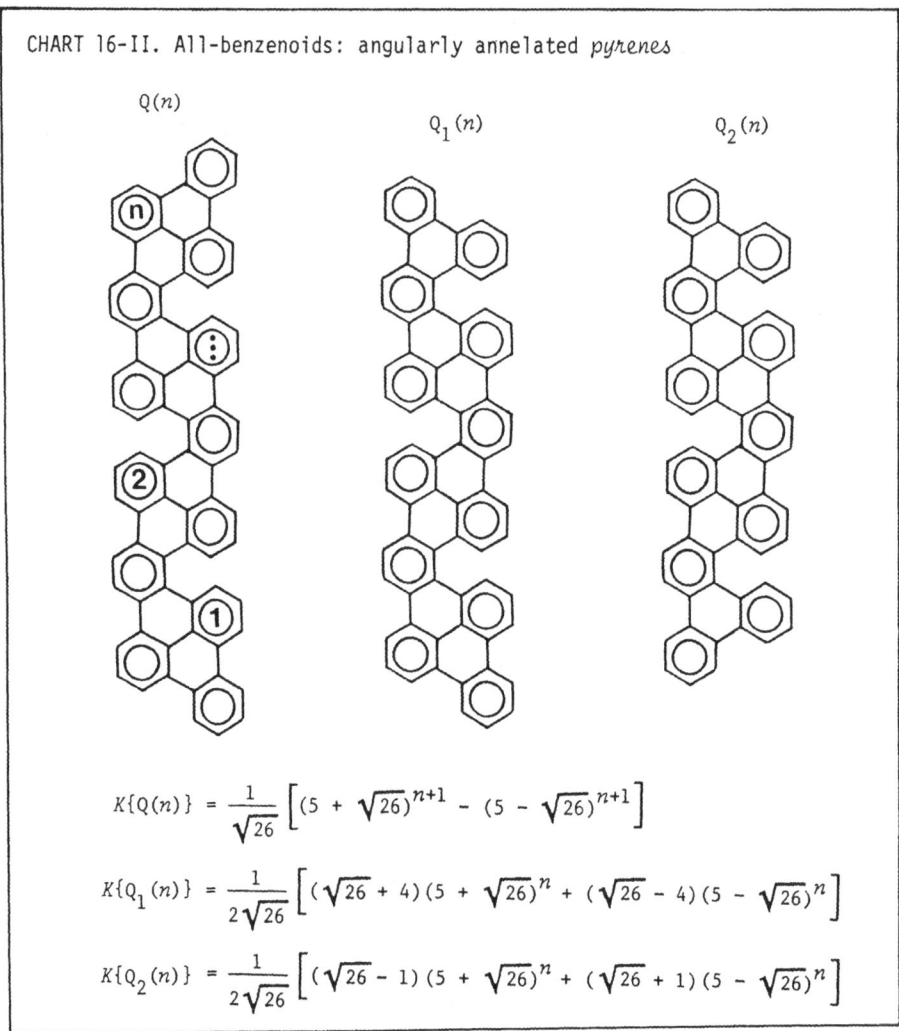

CHART 16-II. All-benzenoids: angularly annelated *pyrenes*

Q(n) $Q_1(n)$ $Q_2(n)$

$$K\{Q(n)\} = \frac{1}{\sqrt{26}}\left[(5 + \sqrt{26})^{n+1} - (5 - \sqrt{26})^{n+1}\right]$$

$$K\{Q_1(n)\} = \frac{1}{2\sqrt{26}}\left[(\sqrt{26} + 4)(5 + \sqrt{26})^{n} + (\sqrt{26} - 4)(5 - \sqrt{26})^{n}\right]$$

$$K\{Q_2(n)\} = \frac{1}{2\sqrt{26}}\left[(\sqrt{26} - 1)(5 + \sqrt{26})^{n} + (\sqrt{26} + 1)(5 - \sqrt{26})^{n}\right]$$

$$K\{P_1{'}(n)\} = \frac{1}{2}\,[K\{P(n)\} = 5K\{P(n-1)\}]; \qquad n \geq 1 \qquad (16.6)$$

For the sake of brevity the explicit formulas (CHART I) and numerical K values (Table 1) are given only for the three all-benzenoid classes in question. The table includes nominal K values (for $n=0$).

Angular Annelation. *Pyrene* units may be annelated not only linearly, but also in a zigzag manner, in order to produce additional all-benzenoid systems; cf. CHART II. The derived recurrence relation for Q(n) reads

$$K\{Q(n)\} = 10K\{Q(n-1)\} + K\{Q(n-2)\}; \qquad n \geq 2 \qquad (16.7)$$

Notice the similarity between (1) and (7).

Five additional classes were defined in the same way as was done in the case of linear annelation. The three (not all-benzenoid) classes $Q'(n)$, $Q''(n)$ and $Q_1'(n)$ are shown in Fig. 3, while the two (all-benzenoid) classes $Q_1(n)$ and $Q_2(n)$ are included in CHART II. All the six classes considered here obey the recurrence relation of the form (7). Linear dependencies for K numbers:

$$K\{Q'(n)\} = \frac{1}{2} [K\{Q(n)\} + K\{Q(n-1)\}]; \qquad n \geq 1 \tag{16.8}$$

$$K\{Q''(n)\} = \frac{1}{4} [K\{Q(n)\} + 2K\{Q(n-1)\} + K\{Q(n-2)\}]; \qquad n \geq 2 \tag{16.9}$$

$$K\{Q_1(n)\} = \frac{1}{2} [K\{Q(n)\} - K\{Q(n-1)\}]; \qquad n \geq 1 \tag{16.10}$$

$$K\{Q_2(n)\} = \frac{1}{4} [K\{Q(n)\} - 2K\{Q(n-1)\} + K\{Q(n-2)\}]; \qquad n \geq 2 \tag{16.11}$$

$$K\{Q_1'(n)\} = \frac{1}{4} [K\{Q(n)\} - K\{Q(n-2)\}]; \qquad n \geq 2 \tag{16.12}$$

Arbitrary Annelation. The algebra of all-benzenoid annelated *pyrenes* resembles very much the algebra of Section 6.5 for catacondensed all-benzenoids. Also in the present case there is much room for isoarithmicity inasmuch as the way of kinks may change without affecting the K numbers. Figure 4 shows a benzenoid which is iso-

$Q'(n)$ $\qquad\qquad$ $Q''(n)$ $\qquad\qquad$ $Q_1'(n)$

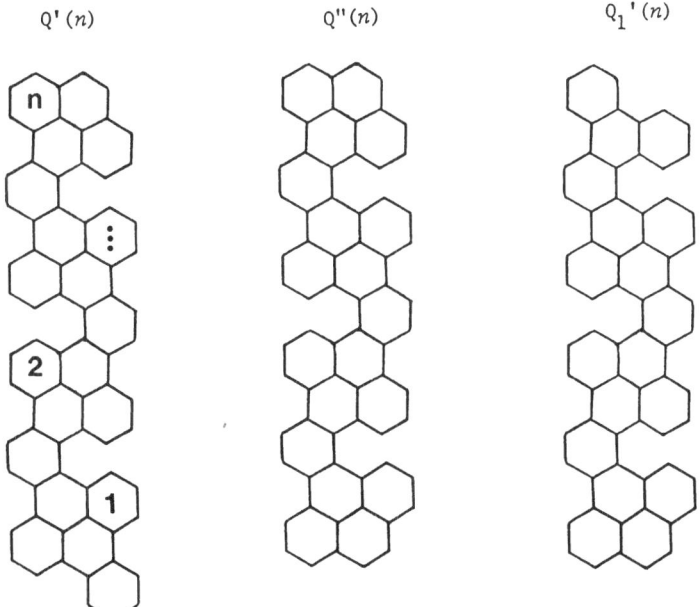

Fig. 16.3. Definition of three benzenoid classes related to $Q(n)$.

$K = 208060$

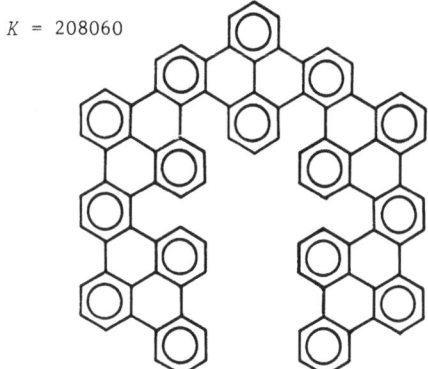

Fig. 16.4. An isoarithmic benzenoid of Q(5).

arithmic with Q(5). Linear and angular annelation may also be combined. In Fig. 5 an example is depicted, in which three all-benzenoid systems meet at a branching hexagon so that the whole system also is all-benzenoid.

16.2.3 *Pericondensed All-Benzenoid Ladder*

Although no precise definition has been given to distinguish the different classes of ladders, we can see a similarity between the class of CHART III and those of CHART 6-IV and CHART 6-VI. There is also a great similarity with T_m of CHART 12-VIII. In the present case of $K(m)$ one has the exceedingly simple recurrence relation

$$K(m) = 10 \; K(m-1); \qquad m \geq 2 \tag{16.13}$$

Here again P(1) is a member of the class (see also Fig. 1). Specifically $K(2) = P(1)$, while the system degenerates to *benzene* for $m=1$: $K(1) = L(1)$. In conclusion, one obtains the explicit formula of CHART III.

16.2.4 *The Class* S(n)

For the class S(n) – see Fig. 1 – Randić (1980) derived a recurrence relation, which also is implied in the work of Ohkami and Hosoya (1983);

$$K\{S(n)\} = 2K\{S(n-1)\} + K\{S(n-2)\} - K\{S(n-3)\}; \qquad n \geq 3 \tag{16.14}$$

It is a four-term relation, and no explicit equation is available. Table 2 gives the numerical values of $K\{S(n)\}$ up to $n=10$.

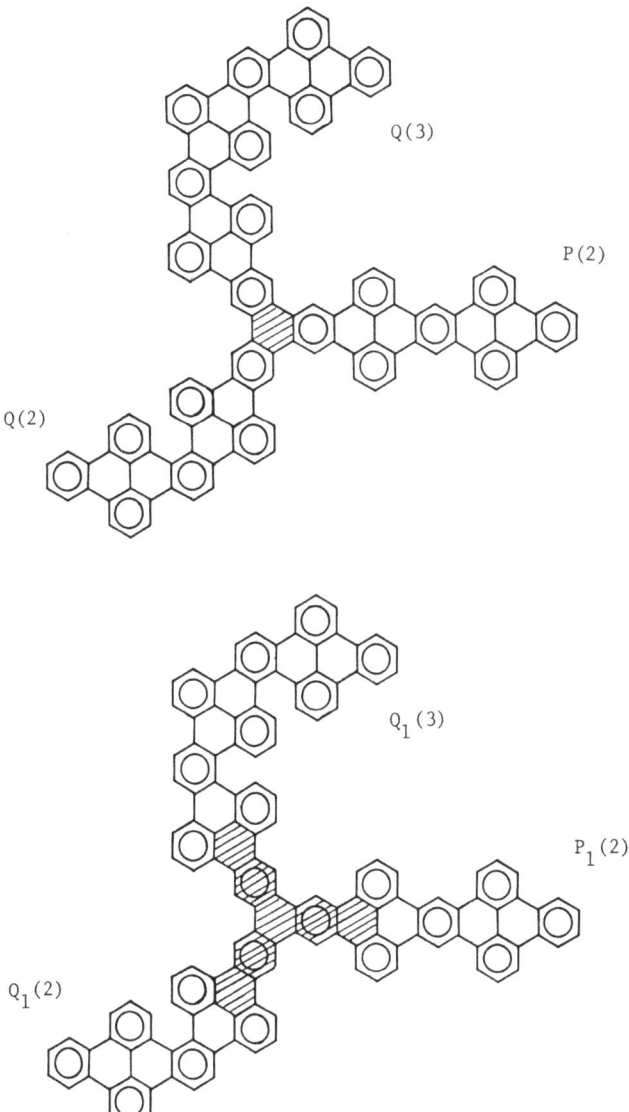

Fig. 16.5. The method of fragmentation applied to an all-benzenoid system with annelated *pyrenes* and a branching hexagon.

$$K = K\{P(2)\} \cdot K\{Q(2)\} \cdot K\{Q(3)\} + K\{P_1(2)\} \cdot K\{Q_1(2)\} \cdot K\{Q_1(3)\} = 89034821$$

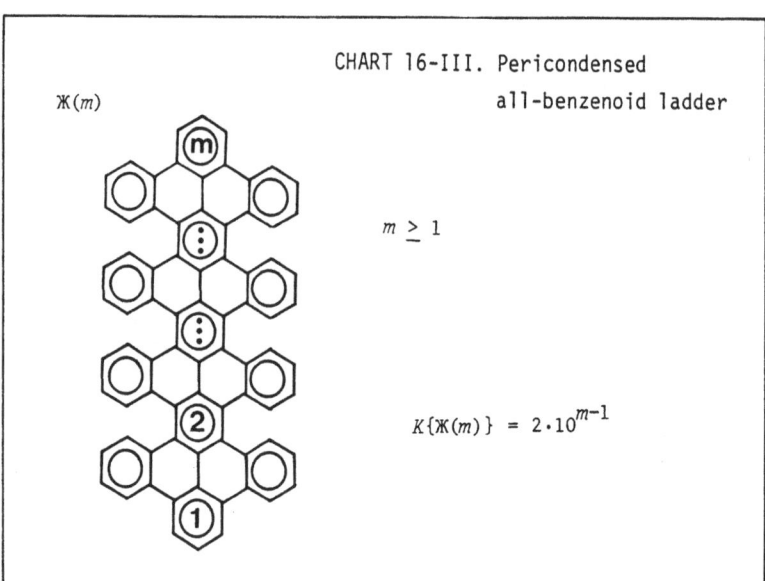

CHART 16-III. Pericondensed all-benzenoid ladder

$\mathbb{K}(m)$

$m \geq 1$

$K\{\mathbb{K}(m)\} = 2 \cdot 10^{m-1}$

Table 16.2. Numerical values of K for S(n).

n	$K\{S(n)\}$
0	4
1	9
2	20
3	45
4	101
5	227
6	510
7	1146
8	2575
9	5786
10	13001

S(n)

16.3 RETICULAR ALL-BENZENOIDS

Figure 6 shows the fragmentation method applied to *tribenzo*[a,g,m]*coronene*.
This system is also found in Fig. 7 as a member of one of three classes. Not a single
combinatorial *K* formula is known for any class of such all-benzenoids, which expand
in two dimensions. The bottom class (with regular hexagonal symmetry) was considered
by Stein and Brown (1985). The *K* number for its second member (*hexabenzocoronene*-A)
has been reported previously (Hosoya 1986a; Cyvin BN and Cyvin 1986).

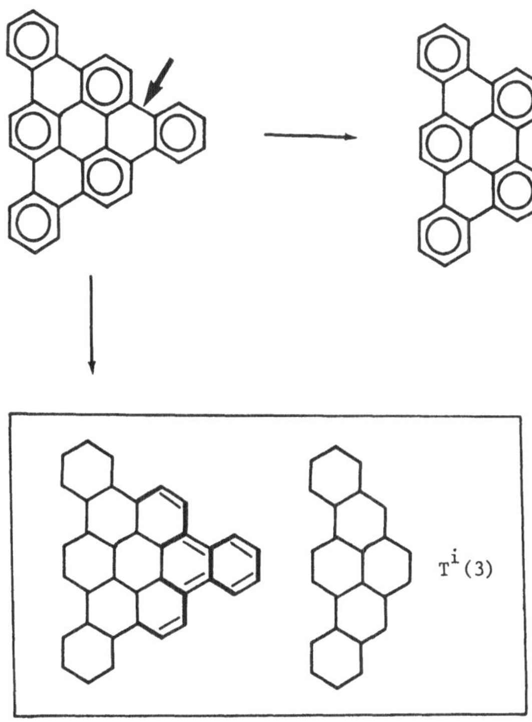

L(1)·S(3)

T^i(3)

Fig. 16.6. The method of fragmentation applied to *tribenzo*[a,g,m]*coronene*. For S(n),
see Table 16.2. T^i(3) is a prolate triangle; see CHART 11-VIII and Table 11.3. The
total number of Kekulé structures: $K = 2 \cdot 45 + 14 = 104$.

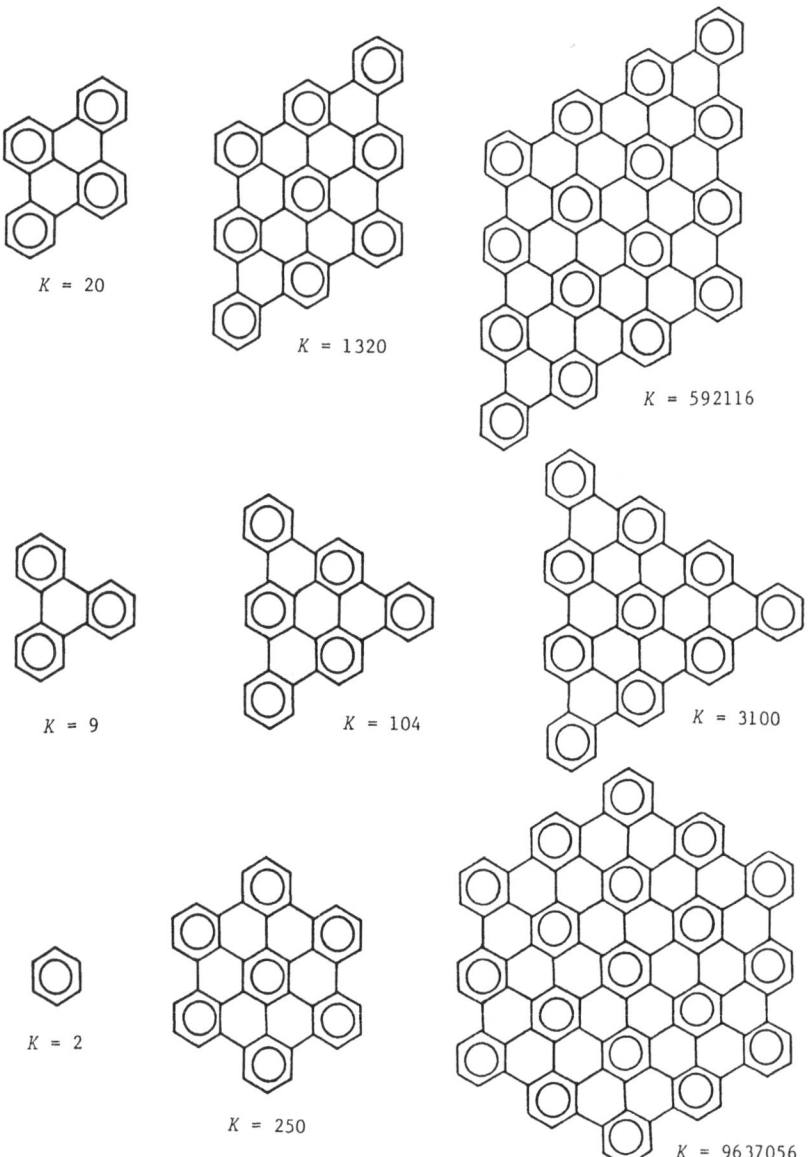

$K = 20$

$K = 1320$

$K = 592116$

$K = 9$

$K = 104$

$K = 3100$

$K = 2$

$K = 250$

$K = 9637056$

Fig. 16.7. Three classes of reticular all-benzenoids.

CHAPTER 17
BENZENOIDS WITH REPEATED UNITS

17.1 INTRODUCTION

17.1.1 *Previous Work and Definitions*

Benzenoids with repeated (identical) units have been encountered in several places of the preceding chapters. Already the single linear and zigzag chains (see, e.g. Fig. 4.9 or CHART 6-II) may be considered as (trivial) benzenoids with repeated units. A less trivial example is the single chain with equal segments in a zigzag arrangement (CHART 6-II). Several typical examples are found among the all-benzenoid classes, both catacondensed (Section 6.5) and pericondensed (Chapter 16).

We are inclined to give Randić (1980) the credit for a first systematic treatment of the number of Kekulé structures (K) for benzenoid systems with repeated units. The recurrence properties are a crucial point in such studies; see, e.g. Hosoya and Ohkami 1983; Ohkami and Hosoya 1983. Especially for some treatments of catacondensed systems it is referred to: Balaban and Tomescu (1985); El-Basil (1986a); Balaban, Artemi and Tomescu (1987). Some general formulations with special applications are also available (Graovac et al. 1986; Křivka et al. 1986; Babić and Graovac 1986; Klein, Hite and Schmalz 1986; Klein, Hite, Seitz and Schmalz 1986; Cyvin SJ, Cyvin and Gutman 1987). Finally we give references to treatments of some special systems: Cyvin SJ, Cyvin and Gutman 1986; Cyvin SJ and Cyvin 1986a.

We shall restrict the considerations to the following kinds of benzenoid systems with repeated units.

(a) A number of benzenoid units fused together. In all our examples we shall consider a system of identical units, occasionally modified at one or both ends.

According to a previous definition (cf. Paragraph 2.2.9 and Section 7.1) two benzenoid units are fused when they share exactly one edge. If at least one of the units is pericondensed the resulting benzenoid is a composite system as defined in Paragraph 2.3.1.

The definition covers cases like $P_1:C_1:P_2:P_3: \ldots$, which is a generalization of the two-sided annelation defined in Section 7.1. Here C_1, C_2, $\ldots$ are catacondensed systems. Also in the general case we may speak about an annelated benzenoid, or we say that the units P_1, P_2, P_3, $\ldots$ are annelated. Assume now the special case of $P_1:L(1):P_2$, where the catacondensed system is a single hexagon (*benzene*). Then we speak about linear (resp. angular) annelation if L(1) designates a hexagon of the mode L_2 (resp. A_2).

(b) A number of identical benzenoid units condensed together, occasionally modified at one or both ends.

By definition (cf. Paragraph 2.2.9) two benzenoid units are condensed when they share exactly two neighbouring edges. In our examples we shall only consider systems where two units are linked by a hexagon of the mode A_4. As the other possibility two condensed units may be linked by a P_3-mode hexagon. By a re-definition of the units, however, this case may be considered as two units which are fused.

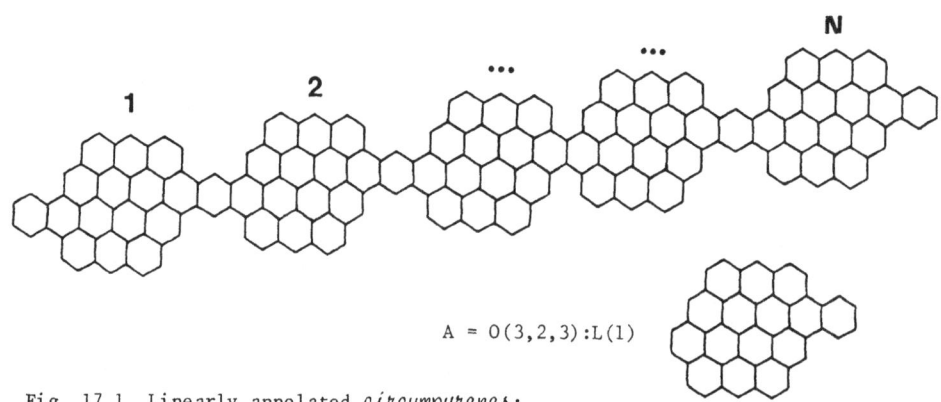

$$A = O(3,2,3):L(1)$$

Fig. 17.1. Linearly annelated *circumpyrenes*:
L(1):O(3,2,3):L(1):O(3,2,3): .... :L(1). The
number of Kekulé structures:

$$^aK = \frac{1}{\sqrt{595}} [(\sqrt{595} + 26)(125 + 5\sqrt{595})^N + (\sqrt{595} - 26)(125 - 5\sqrt{595})^N]$$

[a]Cyvin SJ, Cyvin BN, Gutman I (1986). Coll Sci Papers Fac Sci Kragujevac 7: 5

17.1.2 *Introductory Examples*

Figure 1 shows an example from Cyvin SJ, Cyvin and Gutman (1986). The system could be termed "*circumpyrenes* on a string". It is a system of fused repeated units. In order to fit the definition (a) of the preceding paragraph we may define the unit A as shown in Fig. 1 and assume a modification at the left end. Then the system may be symbolized L(1):A:A: .... :A.

Figure 2 exemplifies systems of condensed benzenoid units.

The left-hand system (Fig. 2) is a rather trivial case of essentially disconnected benzenoids, viz. L(n)·L(n)· .... (m times). Hence $K = (n+1)^m$, exactly as in the case of the prolate rectangle (CHART 12-I).

When the considered system of essentially disconnected benzenoids is modified at the ends as shown in the right-hand part of Fig. 2, we arrive at a non-trivial class here designated by $\mathbb{K}(k,m)$. The K numbers obey the simple recurrence relation

$$K\{\mathbb{K}(k,m)\} = k(2k + 1)K\{\mathbb{K}(k, m-1)\}; \qquad m \geq 2 \tag{17.1}$$

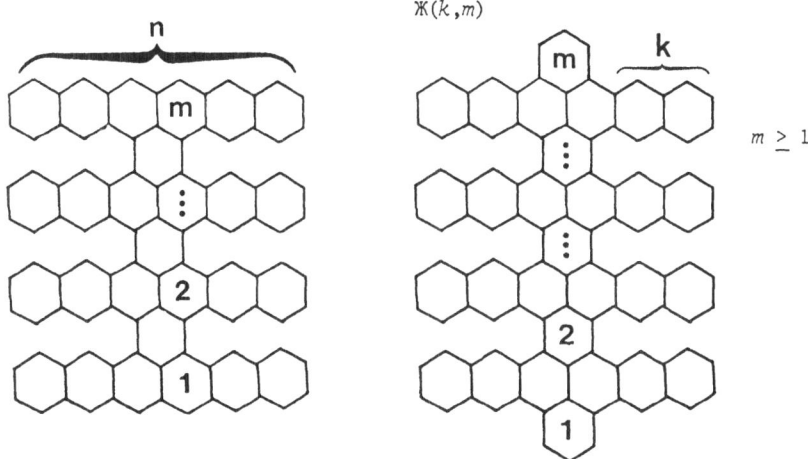

Fig. 17.2. Two classes of benzenoids with condensed repeated units: essentially disconnected benzenoid (left) and $\mathbb{K}(k,m)$, the pericondensed ladder (right). The parameters m are defined differently in the two cases, so that $m=4$ and $m=5$ in the left-hand and right-hand drawing, respectively.

The system degenerates to $\mathbb{K}(k,1) = L(1)$ independently of k. Consequently one arrives at the explicit equation:

$$K\{\mathbb{K}(k,m)\} = 2(2k^2 + k)^{m-1} \qquad (17.2)$$

This class, the pericondensed ladder, is a generalization of the pericondensed all-benzenoid ladder (CHART 16-III), which emerges for $k=1$; $\mathbb{K}(1,m) = \mathbb{K}(m)$. In the special case of $k=0$ the pericondensed ladder coincides with the oblate rectangle for $n=1$; $\mathbb{K}(0,m) = R^j(m,1)$; cf. CHART 12-III. This class has the members *pyrene* ($m=2$), *peropyrene* ($m=3$), *teropyrene* ($m=4$), etc. In our definition of benzenoids with re-peated units this class should be interpreted as condensed *phenalenes*. However, the *pyrene* units are an obvious characteristic of the class. Then the A_4-mode hexagons are reckoned to both of the neighbouring units. Under this scope the system may be termed as "compressed" *pyrenes*.

17.2 FUSED REPEATED UNITS

17.2.1 *General Formulation*

Here we give an account of a general formulation (Cyvin SJ, Cyvin and Gutman 1987) for the K numbers of benzenoids with fused units.

Consider a system symbolized by

$$C_N = A_0 : A_1 : A_2 : \dots : A_N : A_\omega$$

and similarly:

$$B_N = A_0 : A_1 : A_2 : \ldots : A_N$$

For $N=0$ the systems degenerate into:

$$C_0 = A_0 : A_\omega, \qquad B_0 = A_0$$

We define symbols of the type A_i', $'A_i$ and $'A_i'$ as the pertinent unit with one or both edges of fusion removed in the same way as was done in Chapter 7; cf. especially Figs. 7.1, 7.2 and 7.12. The whole theory developed below is based on eqn. (7.20) for $K\{P:Q\}$. Therefore we assume that all the units A_0, A_1, A_2, ..., A_N, A_ω are even (i.e. have an even number of vertices). If some of the units were odd, then eqn. (7.24) would have to be applied, leading to substantially different final results. We shall not pursue this option.

Let us first consider the number of Kekulé structures of B_N, viz. $K\{B_N\}$. In addition to B_N (see above), define also

$$B_N' = A_0 : A_1 : A_2 : \ldots : A_N'$$

Application of (7.20) gives readily

$$K\{B_N\} = K\{A_N\}K\{B_{N-1}\} - K\{'A_N\}K\{B_{N-1}'\} \tag{17.3}$$

$$K\{B_N'\} = K\{A_N'\}K\{B_{N-1}\} - K\{'A_N'\}K\{B_{N-1}'\} \tag{17.4}$$

or in matrix notation:

$$\begin{bmatrix} K\{B_N\} \\ K\{B_N'\} \end{bmatrix} = M_N \begin{bmatrix} K\{B_{N-1}\} \\ K\{B_{N-1}'\} \end{bmatrix} \tag{17.5}$$

where the matrices M_i; $i = 1, 2, \ldots, N$ are defined as

$$M_i = \begin{bmatrix} K\{A_i\} & -K\{'A_i\} \\ K\{A_i'\} & -K\{'A_i'\} \end{bmatrix} \tag{17.6}$$

Repeating the argument additional $N-1$ times we arrive at

$$\begin{bmatrix} K\{B_N\} \\ K\{B_N'\} \end{bmatrix} = M_N M_{N-1} \ldots M_2 M_1 \begin{bmatrix} K\{A_0\} \\ K\{A_0'\} \end{bmatrix} \tag{17.7}$$

Eqns. (3) and (4) present a system of linearly coupled recurrence relations. They can be de-coupled into

$$K\{B_N\} = \alpha_N K\{B_{N-1}\} + \beta_N K\{B_{N-2}\}; \qquad N \geq 2 \qquad (17.8)$$

$$K\{B_N{}'\} = \alpha_N K\{B_{N-1}'\} + \beta_N K\{B_{N-2}'\}; \qquad N \geq 2 \qquad (17.9)$$

where

$$\alpha_N = K\{A_N\} - \frac{K\{'A_N\}K\{'A_{N-1}'\}}{K\{'A_{N-1}\}} \qquad (17.10)$$

$$\beta_N = - \frac{K\{'A_N\}}{K\{'A_{N-1}\}} \det M_{N-1} \qquad (17.11)$$

It is emphasized that B_N and $B_N{}'$ obey the same form of the recurrence relation.

Let us now consider the more general case of C_N. One finds

$$K\{C_N\} = K\{A_\omega\}K\{B_N\} - K\{'A_\omega\}K\{B_N{}'\} \qquad (17.12)$$

By virtue of this linear dependency it is clear that also C_N has the same recurrence properties as B_N and $B_N{}'$, viz.

$$K\{C_N\} = \alpha_N K\{C_{N-1}\} + \beta_N K\{C_{N-2}\}; \qquad N \geq 2 \qquad (17.13)$$

where α_N and β_N are given by (10) and (11), respectively. It was inferred that C_N is a generalization of B_N. The latter case (B_N), where A_ω is absent, is formally derived from the case of C_N by setting $K\{A_\omega\} = 1$ and $K\{'A_\omega\} = 0$.

We have reached the conclusion that the nature of the terminal fragments, viz. A_0 and A_ω, does not influence the form of the recurrence relation for $K\{C_N\}$. The numerical values of $K\{C_N\}$, however, depend on A_0 and A_ω. In other words: the recurrence relations have different initial conditions depending on A_0 and A_ω, leading to different combinatorial K formulas as functions of N. It can also be shown that provided all the units A_0, A_1, A_2, ..., A_N, A_ω are even systems, C_N is Kekuléan if and only if all the units are Kekuléan.

The most general formulation of the problem as described above leads usually to a very complicated analysis. Therefore some simplifying specializations are warranted.

Let us consider repeated units in the sense that $A_i = A$ for all $i = 1, 2, ..., N$. It is also implied that the edges of fusion in these (identical) units conform to each other. In this case eqn. (7) becomes simplified into

$$\begin{bmatrix} K\{B_N\} \\ K\{B_N{}'\} \end{bmatrix} = \begin{bmatrix} K\{A\} & -K\{'A\} \\ K\{A'\} & -K\{'A'\} \end{bmatrix}^N \begin{bmatrix} K\{A_0\} \\ K\{A_0'\} \end{bmatrix} \qquad (17.14)$$

The recurrence relations (8), (9) and (13) become under the simplifying conditions:

$$K\{B_N\} = \alpha K\{B_{N-1}\} + \beta K\{B_{N-2}\}; \qquad N \geq 2 \qquad (17.15)$$

$$K\{B_N'\} = \alpha K\{B_{N-1}'\} + \beta K\{B_{N-2}'\}; \qquad N \geq 2 \qquad (17.16)$$

$$K\{C_N\} = \alpha K\{C_{N-1}\} + \beta K\{C_{N-2}\}; \qquad N \geq 2 \qquad (17.17)$$

where

$$\alpha = K\{A\} - K\{'A'\} \qquad (17.18)$$

$$\beta = K\{A\}K\{'A'\} - K\{A'\}K\{'A\} \qquad (17.19)$$

The explicit (combinatorial) K formulas are obtained from the respective recurrence relations by standard methods (Spiegel 1971); cf. also Section 4.7 and other places in the preceding chapters. In practice this analysis may still be a cumbersome task for eqns. (15)-(17). The case of B_N (i.e. when A_ω is absent) becomes manageable if we also assume that A_0 is absent. Formally this special case is effectuated by setting $K\{A_0\} = 1$ and $K\{A_0'\} = 0$. Then it was found for the system of N idential units:

$$K\{A:A: \ \ldots \ :A\} = \frac{1}{2R}\left[(K\{A\} + K\{'A'\} + R)\left(\frac{K\{A\} - K\{'A'\} + R}{2}\right)^N\right.$$
$$\left. - (K\{A\} + K\{'A'\} - R)\left(\frac{K\{A\} - K\{'A'\} - R}{2}\right)^N\right] \qquad (17.20)$$

where

$$R = \sqrt{(K\{A\} + K\{'A'\})^2 - 4K\{'A\}K\{A'\}} \qquad (17.21)$$

Much of the above theory may be formulated in terms of fragments of the type A", "A and "A" rather than A', 'A and 'A'. The new symbols designate the unit with one or two pairs of vertices removed (cf. Chapter 7 as above). We shall only give the alternative equations for the quantities of (18) and (19).

$$\alpha = K\{A''\} + K\{''A\} - K\{''A''\} \qquad (17.22)$$

$$\beta = K\{A\}K\{''A\} - K\{A''\}K\{''A\} \qquad (17.23)$$

It is especially interesting to notice that

$$\begin{vmatrix} K\{A\} & K\{'A\} \\ K\{A'\} & K\{'A'\} \end{vmatrix} = \begin{vmatrix} K\{A\} & K\{''A\} \\ K\{A''\} & K\{''A''\} \end{vmatrix} \qquad (17.24)$$

17.2.2 *Special Formulas*

It is useful to show some special cases of the general formulas of the preceding paragraph.

Assume a Kekuléan benzenoid unit, U (which necessarily is even). It should be symmetrical in the sense that "U = U" (and therefore also 'U = U'), where the notation conforms the previous usage. For the K numbers we introduce the abbreviations

$$K\{U\} = U, \qquad K\{"U"\} = K\{"U\} = U", \qquad K\{"U"\} = U* \qquad (17.25)$$

Consider the three systems $L(a):U:L(a):U:L(a): \ldots U:L(a), U:L(a):U:L(a): \ldots :L(a):U$ and "U:L(a):U:L(a): \ldots :L(a):U". In all cases it is assumed that the number of the units U, occasionally together with "U and U", is N. We symbolize the three systems by

$$B_N = L(a):[U:L(a)]^N, \qquad C_N = [U:L(a)]^{N-1}:U, \qquad D_N = "U:L(a):[U(L(a)]^{N-2}:U"$$

respectively. Figure 1 shows an example of B_N (for $a=1$). If the two terminal hexagons are deleted, an example of C_N emerges. For the K numbers we shall use the notation:

$$K\{B_N\} = B_N, \qquad K\{C_N\} = C_N, \qquad K\{D_N\} = D_N \qquad (17.26)$$

The K numbers for all three classes (26) are found to obey the same type of recurrence relation:

$$K_N = [2U" + (a-1)U*]K_{N-1} + [U*U - (U")^2]K_{N-2}; \qquad N \geq 2 \qquad (17.27)$$

Here either B, C or D may be substituted for K throughout.

For $a=1$ as a further specialization one attains at the following explicit formulas.

$$B_{N(a=1)} = \frac{1}{2\sqrt{U*U}} \left[(2\sqrt{U*U} + U + U*)(U" + \sqrt{U*U})^N \right.$$
$$\left. + (2\sqrt{U*U} - U - U*)(U" - \sqrt{U*U})^N \right] \qquad (17.28)$$

$$C_{N(a=1)} = \frac{1}{2}\sqrt{\frac{U}{U*}} \left[(U" + \sqrt{U*U})^N - (U" - \sqrt{U*U})^N \right] \qquad (17.29)$$

$$D_{N(a=1)} = \frac{1}{2}\sqrt{\frac{U*}{U}} \left[(U" + \sqrt{U*U})^N - (U" - \sqrt{U*U})^N \right] \qquad (17.30)$$

Notice that

$$U* C_{N(a=1)} = U D_{N(a=1)} \qquad (17.31)$$

17.2.3 *Formulas when* U = L(k,m)

Consider the three systems B_N, C_N and D_N of the preceding paragraph when U is a parallelogram, L(k,m), while a is arbitrary. It is assumed that the parallelograms are fused by the (opposite) edges belonging to L_3-mode hexagons. Figure 3 exemplifies the three benzenoid classes under consideration. In this analysis it is expedient to introduce the symbol

$$r = (a+1)\left[\binom{k+m}{m} - 2\right] \qquad (17.32)$$

Then one has, with the aid of the formulas of CHART 8-IV:

$$U = K\{L(k,m)\} = \frac{r}{a+1} + 2 \qquad (17.33)$$

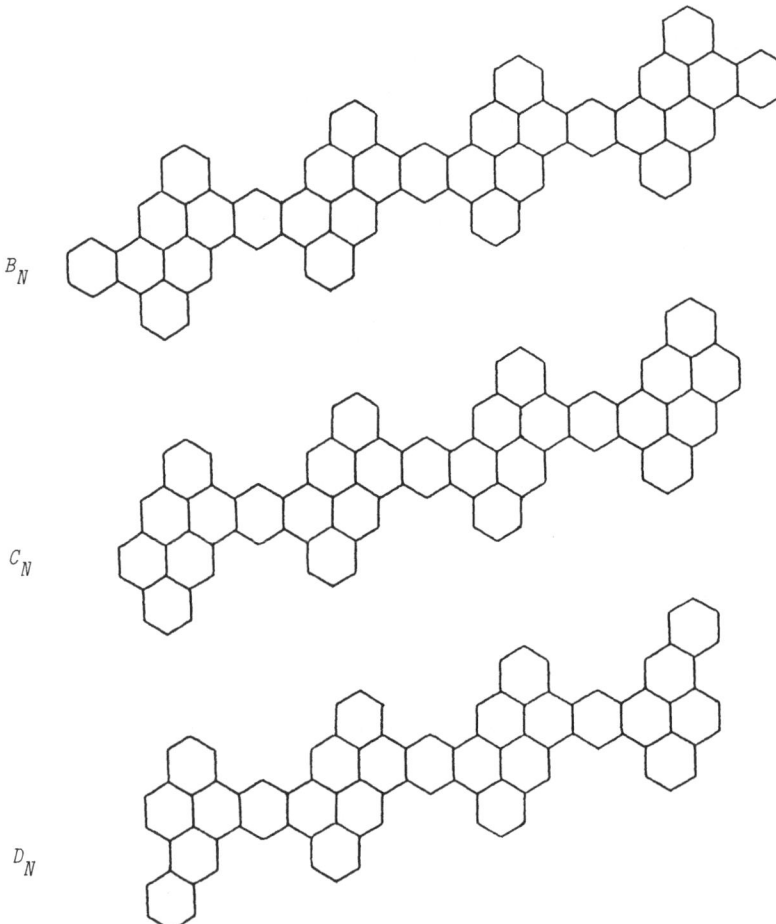

B_N

C_N

D_N

Fig. 17.3. Exemplification of the benzenoid systems B_N, C_N and D_N with fused units. Here $N=4$, $a=1$, and U = L(2,3) = *anthanthrene*.

$$U'' = K\{La(k,m)\} = \frac{r}{a+1} + 1 \tag{17.34}$$

$$U* = K\{Lb(k,m)\} = \frac{r}{a+1} \tag{17.35}$$

The recurrence relation becomes

$$K_N = (r+2)K_{N-1} - K_{N-2}; \qquad N \geq 2 \tag{17.36}$$

where K may be replaced by either B, C or D. The explicit K formulas read:

$$B_N = \frac{a+1}{\sqrt{r(r+4)}} \left\{ \left[\frac{r + 2 + \sqrt{r(r+4)}}{2} \right]^{N+1} - \left[\frac{r + 2 - \sqrt{r(r+4)}}{2} \right]^{N+1} \right\} \tag{17.37}$$

$$C_N = \frac{1}{2(a+1)\sqrt{r(r+4)}} \left\{ \left[a^2(r+2) + 4a + r + 2 - (a^2-1)\sqrt{r(r+4)} \right] \left[\frac{r + 2 + \sqrt{r(r+4)}}{2} \right]^{N} \right.$$

$$\left. - \left[a^2(r+2) + 4a + r + 2 + (a^2-1)\sqrt{r(r+4)} \right] \left[\frac{r + 2 - \sqrt{r(r+4)}}{2} \right]^{N} \right\} \tag{17.38}$$

$$D_N = \frac{1}{a+1} \sqrt{\frac{r}{r+4}} \left\{ \left[\frac{r + 2 + \sqrt{r(r+4)}}{2} \right]^{N} - \left[\frac{r + 2 - \sqrt{r(r+4)}}{2} \right]^{N} \right\} \tag{17.39}$$

It is instructive to show the connection with the general formulation of Paragraph 17.2.1. We set A = $L(k,m):L(a)$ and demonstrate that the coefficients α and β as given by eqns. (18), (19), (22) and (23) are consistent with the relation (36). One has A' = $L(k,m):L(a-1)$, 'A = $L(a)$ and 'A' = $L(a-1)$. Consequently:

$$K\{A\} = (a+1)\binom{k+m}{m} - a = r + a + 2, \qquad K\{A'\} = a\binom{k+m}{m} - (a-1) = \frac{a(r+2) + 2}{a+1} - (a-1),$$

$$K\{'A\} = a+1, \qquad K\{'A'\} = a$$

On inserting these expressions into (18) and (19) one obtains α = $r+2$ and β = -1, respectively, which is consistent with (36). On the other hand one has A" = $La(k,m)$, "A = $La(k,m):L(a)$ and "A" = $Lb(k,m)$. Consequently:

$$K\{A''\} = \binom{k+m}{m} - 1 = \frac{r}{a+1} + 1, \quad K\{''A\} = (a+1)\left[\binom{k+m}{m} - 1\right] - a = r+1, \quad K\{''A''\} = \binom{k+m}{m} - 2 = \frac{r}{a+1}$$

On inserting these expressions - together with $K\{A\}$ from above - into (22) and (23), one obtains the same result (α = $r+2$, β = -1).

We shall show one further specialization, viz. B_N when U = $L(2,3)$ = *anthanthrene* (as in Fig. 3). In this case $r = 8(a+1)$, and:

$$K\{L(a):[L(2,3):L(a)]^N\} = \frac{1}{4} \sqrt{\frac{a+1}{2(2a + 3)}} \left\{ \left[4a + 5 + 2\sqrt{2(a+1)(2a + 3)} \right]^{N+1} \right.$$

$$\left. - \left[4a + 5 - 2\sqrt{2(a+1)(2a + 3)} \right]^{N+1} \right\} \tag{17.40}$$

Especially for $a=1$ (as in Fig. 3):

$$K\{L(1):[L(2,3):L(1)]^{N}\} = \frac{1}{4\sqrt{5}}\left[(9 + 4\sqrt{5})^{N+1} - (9 - 4\sqrt{5})^{N+1}\right] \qquad (17.41)$$

There are usually many isoarithmic variants for benzenoids with fused units. The considered cases of $U = L(k,m)$, for instance, give always rise to isoarithmicity whenever $k \neq m$. An example is shown in Fig. 4.

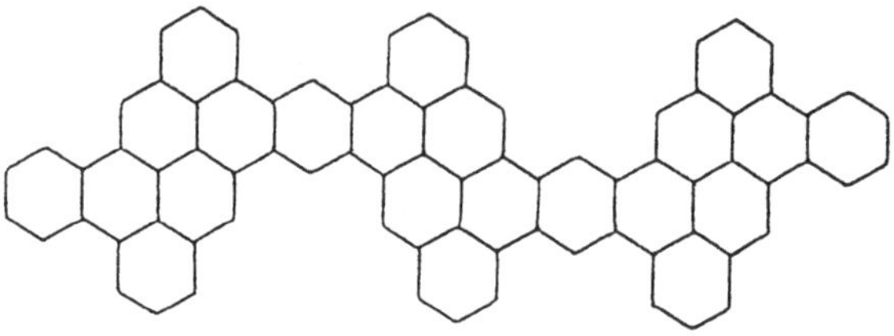

$K = 11592$

Fig. 17.4. An isoarithmic system to $B_3 = L(1):[L(2,3):L(1)]^3$ with reference to Fig. 17.3.

17.2.4 *Formulas for Fused and Annelated Pyrenes*

Consider first the classes B_N, C_N and D_N of the preceding paragraph when $U = L(2,2) = pyrene$. Then $r = 4(a+1)$, $U=6$, $U'' = 5$ and $U* = 4$. The recurrence relation (27) or (36) becomes

$$K_N = 2(2a + 3)K_{N-1} - K_{N-2}; \qquad N \geq 2 \qquad (17.42)$$

Explicit formulas are obtained from (37)-(39).

$$K\{L(a):[L(2,2):L(a)]^{N}\} = \frac{1}{4}\sqrt{\frac{a+1}{a+2}}\left\{\left[2a + 3 + 2\sqrt{(a+1)(a+2)}\right]^{N+1}\right.$$
$$\left. - \left[2a + 3 - 2\sqrt{(a+1)(a+2)}\right]^{N+1}\right\} \qquad (17.43)$$

$$K\{[L(2,2):L(a)]^{N-1}:L(2,2)\}$$
$$= \frac{1}{4\sqrt{(a+1)(a+2)}}\left\{\left[2a^2 + a + 3 - 2(a-1)\sqrt{(a+1)(a+2)}\right]\left[2a + 3 + 2\sqrt{(a+1)(a+2)}\right]^{N}\right.$$
$$\left. - \left[2a^2 + a + 3 + 2(a-1)\sqrt{(a+1)(a+2)}\right]\left[2a + 3 - 2\sqrt{(a+1)(a+2)}\right]^{N}\right\} \qquad (17.44)$$

$$K\{A(3):L(a):[L(2,2):L(a)]^{N-2}:A(3)\} = \frac{1}{\sqrt{(a+1)(a+2)}}\left\{\left[2a + 3 + 2\sqrt{(a+1)(a+2)}\right]^{N}\right.$$
$$\left. - \left[2a + 3 - 2\sqrt{(a+1)(a+2)}\right]^{N}\right\} \qquad (17.45)$$

Here A(3) = *phenanthrene*.

For $a=1$ the three considered classes coincide with some of the classes encountered in Chapter 16. The recurrence relation (42) becomes compatible with
(16.1). Explicit formulas are obtainable either from (28)-(30) or (43)-(45). The
results for $K\{L(1):[L(2,2):L(1)]^N\}$ (*pyrenes* on a string) and $K\{A(3):L(1):[L(2,2):$
$L(1)]^{N-2}:A(3)\}$ coincide with $K\{P(N)\}$ and $K\{P_2(N)\}$, respectively; cf. CHART 16-I.
The remaining class, viz. $[L(2,2):L(1)]^{N-1}:L(2,2)$, is identical with P''(N) in the
terminology of Fig. 16.2. For numerical K values, see Table 16.1 and Table 1.

CHART I is inserted as a continuation of CHART 7-III. Recurrence relations
and explicit formulas for some fused and annelated *pyrenes* are given therein. Cases
are considered where the edge of fusion belongs to a P_2-mode hexagon of *pyrene* and
not only L_3. In the former case (P_2) there are many possibilities of isoarithmicity.
Figure 5 shows two examples.

Fused pyrenes, viz. P* = L(2,2):L(2,2): .... :L(2,2), is a popular class in
the studies of Kekulé structures. The pertinent recurrence relation (cf. CHART I) is
obtained by setting $a=0$ into (42). It is implied in the work of Ohkami and Hosoya
(1983), and has also been reported by Křivka et al. (1986) together with the corresponding relation for L(3,3):L(3,3): .... :L(3,3). The explicit formula for $K\{P*\}$ is
easily obtained on inserting $a=0$ into either (43) or (44). The same result (see
$K\{P_{13}\}$ of CHART I) is also obtainable from (20). The K formula for P* = P_{13}, as derived by the matrix-transfer method, has been reported by Klein, Hite and Schmalz
(1986).

Table 1 gives numerical K values for the classes P''(n) and P*(n).

Additional classes of annelated *pyrenes* are found in Paragraph 16.2.2.

Table 17.1. Numerical values of K for the benzenoids P''(n) and P*(n).

n	$K\{P''(n)\}$	[a]$K\{P*(n)\}$
0	0	1
1	6	6
2	60	35
3	594	204
4	5880	1189
5	58206	6930
6	576180	40391
7	5703594	235416
8	56459760	1372105
9	558894006	7997214
10	5532480300	46611179

[a]Křivka P, Nikolić S, Trinajstić N (1986). Croat Chem Acta 59: 659;
an error is corrected for $n=6$

CHART 17-I. Benzenoids with repeated units: fused and annelated *pyrenes*

P₁₁

$$K\{P_{11}(n)\} = 5K\{P_{11}(n-1)\} - 3K\{P_{11}(n-2)\}$$

P₁₂

$$K\{P_{12}(n)\} = 4K\{P_{12}(n-1)\} + 3K\{P_{12}(n-2)\}$$

P₁₃

$$^{a,b}K\{P_{13}(n)\} = 6K\{P_{13}(n-1)\} - K\{P_{13}(n-2)\}$$

$$K\{P_{11}\} = \frac{1}{2\sqrt{13}}\left[(\sqrt{13}+7)\left(\frac{5+\sqrt{13}}{2}\right)^n + (\sqrt{13}-7)\left(\frac{5-\sqrt{13}}{2}\right)^n\right]$$

$$K\{P_{12}\} = \frac{1}{2\sqrt{7}}[(\sqrt{7}+4)(2+\sqrt{7})^n + (\sqrt{7}-4)(2-\sqrt{7})^n]$$

$$^cK\{P_{13}\} = \frac{1}{2\sqrt{8}}[(3+\sqrt{8})^{n+1} - (3-\sqrt{8})^{n+1}]$$

P₁₄

$$K\{P_{14}(n)\} = 6K\{P_{14}(n-1)\} - 3K\{P_{14}(n-2)\}$$

P₁₅

$$K\{P_{15}(n)\} = 6K\{P_{15}(n-1)\} + 3K\{P_{15}(n-2)\}$$

P₁₆

$$^aK\{P_{16}(n)\} = 10K\{P_{16}(n-1)\} - K\{P_{16}(n-2)\}$$

$$K\{P_{14}\} = \frac{1}{2\sqrt{6}}[(2\sqrt{6}+7)(3+\sqrt{6})^n + (2\sqrt{6}-7)(3-\sqrt{6})^n]$$

$$K\{P_{15}\} = \frac{1}{3}[(3+2\sqrt{3})^{n+1} + (3-2\sqrt{3})^{n+1}]$$

$$^dK\{P_{16}\} = \frac{1}{2\sqrt{6}}[(5+2\sqrt{6})^{n+1} - (5-2\sqrt{6})^{n+1}]$$

[a] Ohkami N, Hosoya H (1983). Theor Chim Acta 64: 153
[b] Křivka P, Nikolić S, Trinajstić N (1986). Croat Chem Acta 59: 659
[c] Klein DJ, Hite GE, Schmalz TG (1986). J Comput Chem 7: 443
[d] $P_{16}(n) = P(n)$; see CHART 16-I

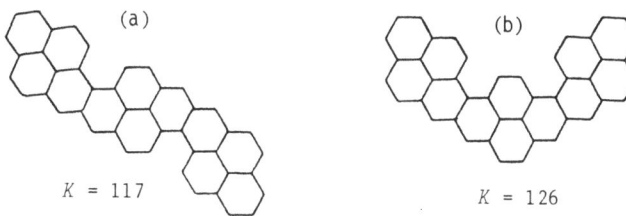

$K = 117$

$K = 126$

Fig. 17.5. An isoarithmic
system to (a) $P_{11}(3)$ and
to (b) $P_{12}(3)$ with reference to CHART 17-I.

17.3 CONDENSED REPEATED UNITS

17.3.1 *Compressed Parallelograms*

Parallel Condensation. Cyvin SJ and Cyvin (1986a) studied the number of Kekulé
structures for a class of compressed parallelograms as defined in Fig. 6. Their
studies include the additional system obtained by a modification at one end (see
Fig. 6, which gives the notation for K numbers). We quote the recurrence relation:

$$P_N = \binom{k+m-2}{m-1}\left[2P_{N-1} - \frac{1}{k}\binom{k+m-1}{m}P_{N-2}\right]; \qquad N \geq 2 \qquad (17.46)$$

The same form holds for $P_N^{(k)}$ and $P_N^{(m)}$ as well. With the aid of the initial conditions $P_0 = 2$ and $P_1 = \binom{k+m}{m}$ the following explicit formula was derived for P_N.

$$P_N = \binom{k+m-2}{m-1}^N \left\{ \left[1 + \frac{k(k-1)+m(m-1)}{2\sqrt{km(k-1)(m-1)}}\right]\left[1 + \sqrt{\frac{(k-1)(m-1)}{km}}\right]^N \right.$$

$$\left. + \left[1 - \frac{k(k-1)+m(m-1)}{2\sqrt{km(k-1)(m-1)}}\right]\left[1 - \sqrt{\frac{(k-1)(m-1)}{km}}\right]^N \right\} \qquad (17.47)$$

Below we give the linear combinations for $P_N^{(k)}$ and $P_N^{(m)}$ in terms of P_N and P_{N-1}.

$$P_N^{(k)} = \frac{k}{(k+m-1)(k-m)}\left[(k-m-1)P_N + \binom{k+m-1}{m-1}P_{N-1}\right]; \qquad k \neq m \qquad (17.48)$$

$$P_N^{(m)} = \frac{m}{(k+m-1)(m-k)}\left[(m-k-1)P_N + \binom{k+m-1}{m}P_{N-1}\right]; \qquad k \neq m \qquad (17.49)$$

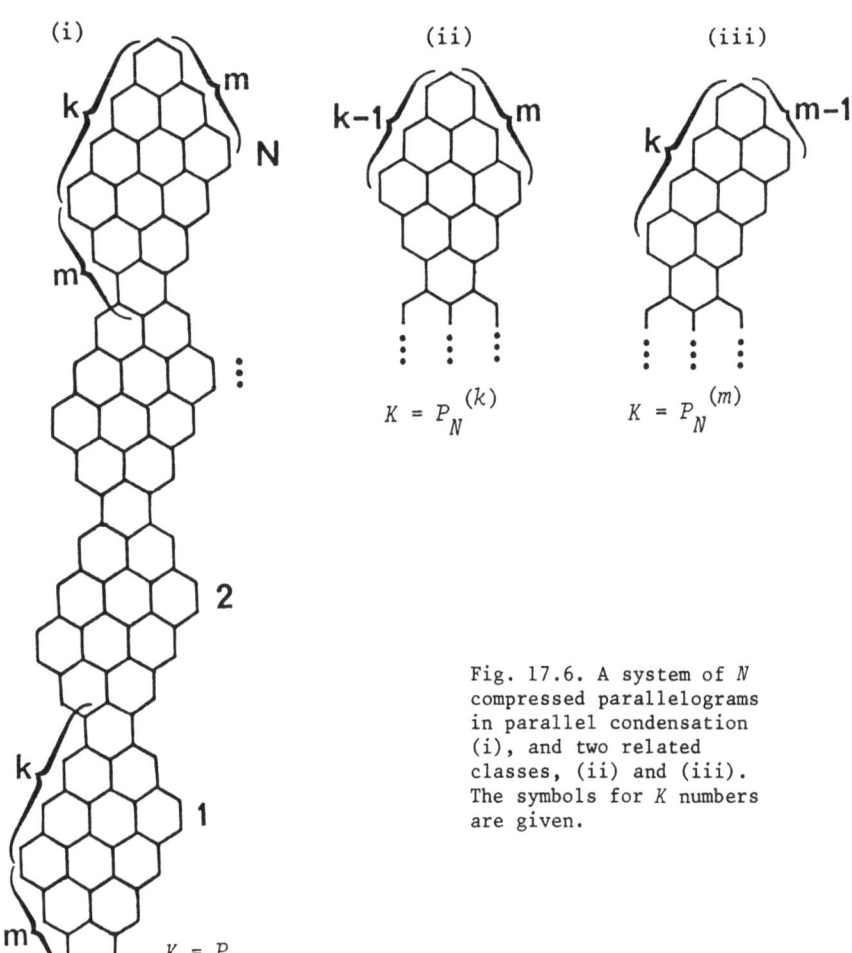

$$K = P_N^{(k)} \qquad K = P_N^{(m)}$$

$$K = P_N$$

Fig. 17.6. A system of N compressed parallelograms in parallel condensation (i), and two related classes, (ii) and (iii). The symbols for K numbers are given.

Antiparallel Condensation. Figure 7 shows another version of compressed parallelograms. If the system (i) of Fig. 6 is characterized as "parallel condensation", it is natural to use the term "antiparallel condensation" for the present system. The pertinent recurrence relation reads

$$Q_N = \left[\binom{k+m-2}{m} + \binom{k+m-2}{m-2} \right] Q_{N-1} + \frac{1}{k} \binom{k+m-1}{m} \binom{k+m-2}{m-1} Q_{N-2} ; \qquad N \geq 2 \qquad (17.50)$$

The initial conditions are $Q_0 = 2$, $Q_1 = P_1 = \binom{k+m}{m}$.

Connection Between P_2 and Q_2. The relations (46) and (50) along with the initial conditions give

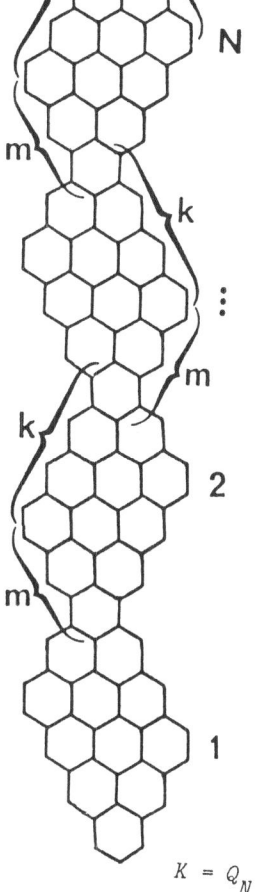

Fig. 17.7. A system of N compressed parallelograms in antiparallel condensation.

$$P_2 = 2 \binom{k+m-1}{m}\binom{k+m-1}{m-1} \qquad (17.51)$$

and

$$Q_2 = \binom{k+m-1}{m}^2 + \binom{k+m-1}{m-1}^2 \qquad (17.52)$$

respectively. Consequently

$$P_2 + Q_2 = \left[\binom{k+m-1}{m} + \binom{k+m-1}{m-1}\right]^2 = \binom{k+m}{m}^2 \qquad (17.53)$$

$$K = Q_N$$

17.3.2 *Compressed Rhombs*

The problem of the preceding paragraph becomes considerably simpler when $k=m$; see CHART II. A simpler recurrence relation than (46) is valid, viz.

$$R_N = \binom{2m-1}{m} R_{N-1}, \qquad R_N' = \binom{2m-1}{m} R_{N-1}'; \qquad N \geq 1 \qquad (17.54)$$

The explicit formulas for R_N and R_N' are immediately obtained from (54) by means of the initial conditions $R_0 = 2$ and $R_0' = 1$, respectively. The results are entered into CHART II. The linear dependence between R_N and R_N' is explicitly:

$$R_N' = \frac{1}{2} R_N \qquad (17.55)$$

The first formula of CHART 12-III pertains to oblate rectangles $R^j(m', 1)$ or,

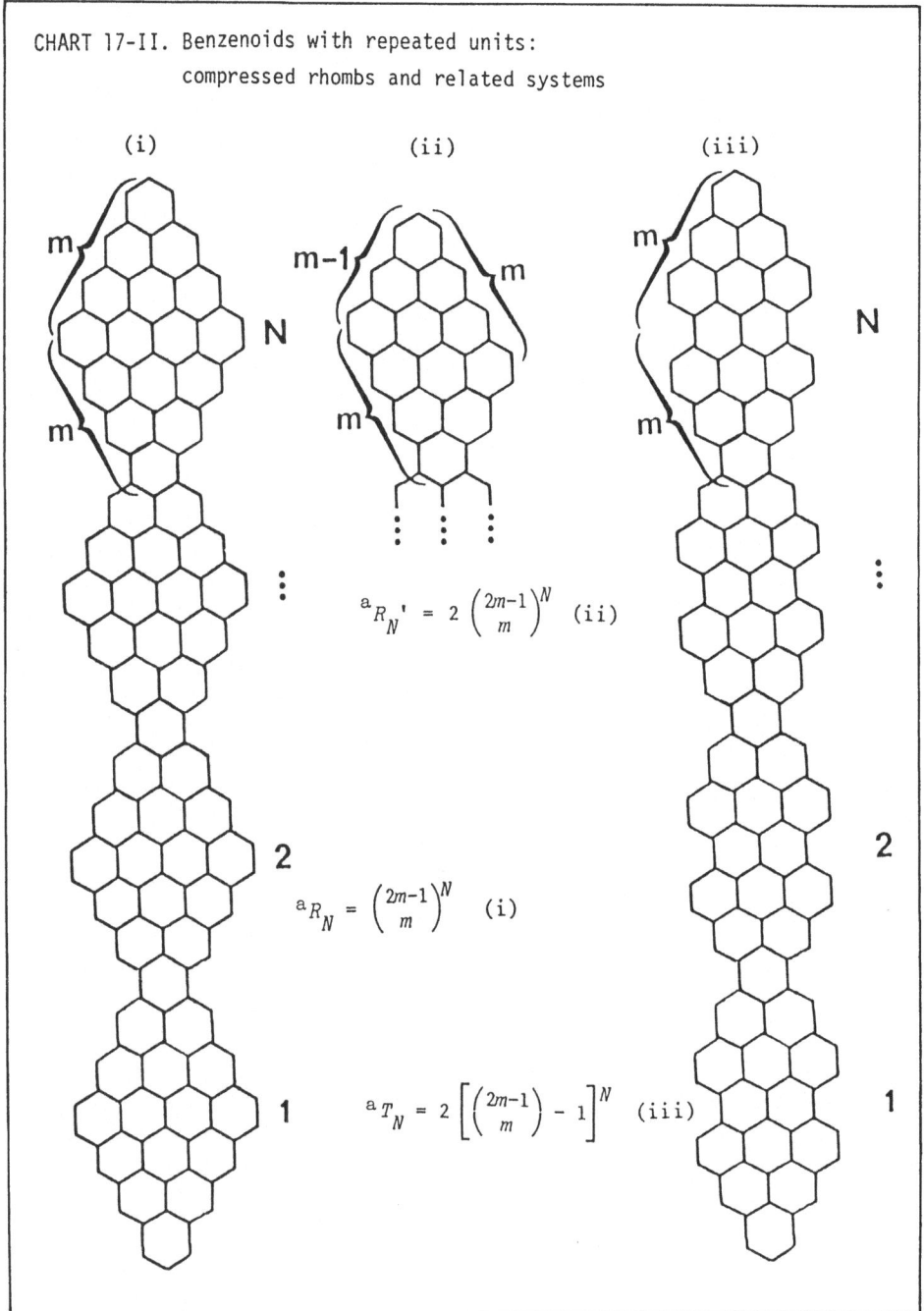

CHART 17-II. Benzenoids with repeated units:
compressed rhombs and related systems

(i)   (ii)   (iii)

$$^{a}R_{N}{'} = 2\left(\frac{2m-1}{m}\right)^{N} \quad (ii)$$

$$^{a}R_{N} = \left(\frac{2m-1}{m}\right)^{N} \quad (i)$$

$$^{a}T_{N} = 2\left[\left(\frac{2m-1}{m}\right) - 1\right]^{N} \quad (iii)$$

[a]Cyvin SJ, Cyvin BN (1986). Match 20: 181

in other words, compressed *pyrenes*. It emerges from R_N by inserting $m=2$ and $N = m' - 1$.

17.3.3 *Compressed Rhombs without Corners*

Cyvin SJ and Cyvin (1986a) considered also the system of compressed rhombs without corners. Here the appropriate unit is $Lb(m,m)$ in the notation of CHART 8-IV. The explicit K formula is included in CHART II.

17.3.4 *General Formulation*

The considerations at the end of Paragraph 17.3.1 may be generalized in the following way. Assume two benzenoids, P and Q, which each have (at least) one hexagon of the mode P_2. Compress P with Q into a new benzenoid by identifying one P_2 hexagon from each of them. This can in general be done in two ways (parallel and antiparallel). We denote the two new benzenoids by P↑Q and P↓Q (without necessarily specifying which is which).

If P is even and Q odd (or vice versa), then P↑Q and P↓Q are both obvious non-Kekuléan. If both P and Q are odd, the compressed system may be Kekuléan, but always essentially disconnected in that case. It may also be non-Kekuléan. The cases of most interest are those when both P and Q are even. This is assumed to be fulfilled in the following.

The below peculiar identity was verified.

$$K\{P\uparrow Q\} + K\{P\downarrow Q\} = K\{P\}K\{Q\} \qquad (17.56)$$

Eqn. (53) is an example. It relates to (56) by the interpretation $P_2 = K\{L\uparrow L\}$, $Q_2 = K\{L\downarrow L\}$, where $L = L(k,m)$.

If for some reason

$$K\{P\uparrow Q\} = K\{P\downarrow Q\} \qquad (17.57)$$

we get

$$K\{P\uparrow Q\} = \frac{1}{2} K\{P\}K\{Q\} \qquad (17.58)$$

The condition (57) is fulfilled when either P or Q is symmetric in the sense that the unit has a vertical plane of symmetry through the P_2-mode hexagon used in the compression. Such a case is illustrated in Fig. 8.

Assume now a system A_N of N identical compressed units A, where A is even and mirror-symmetrical as described above. Then eqn. (58) gives the recurrence relation

$$K\{A_N\} = \frac{1}{2} K\{A_1\}K\{A_{N-1}\}; \qquad N \geq 1 \qquad (17.59)$$

For the degenerate case of $N=0$ one has $K\{A_0\} = 2$. Repeated application of (59) leads to the explicit formula

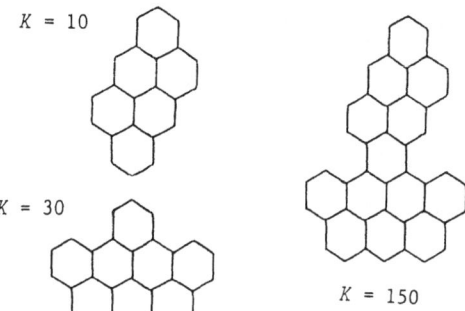

$K = 10$

$K = 30$

$K = 150$

Fig. 17.8. Illustration of eqn. (17.58).

$$K\{A_N\} = 2 \left(\frac{1}{2} K\{A_1\} \right)^N \tag{17.60}$$

An example is furnished by A = L(m,m), the rhomb of CHART II. Eqn. (60) gives

$$R_N = 2 \left[\frac{1}{2} \left(\frac{2m}{m} \right) \right]^N \tag{17.61}$$

which is equivalent to the corresponding formula of CHART II.

For cases where the unit A is not symmetrical (e.g. Figs. 6 and 7) no general formulation is available.

309

17.4 BENZENOIDS WITH HEXAGONAL AND TRIGONAL SYMMETRIES

17.4.1 *Hexagonal Symmetry*

Benzenoid systems of hexagonal symmetry (D_{6h} or C_{6h}) may be interpreted as consisting of six repeated units in a circular arrangement. Extensive computer-aided generations of such systems have been performed (Brunvoll, Cyvin BN and Cyvin

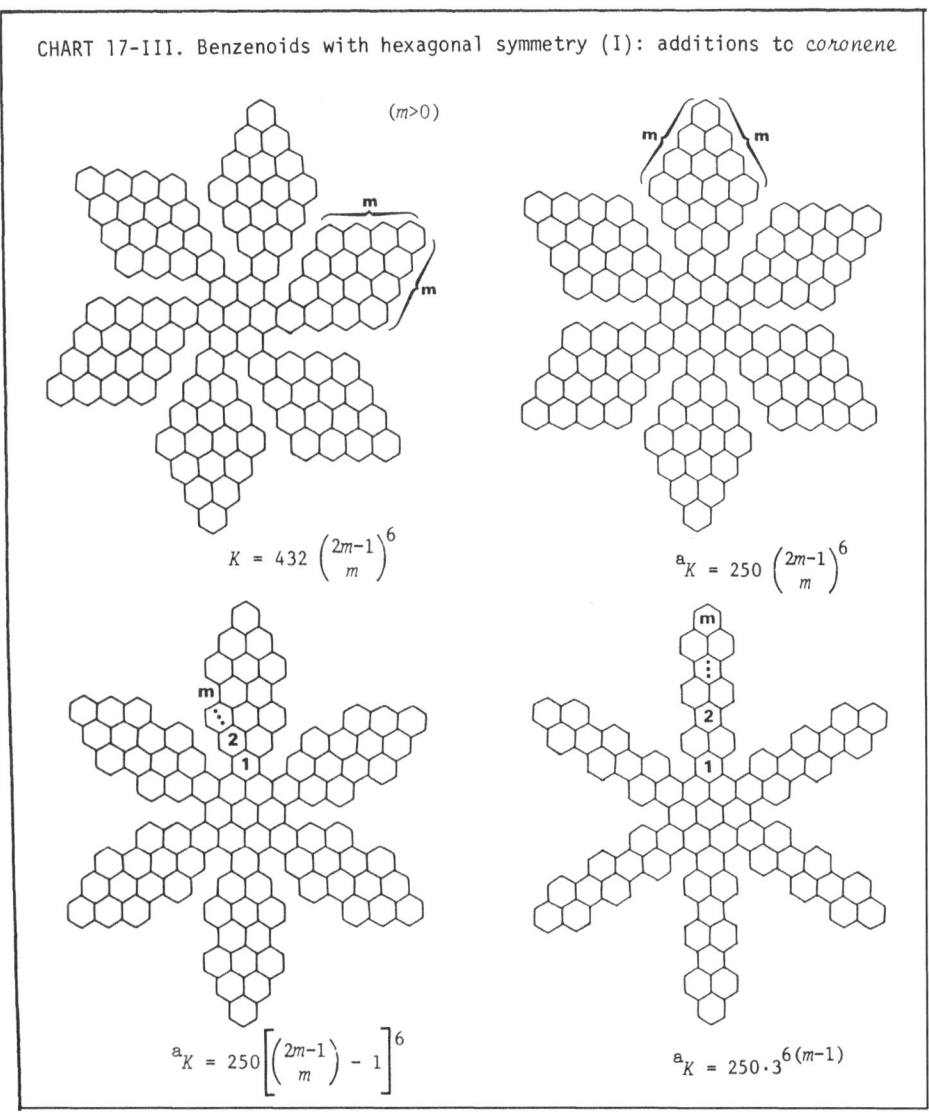

CHART 17-III. Benzenoids with hexagonal symmetry (I): additions to *coronene*

$(m>0)$

$$K = 432 \left(\frac{2m-1}{m}\right)^6$$

$$^a K = 250 \left(\frac{2m-1}{m}\right)^6$$

$$^a K = 250 \left[\left(\frac{2m-1}{m}\right) - 1\right]^6$$

$$^a K = 250 \cdot 3^{6(m-1)}$$

[a]Cyvin SJ, Bergan JL, Cyvin BN (1987). Acta Chim Hung 124: 691

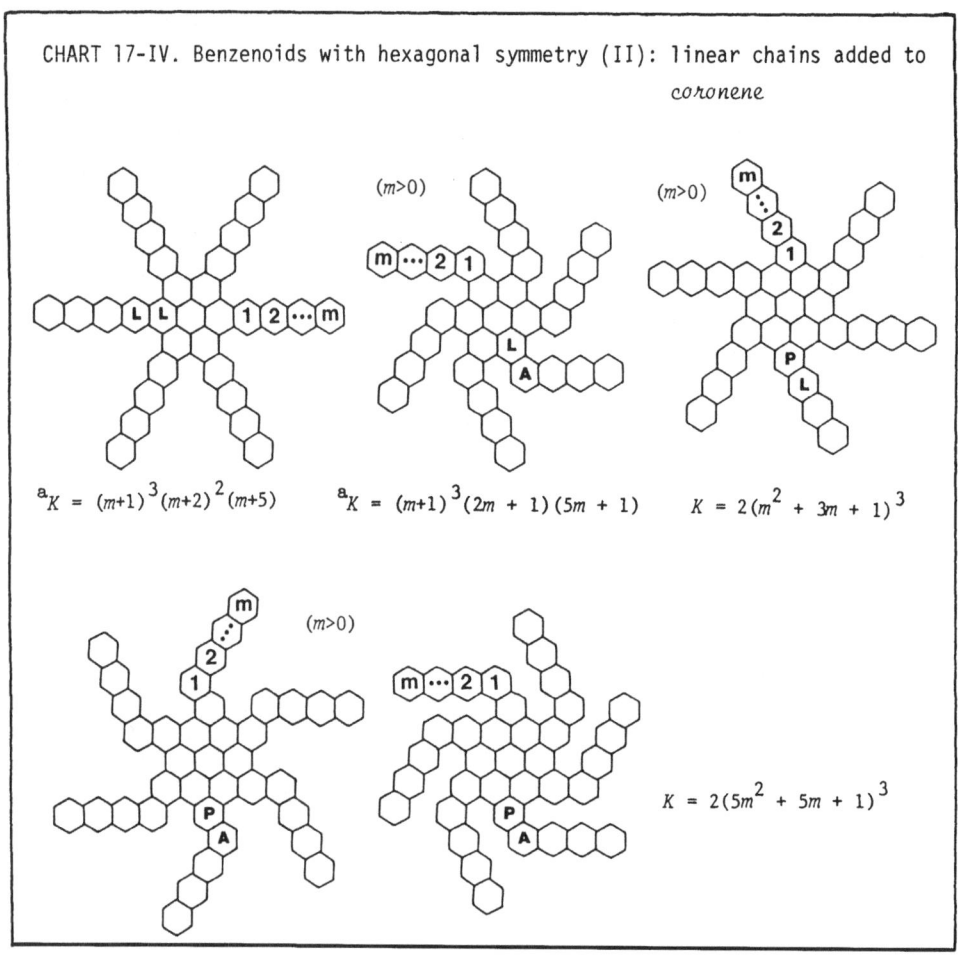

CHART 17-IV. Benzenoids with hexagonal symmetry (II): linear chains added to *coronene*

$^a K = (m+1)^3 (m+2)^2 (m+5)$

$^a K = (m+1)^3 (2m + 1)(5m + 1)$

$K = 2(m^2 + 3m + 1)^3$

$K = 2(5m^2 + 5m + 1)^3$

^{a}Cyvin SJ, Bergan JL, Cyvin BN (1987). Acta Chim Hung 124: 691

1987) combined with computations of K numbers. Some combinatorial K formulas are also available (Cyvin SJ, Bergan and Cyvin 1987). They are collected in CHARTS III and IV, supplemented by some new formulas. The formulas of CHART V are all new. In the cases of additions of linear chains (CHARTS IV and V) some hexagon modes are given for the sake of clarity; L indicates L_2 or L_3, A indicates A_2, and P indicates P_3.

CHART 17-V. Benzenoids with hexagonal symmetry (III): linear chains added to
circumcoronene

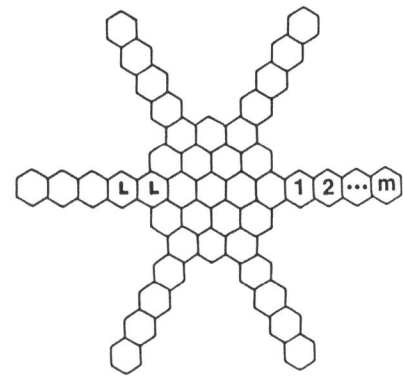

$$K = 10(m+1)^3(m+2)(5m + 7)^2$$

(*m*>0)

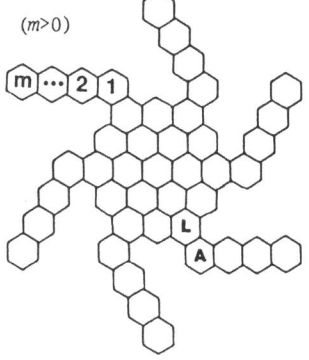

$$K = 10(m+1)^3(2m + 1)(7m + 5)^2$$

(*m*>0)

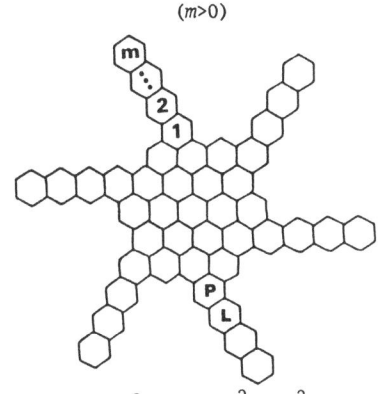

$$K = 4(m+1)^2(3m + 1)^2(12m^2 + 20m + 5)$$

(*m*>0)

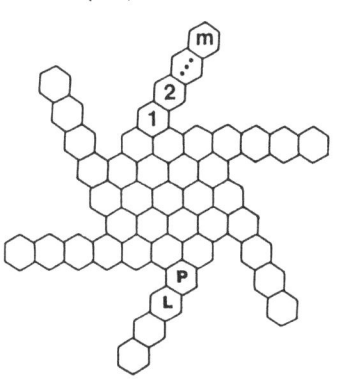

$$K = 4(m+1)^2(m+3)^2(5m^2 + 20m + 12)$$

CHART 17-VI. Benzenoids of trigonal symmetry (I)

$(m>0)$

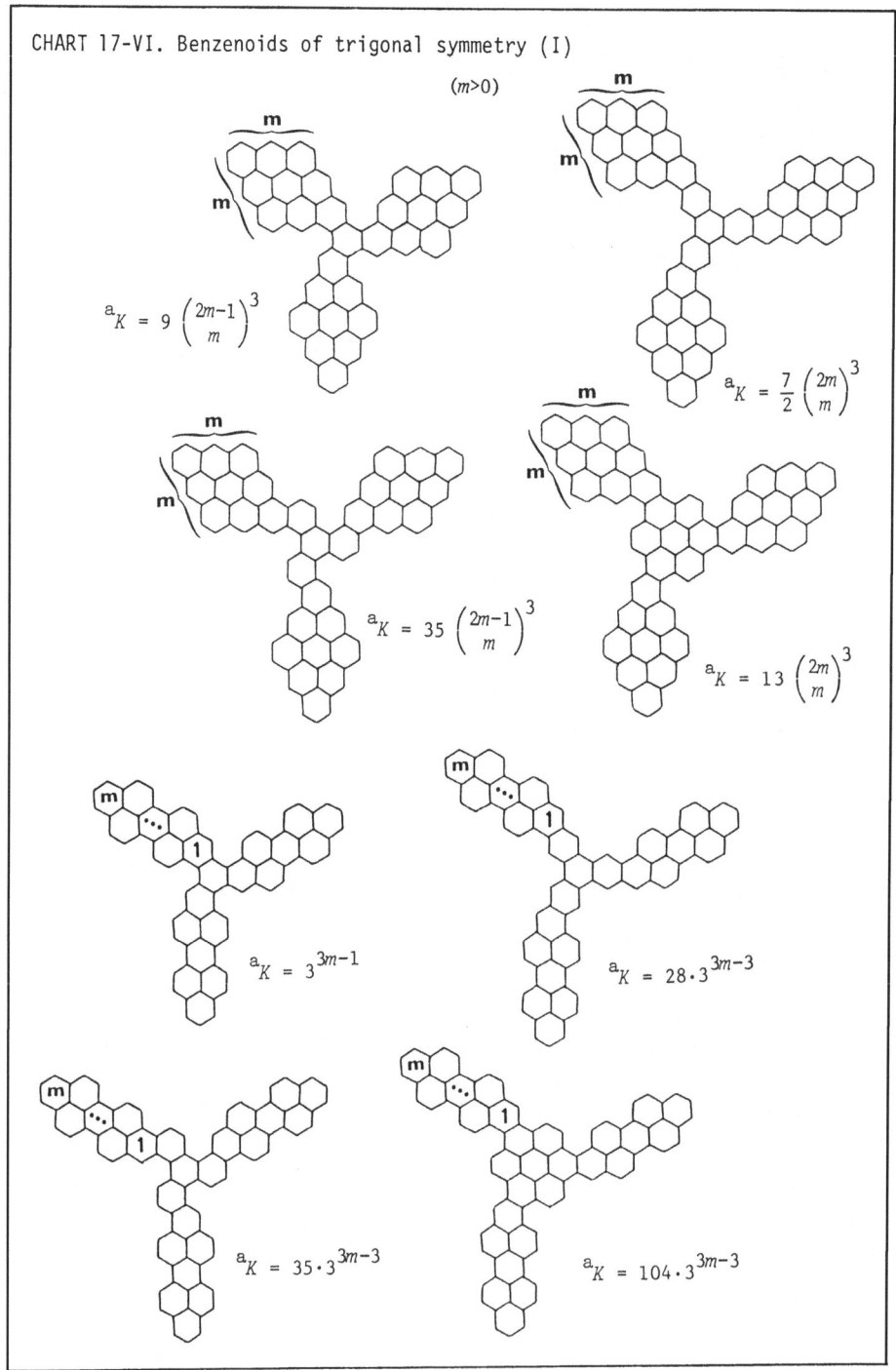

$$^aK = 9 \left(\frac{2m-1}{m}\right)^3$$

$$^aK = \frac{7}{2} \left(\frac{2m}{m}\right)^3$$

$$^aK = 35 \left(\frac{2m-1}{m}\right)^3$$

$$^aK = 13 \left(\frac{2m}{m}\right)^3$$

$$^aK = 3^{3m-1}$$

$$^aK = 28 \cdot 3^{3m-3}$$

$$^aK = 35 \cdot 3^{3m-3}$$

$$^aK = 104 \cdot 3^{3m-3}$$

[a]Cyvin SJ, Cyvin BN, Brunvoll J (1987). J Mol Struct (Theochem) 151: 271

17.4.2 *Trigonal Symmetry*

Benzenoid systems of trigonal symmetry (D_{3h} or C_{3h}) may be interpreted as consisting of three repeated units in a circle. Some combinatorial K formulas from an analysis by Cyvin SJ, Cyvin and Brunvoll (1987) are given in CHART VI and CHART VII. The mentioned authors have also reported some more general K formulas, of which those in CHARTS VI and VII are special cases.

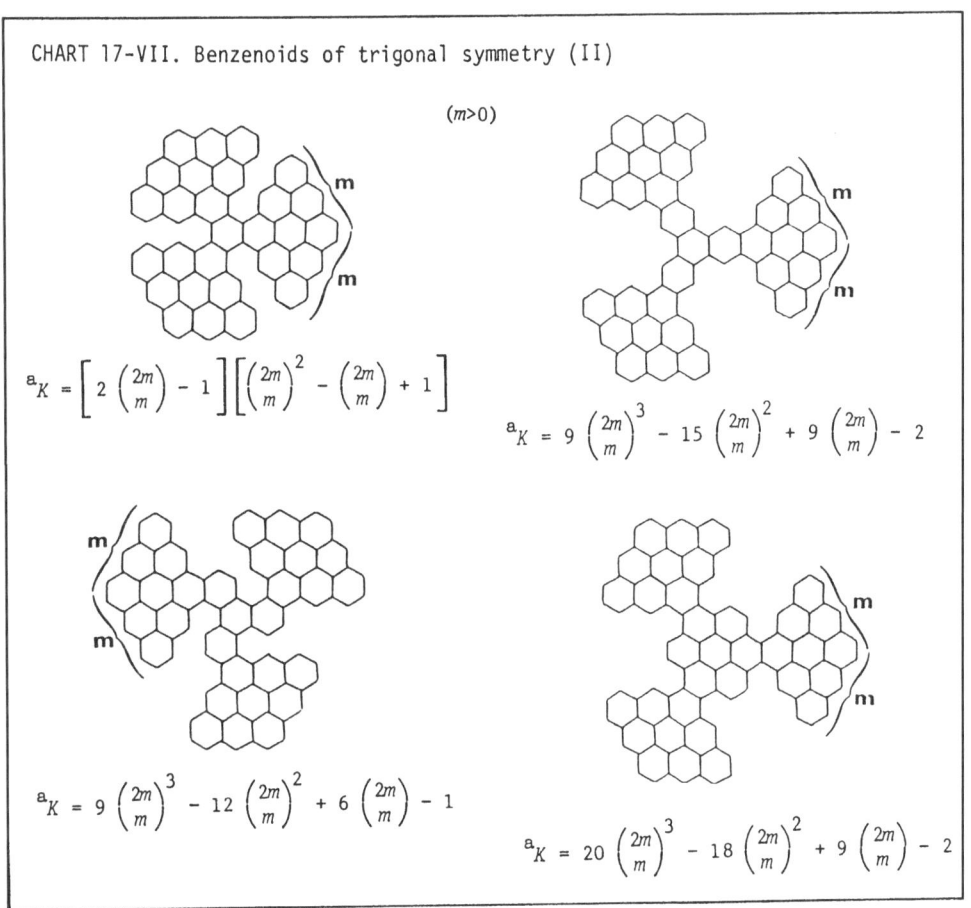

CHART 17-VII. Benzenoids of trigonal symmetry (II)

$(m>0)$

$${}^aK = \left[2 \binom{2m}{m} - 1 \right] \left[\binom{2m}{m}^2 - \binom{2m}{m} + 1 \right]$$

$${}^aK = 9 \binom{2m}{m}^3 - 15 \binom{2m}{m}^2 + 9 \binom{2m}{m} - 2$$

$${}^aK = 9 \binom{2m}{m}^3 - 12 \binom{2m}{m}^2 + 6 \binom{2m}{m} - 1$$

$${}^aK = 20 \binom{2m}{m}^3 - 18 \binom{2m}{m}^2 + 9 \binom{2m}{m} - 2$$

[a]Cyvin SJ, Cyvin BN, Brunvoll J (1987). J Mol Struct (Theochem) 151: 271

CHAPTER 18

DISTRIBUTION OF K, AND KEKULÉ STRUCTURE STATISTICS

18.1 INTRODUCTION AND PREVIOUS WORK

The problem to find all benzenoid systems which have a given number of Kekulé structures (K) for $h \leq 8$ was posed relatively early (Gutman 1982c). On the other hand it was known at that time that infinitely many benzenoids have $K=9$. Until quite recently the only known theorem about this question was: If $K\{B\} = 2$, then B = *benzene* (Gutman 1983). In other words, there exists one benzenoid with $K=2$. Furthermore, it was conjectured that there is one benzenoid with $K=3$ (viz. *naphthalene*), one with $K=4$ (*anthracene*), and that there are two with $K=5$ (*naphthacene* and *phenanthrene*). The distinction between normal and essentially disconnected benzenoids (Paragraph 2.3.2) was a break-through for this problem. Based on *Theorem 13* (see Section 3.4), which was put forward recently by Gutman and Cyvin (1988), it was deduced that there exists a limited number of *normal* benzenoids with any K number ($K > 1$). The infinite number for $K=9$ (and other values of $K > 9$) is due to the essentially disconnected systems. Hence it has sense to ask, for instance: – How many normal benzenoids have $K = 100$? (The answer is 444.)

In Fig. 1 we give a diagram of the numbers of normal benzenoids with $K \leq 50$.

For given numbers of hexagons (h) the number of normal benzenoids with the different K numbers, i.e. the distribution of K, have been studied extensively. The complete distributions have been reported for $h = 1, 2, \ldots, 6$ (Cyvin 1986b; Cyvin and Gutman 1986c), $h=7$ (Cyvin 1986b), $h=8$ and $h=9$ (Cyvin BN, Brunvoll, Cyvin and Gutman 1986). Data for $h=10$ and $h=11$ are also available (Cyvin SJ, Brunvoll and Cyvin 1986). The distributions of K for essentially disconnected benzenoids with $5 \leq h \leq 9$ have also been reported (Cyvin BN, Brunvoll, Cyvin and Gutman 1986).

The "sieve method" was devised in order to determine successively all normal benzenoids with given K (Cyvin and Gutman 1986c). The numbers for $K \leq 24$ were reported, and the actual forms for $K \leq 14$ were depicted in the cited reference.

"Kekulé structure statistics" (Cyvin BN, Brunvoll, Cyvin and Gutman 1986) does not deal with statistics in a strict sense. The distributions of K are mentioned above. In the same connection some average values of K for given h values have been computed. Although these terms are adapted from probability calculus and statistics all the numerical values for the benzenoids are exact quantities (only subjected to round-off errors in the decimals).

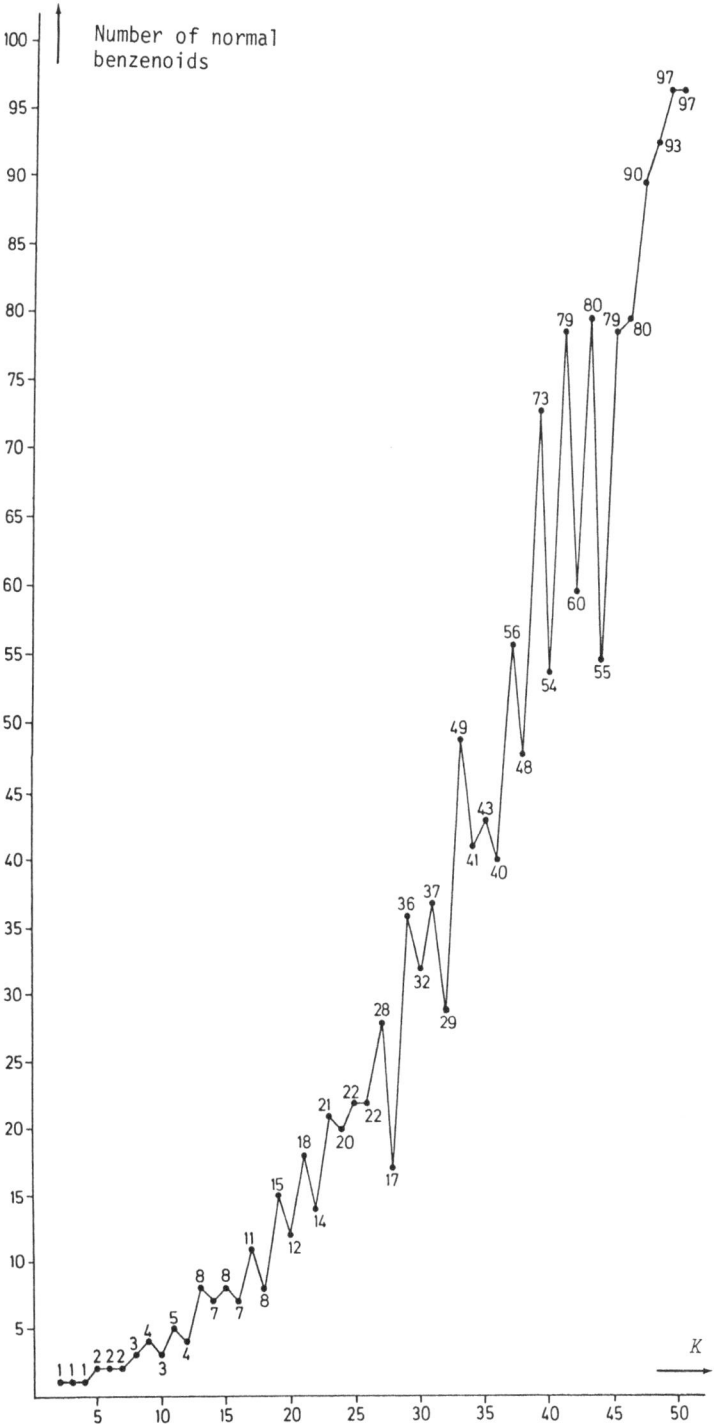

Fig. 18.1. Number of normal benzenoids as a function
of K (the number of Kekulé structures).

316

18.2 DISTRIBUTIONS OF K

18.2.1 *Ranges of K*

For a given h the K numbers, viz. $K(h)$, are limited to certain ranges. For normal benzenoids with $h \leq 10$ they are:

$$K(1) = 2, \quad K(2) = 3, \quad 4 \leq K(3) \leq 5, \quad 5 \leq K(4) \leq 9,$$
$$6 \leq K(5) \leq 14, \quad 7 \leq K(6) \leq 24, \quad 8 \leq K(7) \leq 41,$$
$$9 \leq K(8) \leq 66, \quad 10 \leq K(9) \leq 110, \quad 11 \leq K(10) \leq 189$$

The lower limits are equal to $h+1$ (cf. *Theorem 13*). The upper limits are consistent with Fig. 3.1.

If K is the number of Kekulé structures for an essentially disconnected benzenoid, then $K = p \cdot q$, where p and q are integers greater than 2. Consequently some of the lowest numbers compatible with K for essentially disconnected benzenoids are: 9, 12, 15, 16, 18, 20, 21, 24, 25. The number $K=9$ is represented among essentially disconnected systems with any $h \geq 5$. Furthermore, if $K = K'$ is realized for $h = h'$, then the same K number (viz. K') is also found among the systems with any $h > h'$. It is also true that K for a given h may be any number from the interval $[9, K_{max}^e]$, which may be written in the form $p \cdot q$ (see above). Here K_{max}^e is the maximum, a function of h.

18.2.2 *Distribution of K for Normal Benzenoids*

Figure 2.7 gives the full account of the K numbers for normal benzenoids with $h \leq 6$. The distributions of K for normal benzenoids with $h = 7$, 8 and 9 are shown graphically in Figs. 2, 3 and 4, respectively. The curve for $h=10$ has been reported

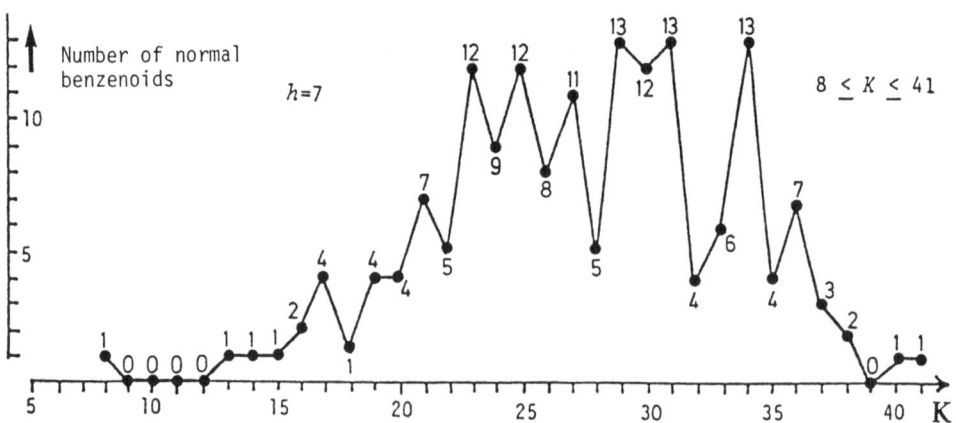

Fig. 18.2. Distribution of K for normal benzenoids with $h=7$.

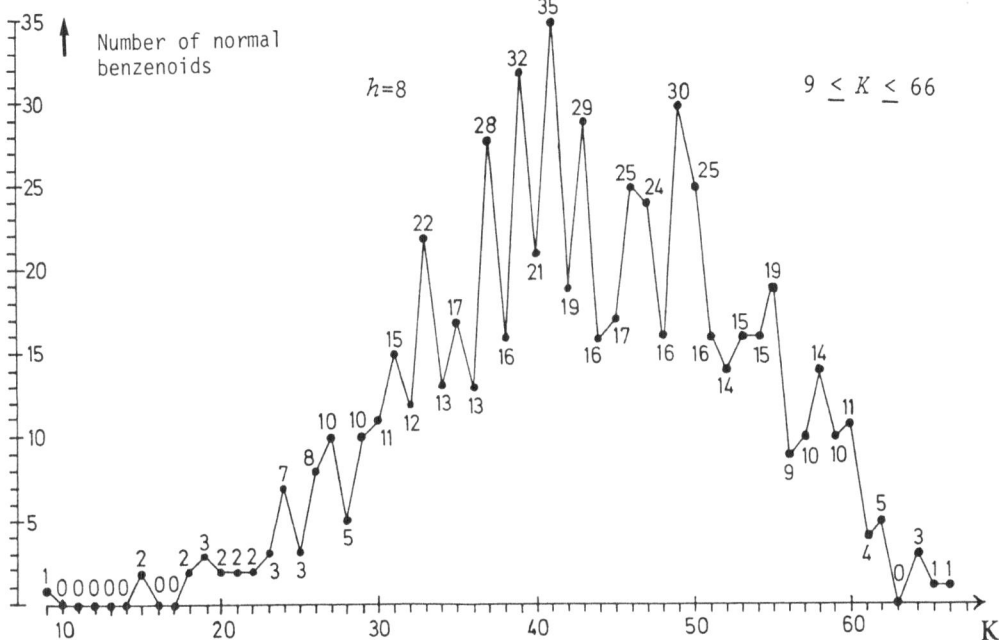

Fig. 18.3. Distribution of K for normal benzenoids with h=8.

in Cyvin SJ, Brunvoll and Cyvin (1986). Here we give the pertinent numbers in a table (see Table 1).

18.2.3 *Distribution of K for Essentially Disconnected Benzenoids*

A full account of the K numbers of essentially disconnected benzenoids with $5 \leq h \leq 7$ is found in Fig. 2.8. The distributions of K for h=8 and h=9 are shown graphically in one figure (Fig. 5). For h=10 the pertinent numbers are included in Table 1.

18.3 AVERAGE VALUES OF K, AND RELATED QUANTITIES

In connection with the distributions of K the average values, viz. $\langle K \rangle$, have been computed for different h values. The results for normal and essentially disconnected benzenoids, viz. $\langle K \rangle_n$ and $\langle K \rangle_e$, respectively, and for $h \leq 10$, are shown in Table 2.

Especially interesting are the quantities $(\ln \langle K \rangle)/h$; they represent some kind of an average resonance energy per hexagon; cf. eqn. (1.2). For the numerical values, see Table 2. The graphical representations (Fig. 6) show that these quantities seem to approach limit values when h increases, but nothing has been proved to this effect.

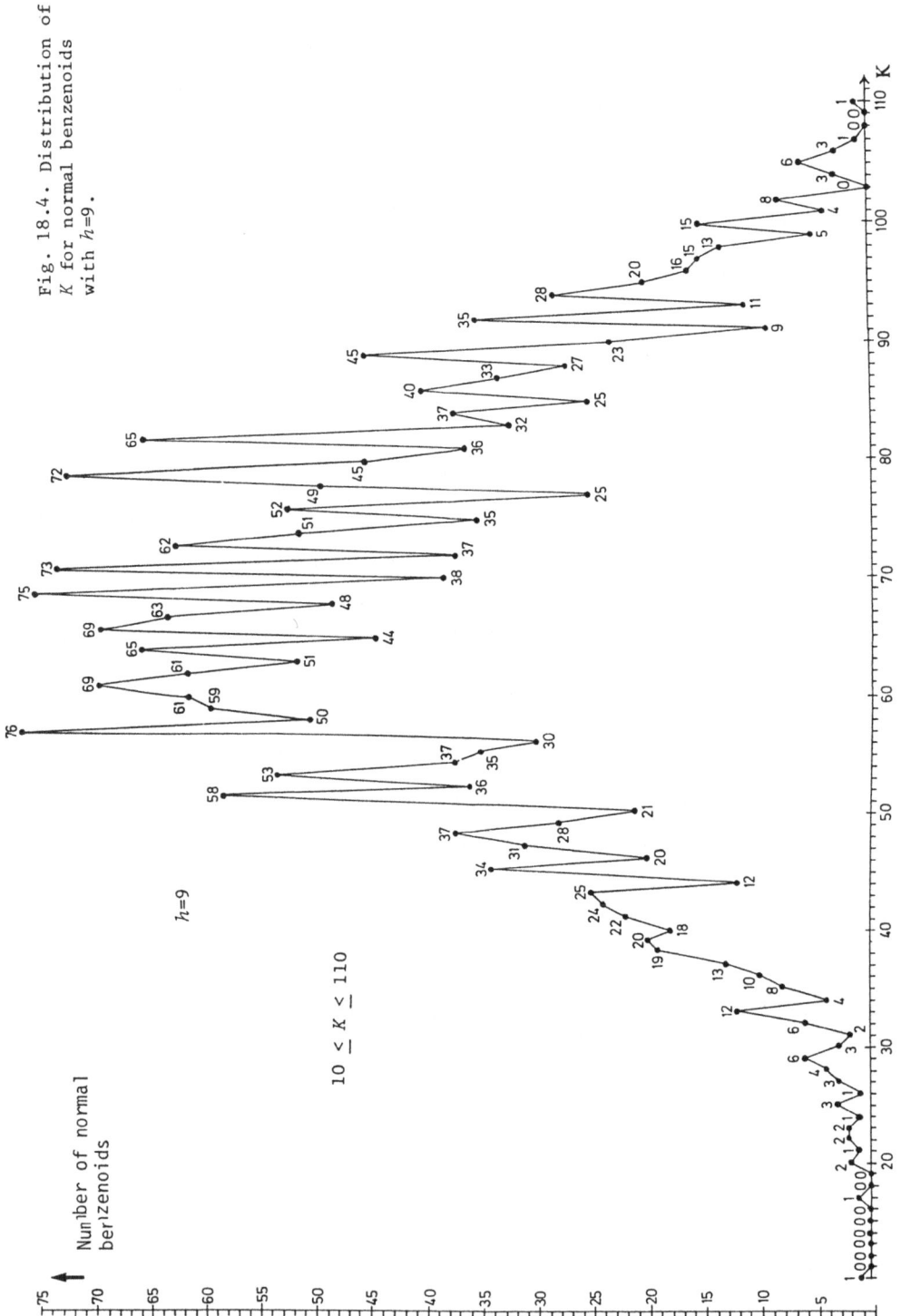

Fig. 18.4. Distribution of
K for normal benzenoids
with h=9.

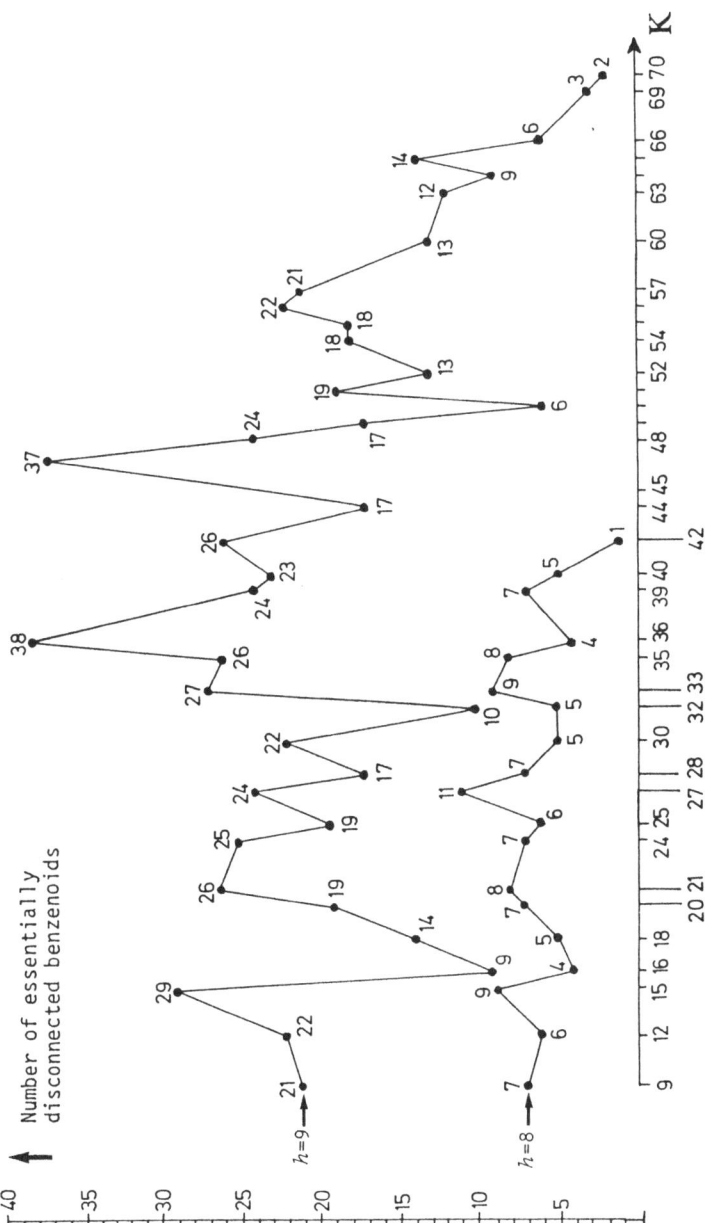

Fig. 18.5. Distributions of K for essentially disconnected benzenoids with $h=8$ and $h=9$. Only those K values are indicated which are realized in essentially disconnected benzenoids.

Table 18.1. Distribution of K for benzenoids with $h=10$: Numbers of normal (n) and essentially disconnected (e) systems.

K	n	e	K	n	e	K	n	e	K	n	K	n
9		64	45	10	132	81	142	46	117	102	153	32
10		0	46	20	0	82	84	0	118	157	154	42
11	1	0	47	14	0	83	134	0	119	123	155	14
12	0	74	48	21	97	84	119	65	120	79	156	40
13	0	0	49	16	37	85	105	37	121	112	157	38
14	0	0	50	27	28	86	87	0	122	132	158	52
15	0	99	51	33	64	87	191	51	123	109	159	14
16	0	28	52	19	46	88	96	56	124	83	160	36
17	0	0	53	32	0	89	133	0	125	79	161	4
18	0	54	54	31	62	90	130	71	126	96	162	17
19	1	0	55	25	79	91	121	52	127	114	163	0
20	0	63	56	27	74	92	119	6	128	152	164	34
21	1	83	57	45	69	93	185	49	129	111	165	6
22	0	0	58	27	0	94	116	0	130	107	166	15
23	0	0	59	38	0	95	129	42	131	110	167	9
24	1	91	60	29	103	96	184	34	132	73	168	9
25	2	50	61	48	0	97	178	0	133	47	169	9
26	4	0	62	48	0	98	113	8	134	125	170	5
27	0	80	63	69	135	99	146	17	135	77	171	4
28	0	53	64	48	34	100	107	10	136	81	172	3
29	3	0	65	57	67	101	166	0	137	97	173	1
30	2	74	66	61	59	102	166	36	138	83	174	4
31	4	0	67	63	0	103	140	0	139	40	175	1
32	3	32	68	58	32	104	114	35	140	67	176	2
33	3	73	69	95	69	105	166	32	141	51	177	0
34	5	0	70	48	58	106	160	0	142	105	178	1
35	5	77	71	68	0	107	135	0	143	50	179	0
36	2	108	72	102	134	108	142	17	144	83	180	1
37	4	0	73	97	0	109	171	0	145	48	181	0
38	6	0	74	73	0	110	101	12	146	68	182	2
39	9	70	75	130	89	111	189	5	147	8	183	0
40	7	66	76	74	37	112	167	5	148	48	184	0
41	7	0	77	78	65	113	109	0	149	39	185	0
42	9	81	78	118	51	114	174	2	150	73	186	0
43	12	0	79	110	0	115	112	6	151	26	187	2
44	12	54	80	62	43	116	107		152	60	188	0
											189	1

Table 18.2. Average K numbers, $\langle K \rangle$, and the quantities $(\ln \langle K \rangle)/h$.

h	Normal		Essentially disconnected	
	$\langle K \rangle_n$	$(\ln \langle K \rangle_n)/h$	$\langle K \rangle_e$	$(\ln \langle K \rangle_e)/h$
1	2	0.6931	–	–
2	3	0.5493	–	–
3	4.5	0.5014	–	–
4	7.17	0.4924	–	–
5	11.21	0.4834	9	0.4394
6	17.5	0.4770	12	0.4142
7	27.46	0.4733	17.70	0.4105
8	42.83	0.4697	25.53	0.4050
9	66.94	0.4671	37.58	0.4029
10	104.40	0.4648	55.00	0.4007

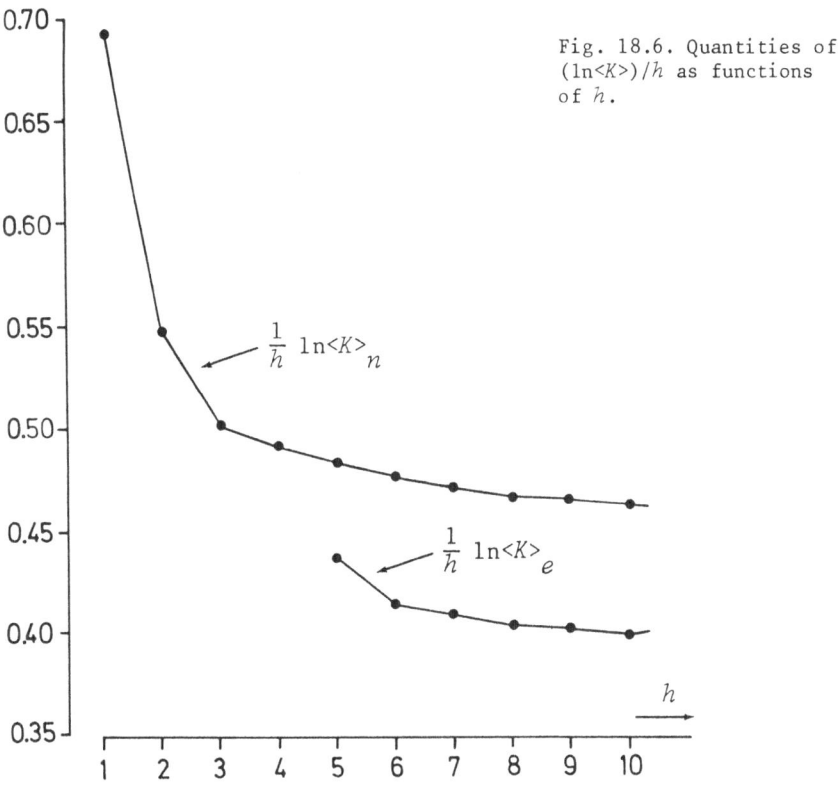

Fig. 18.6. Quantities of $(\ln\langle K\rangle)/h$ as functions of h.

18.4 NUMBER OF NORMAL BENZENOIDS WITH A GIVEN K

18.4.1 *Generation of Normal Benzenoids*

With reference to Section 3.6 we will call it a normal addition when a hexagon is added into a mode L_1, L_3 or L_5 to a normal benzenoid. The new benzenoid is also normal.

It has been conjectured (Cyvin and Gutman 1986c) that any normal benzenoid B_{h+1} may be generated by a normal addition to B_h; here B_{h+1} and B_h denote normal benzenoids with $h+1$ and h hexagons, respectively.

As a corollary of the above conjecture it should be possible to tear down any normal benzenoid, hexagon by hexagon, through normal benzenoids exclusively, right down to one hexagon (*benzene*). This is by definition a normal tearing down. In this process only a hexagon of the mode L_1, L_3 or L_5 is removed every time. The sequence in which the hexagons are removed is not arbitrary. Below we show two examples (left-hand side and middle) how to tear down *coronene* correctly. In the right-hand drawing the (incorrect) tearing down leads to an essentially disconnected benzenoid (*perylene*), which can not be torn down further. The hexagons are supposed to be re-

  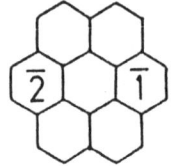

moved in the sequence $\bar{1}$, $\bar{2}$, $\bar{3}$, ...

18.4.2 *Theoretical Solution*

Theorem 13 - eqn. (3.12) - gives information about the minimum of K for a normal benzenoid with h hexagons, viz.

$$K_{min}(h) = h+1 \qquad (18.1)$$

Consequently the h values for the set of normal benzenoids with a given K are limited; the maximum value is clearly

$$h_{max}(K) = K-1 \qquad (18.2)$$

Suppose we wish to deduce the number of normal benzenoids with $K \leq K^*$. In principle this task is accomplished by inspecting all the normal benzenoids with $h = 1$, 2, 3, ..., $K^* - 1$. Hence it is clear that the number in question is limited.

In this way the material of Paragraph 18.2.2 (distributions of K for all normal benzenoids with $h \leq 10$) allows us to determine the numbers of normal benzenoids for $K^* \leq 11$. The analysis may be stretched to $K^* = 12$ by adding one benzenoid to the material, viz. the *polyacene* L(11), which is the only system with $K = 12$ and $h = 11$.

It would be hopeless to continue this analysis to substantially larger values of K^* because of the rapidly increasing numbers of normal benzenoids with increasing h. A practical solution of the problem is furnished by the "sieve method".

18.4.3 *Practical Solution: Sieve Method*

Theory. The sieve method is based on *Theorem 13* and the conjecture about generation of normal benzenoids (cf. Paragraph 18.4.1). We adhere to the notation B_h for a normal benzenoid with h hexagons. We are seeking all the normal benzenoids with $K \leq K^*$. Finally we assume that the distribution of K is known for $h \leq h'$.

1. Consider the complete set of benzenoids $B_{h'}$ with $K \leq K^*$. Here h' is an arbitrarily small number, occasionally $h' = 1$.

2. Execute all normal additions to the $B_{h'}$ systems with $K < K^*$, thus generating $B_{h'+1}$.

3. Discard all the generated benzenoids (from $B_{h'+1}$) which have $K > K^*$.

323

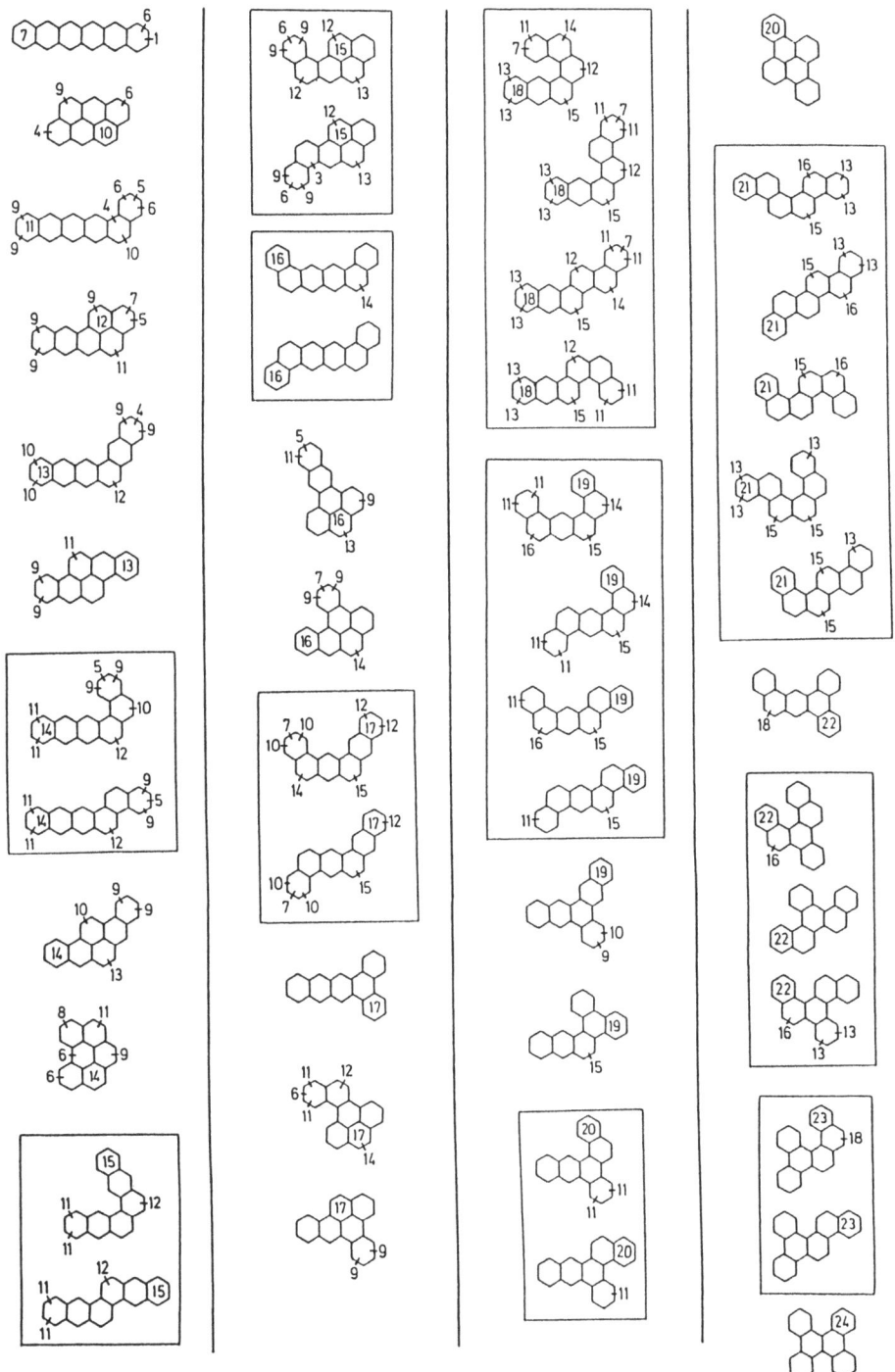

Fig. 18.7. Mapping of the normal additions to all the 48 normal benzenoids with h=6. Framed systems are isoarithmic. K numbers are inscribed. The figure contains information about the 167 existing normal benzenoids with h=7 and their K numbers (given in terms of increments).

4. Continue the process (if necessary) from $B_{h'+1}$ to $B_{h'+2}$, again by normal additions to the systems with $K < K^*$.

5. The process stops automatically by generating $L(K^*-1)$ from $L(K^*-2)$.

Practical Examples. First we give an exemplification of *Theorem 16* (3.18). Consider three additions to B_4 = *chrysene*, A(4), as shown below. Here $K\{B_4\} = 8$. The added hexagon in the generation $B_4 \to B_5$ is marked by an asterisk. The increment in K, viz. $\Delta K = K\{B_5\} - K\{B_4\}$ is symbolized in each case by the unhatched part. The examples show that ΔK may be (a) unity (when the whole benzenoid B_4 is hatched), (b) K number of a normal benzenoid or (c) K number of a disconnected system.

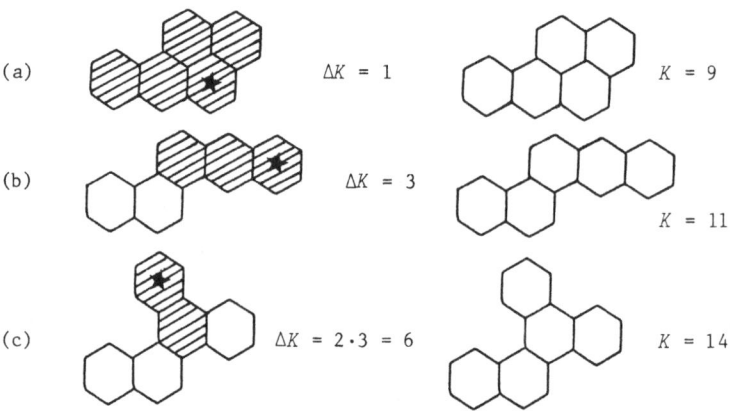

The information of the above figures (a)-(c), in addition to more, may be compressed into one drawing in the style of Fig. 7.11:
In contrast to Chapter 7 we now allow
for not only annelations (L_1), but also
the additions into L_3 and L_5.

   In the following we assume as an example
$K^* = 24$ (which is K_{max} for $h=6$). In other words
we seek all the normal benzenoids with $K \leq 24$. The distributions of K for $h \leq 6$ are assumed to be known (cf. Fig. 2.7). All the *24* normal benzenoids with $h \leq 5$ have $K < 24$.

   1. We take advantage of the totality of the *48* normal benzenoids with $h=6$ (Fig. 2.7). All of them have $K \leq 24$ and therefore constitute B_6.

   2. Only one system in B_6 has $K = 24$. The remaining 47 benzenoids are subjected to normal additions as indicated in Fig. 7. In this process many $h=7$ systems are duplicated. They have been eliminated so that the figure indicates exactly the existing 167 normal benzenoids with $h=7$, viz. B_7.

   3. The 115 systems from B_7 with $K > 24$ are to be discarded.

   4. Among the remaining *52* systems there are 9 with $K = 24$, leaving 43 systems

with $K < 24$. The process is continued by normal additions to the 43 systems from B_7.

The number of systems to be inspected rapidly decreases. Concretely one arrives at *24* systems with $K \leq 24$ from B_8, *10* from B_9, *4* from B_{10}, *2* from B_{11}, *2* from B_{12}, and finally one each from B_{13}, B_{14}, ..., B_{23}. The last *11* systems are the *polyacenes* L(13), L(14), ..., L(23).

Altogether we end up with 177 normal benzenoids, which have $K \leq 24$. The number *177* is the sum of the italicized figures of the above description. The actual forms are depicted in Fig. 8.

Fig. 18.8. The forms of all the normal benzenoids with $K \leq 14$, and (on the following pages) $15 \leq K \leq 24$. Framed systems are isoarithmic. h values are inscribed.

326

Fig. 18.8 (continued)

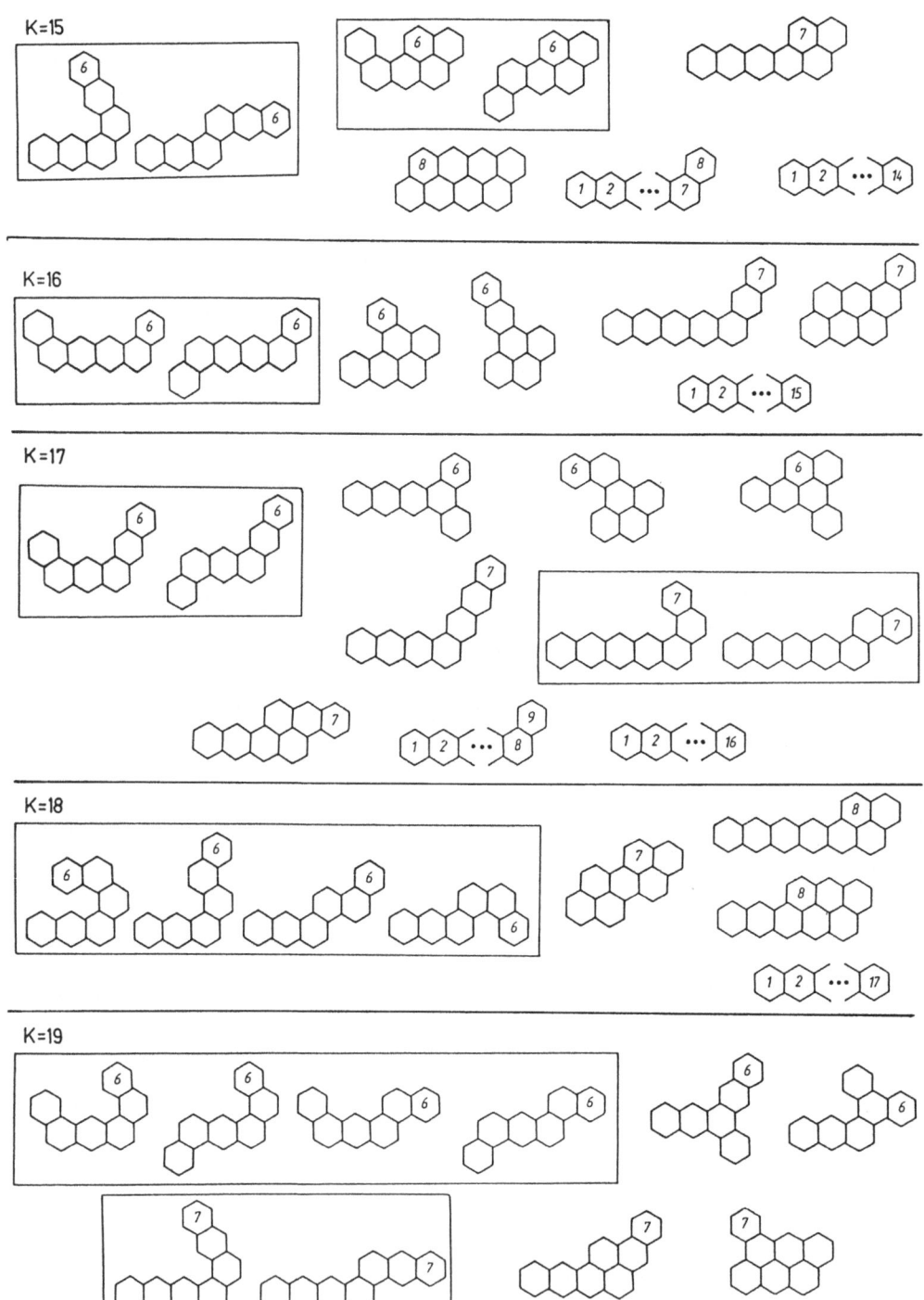

Fig. 18.8 (continued)

Fig. 18.8 (continued)

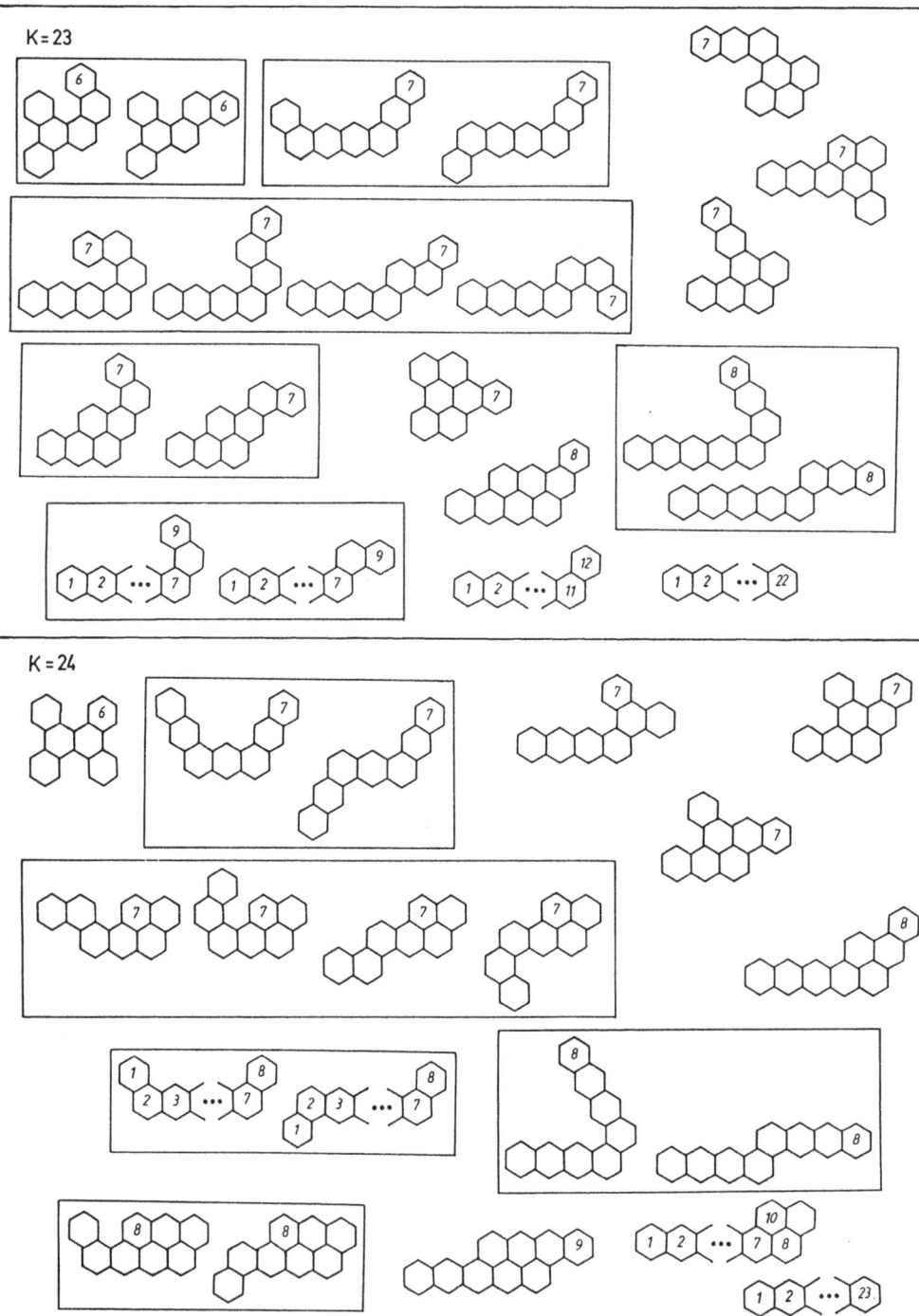

18.4.4 *Listing of Results*

The numbers from the above analysis are found in Table 3. The procedure was adapted to computer programming and carried out by J. Brunvoll for K numbers up to 110. The results for $K \leq 50$ are listed in the table.

The columns of Table 3 may be added to find the total numbers of normal benzenoids with given K. This is clear since eqn. (2) guarantees that there will not appear any nonvanishing numbers below the diagonal of unities. The sums were used to produce Fig. 1.

The fact that the set of normal benzenoids with a given h has a maximum K number implies that every horizontal row of Table 3 (pertaining to a given h) eventually terminates. The system with $K = K_{max}(h)$ indicates the minimum of h for all normal benzenoids with this particular K number. In general there is a lower limit of h

Table 18.3. Numbers of normal benzenoids with given K and different h values. Listing up to $K = 50$.

h	$K=2$	3	4	5	6	7	8	9	10
1	1								
2		1							
3			1	1					
4				1	1	1	2	1	
5					1	0	0	2	1
6						1	0	0	1
7							1	0	0
8								1	0
9									1
Sum	1	1	1	2	2	2	3	4	3

h	$K=11$	12	13	14	15	16	17	18	19	20
5	3	2	4	1						
6	1	1	2	4	4	4	5	4	6	3
7	0	0	1	1	1	2	4	1	4	4
8	0	0	0	0	2	0	0	2	3	2
9	0	0	0	0	0	0	1	0	0	2
10	1	0	0	0	0	0	0	0	1	0
11		1	0	0	0	0	0	0	0	0
12			1	0	0	0	0	0	0	0
13				1	0	0	0	0	0	0
14					1	0	0	0	0	0
15						1	0	0	0	0
16							1	0	0	0
17								1	0	0
18									1	0
19										1
Sum	5	4	8	7	8	7	11	8	15	12

Table 18.3 (continued)

h	$K =$ 21	22	23	24	25	26	27	28	29	30
6	5	4	2	1						
7	7	5	12	9	12	8	11	5	13	12
8	2	2	3	7	3	8	10	5	10	11
9	1	2	2	1	3	1	3	4	6	3
10	1	0	0	1	2	4	0	0	3	2
11	1	0	0	0	0	0	2	1	2	2
12	0	0	1	0	0	0	0	1	0	1
13	0	0	0	0	1	0	0	0	0	0
14	0	0	0	0	0	0	1	0	0	0
15	0	0	0	0	0	0	0	0	1	0
$K-1$	1	1	1	1	1	1	1	1	1	1
Sum	18	14	21	20	22	22	28	17	36	32

h	$K =$ 31	32	33	34	35	36	37	38	39	40
7	13	4	6	13	4	7	3	2	0	1
8	15	12	22	13	17	13	28	16	32	21
9	2	6	12	4	8	10	13	19	20	18
10	4	3	3	5	5	2	4	6	9	7
11	0	1	2	2	2	5	3	0	4	2
12	1	2	1	2	2	0	2	1	4	2
13	0	0	1	1	3	0	0	1	1	1
14	0	0	0	0	0	2	1	2	0	0
15	0	0	0	0	0	0	0	0	1	1
$\frac{1}{2}(K+1)$	1		1		1		1		1	
$K-1$	1	1	1	1	1	1	1	1	1	1
Sum	37	29	49	41	43	40	56	48	73	54

h	$K =$ 41	42	43	44	45	46	47	48	49	50
7	1									
8	35	19	29	16	17	25	24	16	30	25
9	22	24	25	12	34	20	31	37	28	21
10	7	9	12	12	10	20	14	21	16	27
11	5	2	6	4	8	5	9	4	7	13
12	2	2	2	3	1	5	5	9	2	2
13	3	0	2	4	1	2	1	1	5	3
14	0	2	1	0	4	0	2	3	3	1
15	2	0	0	1	0	1	0	0	3	1
16	0	1	1	2	1	0	0	0	0	1
17	0	0	0	0	1	1	2	0	0	0
18	0	0	0	0	0	0	0	1	1	2
$\frac{1}{2}(K+1)$	1		1		1		1		1	
$K-1$	1	1	1	1	1	1	1	1	1	1
Sum	79	60	80	55	79	80	90	93	97	97

associated with every K number. Let it be designated $h_{min}(K)$. It is clearly greater than unity for every $K \geq 3$. For $K = 2, 3, \ldots, 10$ the numbers are $h_{min}(K) = 1, 2, 3, 3, 4, 4, 4, 4$ and 5, respectively. As a general rule the limit $h_{min}(K)$ is the same for $K_{max}(h) < K \leq K_{max}(h+1)$.

18.4.5 *The Fading-Out Phenomenon*

The sieve method (see above) tends to fade out in the sense that for a given K the numbers vanish when h passes a limit, until they terminate with a unity for $h = K-1$. These last numbers constitute the diagonal of unities; cf. Table 3. In the lower part of the table ($K \geq 21$) this diagonal is actually represented (in a space-saving way) by a horizontal row. These numbers, of course, correspond to the *poly-acenes* L($K-1$).

The fading-out phenomenon goes far beyond the linear chains L($K-1$) for sufficiently large h values. Already for all odd K numbers $K \geq 5$ the next-to-largest h value is invariably equal to $\frac{1}{2}(K+1)$. It is realized by the systems of one hexagon angularly annelated to a linear chain (*benzoacenes*). It is seen from Fig. 8 that the same values of h and K are realized in some pericondensed systems for $K=9$ and $K=15$. Otherwise the *benzoacene* is the only system with $h = \frac{1}{2}(K+1)$. Also this feature was exploited in order to save space in Table 3; cf. the entries for $K \geq 31$.

Combinatorial K formulas as functions of h for the two described forms are given in Table 4. They have the limits of K/h equal to 1 and 2, respectively, when $h \to \infty$. Other characteristic forms having large h values in relation to K may be distinguished. They belong to classes where the members only differ by the length of a linear chain. Table 4 includes the formulas for such classes where $\lim(K/h) = 1, 2, 3, 4, 5$ when $h \to \infty$. The class of *aceno*(a)*pyrenes* (3.17) mentioned in Section 3.5 is among them.

Table 18.4. Combinatorial K formulas as functions of the number of hexagons (h) for benzenoids containing a linear chain of arbitrary length.

$\lim(K/h)$ $h \to \infty$	Form*	Formula $K(h) = Ah + B$
1		$K = h + 1 = K\{L(h)\}$
2		$K = 2h - 1; \qquad h \geq 2$
3		$K = 3h - 6; \qquad h \geq 4$
		$K = 3h - 5; \qquad h \geq 3$
		$K = 3h - 4; \qquad h \geq 3$
4		$K = 4h - 14; \qquad h \geq 6$
		$K = 4h - 11; \qquad h \geq 4$
		$K = 4h - 11; \qquad h \geq 5$
		$K = 4h - 9; \qquad h \geq 4$
		$K = 4h - 8; \qquad h \geq 4$

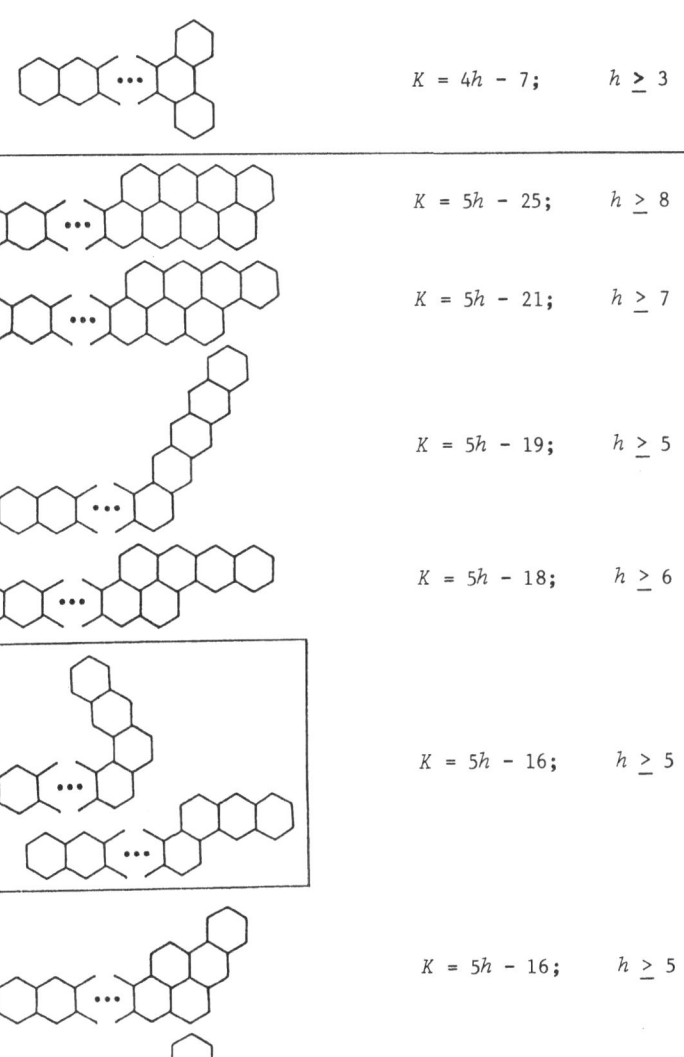

$$K = 4h - 7; \qquad h \geq 3$$

5

$$K = 5h - 25; \qquad h \geq 8$$

$$K = 5h - 21; \qquad h \geq 7$$

$$K = 5h - 19; \qquad h \geq 5$$

$$K = 5h - 18; \qquad h \geq 6$$

$$K = 5h - 16; \qquad h \geq 5$$

$$K = 5h - 16; \qquad h \geq 5$$

$$K = 5h - 14; \qquad h \geq 4$$

$$K = 5h - 14; \qquad h \geq 5$$

$$K = 5h - 13; \qquad h \geq 4$$

Table 18.4 (continued)

$$K = 5h - 12; \qquad h \geq 4$$

*Framed systems are isoarithmic.

BIBLIOGRAPHY

Armitt TW, Robinson R (1925) Polynuclear Heterocyclic Aromatic Types - Part II - Some Anhydronium Bases. J Chem Soc: 1604

Babić D, Graovac A (1986) Enumeration of Kekulé Structures in One-Dimensional Polymers. Croat Chem Acta 59: 731

Balaban AT (1969) Chemical Graphs - VII - Proposed Nomenclature of Branched Cata-Condensed Benzenoid Polycyclic Hydrocarbons. Tetrahedron 25: 2949

Balaban AT (1977) Chemical Graphs - XXVIII - A New Topological Index for Catafusenes - L-Transform of Their Three-Digit Codes. Rev Roum Chim 22: 45

Balaban AT (1981) Chemical Graphs - XXXVII - A Simple Rule for Classifying Peri-Condensed Benzenoid Hydrocarbons as Closed-Shell or Open-Shell (Polyradicalic) Systems Using Dualist (Characteristic) Graphs. Rev Roum Chim 26: 407

Balaban AT (1982) Challenging Problems Involving Benzenoid Polycyclics and Related Systems. Pure & Appl Chem 54: 1075

Balaban AT (1985a) Application of Graph Theory in Chemistry. J Chem Inf Comput Sci 25: 334

Balaban AT (1985b) Graph Theory and Theoretical Chemistry. J Mol Struct (Theochem) 120: 117

Balaban AT, Artemi C, Tomescu I (1987) Algebraic Expressions for Kekulé Structure Counts of Non-Branched Regular Cata-Condensed Benzenoid Hydrocarbons. Match 22: 77

Balaban AT, Brunvoll J, Cioslowski J, Cyvin BN, Cyvin SJ, Gutman I, He WC, He WJ, Knop JV, Kovačević M, Müller WR, Szymanski K, Tošić R, Trinajstić N (1987) Emumeration of Benzenoid and Coronoid Hydrocarbons. Z Naturforsch 42a: 863

Balaban AT, Harary F (1968) Chemical Graphs - V - Enumeration and Proposed Nomenclature of Benzenoid Cata-Condensed Polycyclic Aromatic Hydrocarbons. Tetrahedron 24: 2505

Balaban AT, Tomescu I (1983) Algebraic Expressions for the Number of Kekulé Structures of Isoarithmic Cata-Condensed Benzenoid Polycyclic Hydrocarbons. Match 14: 155

Balaban AT, Tomescu I (1984) Chemical Graphs - XL - Three Relations Between the Fibonacci Sequence and the Numbers of Kekulé Structures for Non-Branched Cata-Condensed Polycyclic Aromatic Hydrocarbons. Croat Chem Acta 57: 391

Balaban AT, Tomescu I (1985) Chemical Graphs - XLI - Numbers of Conjugated Circuits and Kekulé Structures for Zigzag Catafusenes and (J,K)-Hexes - Generalized Fibonacci Numbers. Match 17: 91

Bergan JL, Cyvin BN, Cyvin SJ (1987) The Fibonacci Numbers, and Kekulé Structures of Some Corona-Condensed Benzenoids (Corannulenes). Acta Chim Hung 124: 299

Biermann D, Schmidt W (1980a) Diels-Alder Reactivity of Polycyclic Aromatic Hydrocarbons - III - New Experimental and Theoretical Results. Israel J Chem 20: 312

Biermann D, Schmidt W (1980b) Diels-Alder Reactivity of Polycyclic Aromatic Hydrocarbons - 1 - Acenes and Benzologs. J Am Chem Soc 102: 3163

Biermann D, Schmidt W (1980c) Diels-Alder Reactivity of Polycyclic Aromatic Hydrocarbons - 2 - Phenes and Starphenes. J Am Chem Soc 102: 3173

Bodroža O, Gutman I, Cyvin SJ, Tošić R (1988) Number of Kekulé Structures of Hexagon-Shaped Benzenoids. (To be published)

Bräuchle C, Kabza H, Voitländer J (1980) A Concept to Explain the Variation of the Triplet Zero-Field Splitting Parameters with the Structure of Aromatic Hydrocarbons in Terms of Local Benzenoid Characteristics. Chem Phys 48: 369

Braeuchle C, Kabza H, Voitländer J, Clar E (1978) Phosphorescence and ODMR Studies of Some Coronene Compounds. Chem Phys 32: 63

Bräuchle C, Voitländer J (1982) The Zero-Field Splitting Parameter D of the Triplet State T_1 of Benzenoid Hydrocarbons Described in Terms of Basic Molecular Subunits. Tetrahedron 38: 279

Bräuchle C, Voitländer J, Vogler H (1981) Correlation Between the Ground State and the Triplet State Character Orders of Benzenoid Hydrocarbons. Z Naturforsch 36a: 651

Brown RL (1983) Counting of Resonance Structures for Large Benzenoid Polynuclear Hydrocarbons. J Comput Chem 4: 556

Brunvoll J, Cyvin BN, Cyvin SJ (1987) Enumeration and Classification of Benzenoid Hydrocarbons - Part II - Symmetry and Regular Hexagonal Benzenoids. J Chem Inf Comput Sci (in press)

Brunvoll J, Cyvin SJ, Cyvin BN (1987) Enumeration and Classification of Benzenoid Hydrocarbons. J Comput Chem 8: 189

Brunvoll J, Cyvin SJ, Cyvin BN, Gutman I, He WJ, He WC (1987) There are Exactly Eight Concealed Non-Kekuléan Benzenoids with Eleven Hexagons. Match 22: 105

Chen RS (1986a) Kekulé Structure Count for a Group of Homologous Series of Hexagonal Systems. J Xinjiang Univ 3(2): 13

Chen RS (1986b) On the Number of Kekulé Structures for Rectangle-Shaped Benzenoids. Match 21: 259

Chen RS (1986c) On the Number of Kekulé Structures for Rectangle-Shaped Benzenoids - Part II. Match 21: 277

Chen RS, Cyvin SJ (1987) Distribution of K, the Number of Kekulé Structures, in Benzenoid Hydrocarbons - Part IA - Comments on Upper Bounds of K. Match 22: 175

Chen RS, Cyvin SJ, Cyvin BN (1987) On the Number of Kekulé Structures for Rectangle-Shaped Benzenoids - Part III. Match 22: 111

Cioslowski J (1986) The Generalized Mc Clelland Formula. Match 20: 95

Clar E (1964a) Die Bauprinzipien der aromatischen Kohlenwasserstoffe. Chimia 18: 375

Clar E (1964b) Polycyclic Hydrocarbons - Vols. I, II. Academic Press, London

Clar E (1972) The Aromatic Sextet. Wiley, London

Clar E, Zander M (1958) 1:12-2:3-10:11-Tribenzoperylene. J Chem Soc: 1861

Cotton FA (1971) Chemical Applications of Group Theory (2nd Ed). Wiley-Interscience, New York

Coulson CA (1952) Valence. Oxford University Press, Glasgow

Coulson CA, Longuet-Higgins HC (1948) The Electronic Structure of Conjugated Systems II - Unsaturated Hydrocarbons and Their Hetero-Derivatives. Proc Roy Soc A192: 16

Cvetković D, Doob M, Sachs H (1980) Spectra of Graphs - Theory and Application. Academic Press, New York

Cvetković D, Gutman I (1974) Kekulé Structures and Topology - II - Cata-Condensed Systems. Croat Chem Acta 46: 15

Cvetković D, Gutman I, Trinajstić N (1974) Graph Theory and Molecular Orbitals - VII - The Role of Resonance Structures. J Chem Phys 61: 2700

Cyvin BN, Brunvoll J, Cyvin SJ, Gutman I (1986) Distribution of K, the Number of Kekulé Structures in Benzenoid Hydrocarbons - Part III - Kekulé Structure Statistics. Match 21: 301

Cyvin BN, Cyvin SJ (1986) Molecular Vibrations of Hexabenzocoronenes. Spectroscopy Letters 19: 1161

Cyvin BN, Cyvin SJ (1987) Enumeration of Kekulé Structures - Pentagon-Shaped Benzenoids - Part IV - Seven-Tier Pentagons and Related Classes. Match 22: 157

Cyvin BN, Cyvin SJ, Brunvoll J (1986) Enumeration of Kekulé Structures - Pentagon-Shaped Benzenoids - Part I. Match 21: 291

Cyvin BN, Klaeboe P, Whitmer JC, Cyvin SJ (1982) Condensed Aromatics - Part XV - Chrysene. Z Naturforsch 37a: 251

Cyvin SJ (1982a) Group-Theoretical Treatment of Kekulé Structures. Match 13: 167

Cyvin SJ (1982b) Kekulé Structures of Polyphenes. Monatsh Chem 113: 1127

Cyvin SJ (1983a) Kekulé Structures and the Fibonacci Series. Acta Chim Hung 112: 281

Cyvin SJ (1983b) Symmetry of Kekulé Structures. J Mol Struct 100: 75

Cyvin SJ (1983c) Number of Kekulé Structures of Single-Chain Aromatics. Monatsh Chem 114: 13

Cyvin SJ (1983d) Number of Kekulé Structures for Some Aromatic Chain Molecules. Monatsh Chem 114: 525

Cyvin SJ (1985) Enumeration of Kekulé Structures - Chevrons. J Mol Struct (Theochem) 133: 211

Cyvin SJ (1986a) Number of Kekulé Structures for Rectangle-Shaped Benzenoids - Part II. Match 19: 213

Cyvin SJ (1986b) Distribution of K, the Number of Kekulé Structures in Benzenoid Hydrocarbons - Part I - Upper and Lower Bounds of K. Match 20: 165

Cyvin SJ (1986c) Enumeration of Kekulé Structures - Triangle-Shaped Benzenoids. Match 21: 285

Cyvin SJ (1986d) The Number of Kekulé Structures of Hexagon-Shaped Benzenoids and Members of Other Related Classes. Monatsh Chem 117: 33

Cyvin SJ (1987a) Extended Application of an Algorithm for the Number of Kekulé Structures. Match 22: 101

Cyvin SJ (1987b) Contribution to the Techniques of Enumeration of Kekulé Structures. Monatsh Chem (in press)

Cyvin SJ, Bergan JL, Cyvin BN (1987) Benzenoids and Coronoids with Hexagonal Symmetry ("Snowflakes"). Acta Chim Hung 124: 691

Cyvin SJ, Brunvoll J, Cyvin BN (1986) Distribution of K, the Number of Kekulé Structures in Benzenoid Hydrocarbons - Part IV - Benzenoids with 10 and 11 Hexagons. Z Naturforsch 41a: 1429

Cyvin SJ, Chen RS, Cyvin BN (1987) The Number of Kekulé Structures for Rectangle-Shaped Benzenoids - Part IV. Match 22: 129

Cyvin SJ, Cyvin BN (1986a) Number of Kekulé Structures for Condensed Parallelograms and Related Benzenoids with Repeated Units. Match 20: 181

Cyvin SJ, Cyvin BN (1986b) Enumeration of Kekulé Structures - Pentagon-Shaped Benzenoids - Part II. Match 21: 295

Cyvin SJ, Cyvin BN (1987a) Enumeration of Kekulé Structures - Prolate Pentagons. J Mol Struct (Theochem) 152: 347

Cyvin SJ, Cyvin BN (1987b) Enumeration of Kekulé Structures - Étagères and Related Benzenoid Classes. Monatsh Chem 118: 337

Cyvin SJ, Cyvin BN, Bergan JL (1986) Number of Kekulé Structures for Rectangle-Shaped Benzenoids. Match 19: 189

Cyvin SJ, Cyvin BN, Brunvoll J (1987) Trigonal Benzenoid Hydrocarbons. J Mol Struct (Theochem) 151: 271

Cyvin SJ, Cyvin BN, Brunvoll J, Chen RS, Su LX (1987) The Number of Kekulé Structures for Rectangle-Shaped Benzenoids - Part V. Match 22: 141

Cyvin SJ, Cyvin BN, Brunvoll J, Gutman I (1987) Enumeration of Kekulé Structures for Multiple Zigzag Chains and Related Benzenoid Hydrocarbons. Z Naturforsch 42a: 722

Cyvin SJ, Cyvin BN, Brunvoll J, Whitmer JC, Klaeboe P (1982) Condensed Aromatics - Part XX - Coronene. Z Naturforsch 37a: 1359

Cyvin SJ, Cyvin BN, Chen RS (1987) On the Number of Kekulé Structures for Rectangle-Shaped Benzenoids - Part VI. Match 22: 151

Cyvin SJ, Cyvin BN, Gutman I (1985) Number of Kekulé Structures of Five-Tier Strips. Z Naturforsch 40a: 1253

Cyvin SJ, Cyvin BN, Gutman I (1986) Topological Properties of Benzenoid Systems - XLII - Number of Kekulé Structures of More Peri-Condensed Molecules. Coll Sci Papers Fac Sci Kragujevac 7: 5

Cyvin SJ, Cyvin BN, Gutman I (1987) Number of Kekulé Structures of Systems with Repeated Units. Z Naturforsch 42a: 181

Cyvin SJ, Gutman I (1985) Topological Properties of Benzenoid Systems - XXXIX - The Number of Kekulé Structures of Benzenoid Hydrocarbons Containing a Chain of Hexagons. J Serb Chem Soc 50: 443

Cyvin SJ, Gutman I (1986a) Kekulé Structures and Their Symmetry Properties. Comp & Maths with Appls 12B: 859

Cyvin SJ, Gutman I (1986b) Topological Properties of Benzenoid Systems - Part XXXVI - Algorithm for the Number of Kekulé Structures in Some Peri-Condensed Benzenoids. Match 19: 229

Cyvin SJ, Gutman I (1986c) Number of Kekulé Structures as a Function of the Number of Hexagons in Benzenoid Hydrocarbons. Z Naturforsch 41a: 1079

Cyvin SJ, Gutman I (1987a) Topological Properties of Benzenoid Hydrocarbons - Part XLIV - Obvious and Concealed Non-Kekuléan Benzenoids. J Mol Struct (Theochem) 150: 157

Cyvin SJ, Gutman I (1987b) On Recognizing Kekuléan Benzenoid Molecules. J Mol Struct (Theochem) (in press)

Dewar MJS, Longuet-Higgins HC (1952) The Correspondence Between the Resonance and Molecular Orbital Theories. Proc Roy Soc A214: 482

Dias JR (1984) Isomer Enumeration of Nonradical Strictly Peri-Condensed Polycyclic Aromatic Hydrocarbons. Can J Chem 62: 2914

Dias JR (1987) A Periodic Table for Polycyclic Aromatic Hydrocarbons - Part X - On the Characteristic Polynomial and Other Structural Invariants. J Mol Struct (Theochem) 149: 213

Diederich F, Staab HA (1978) Benzenoid versus Annulenoid Aromaticity - Synthesis and Properties of Kekulene. Angew Chem Int Ed Engl 17: 372

Džonova-Jerman-Blažič B, Trinajstić N (1982a) Computer-Aided Enumeration and Generation of the Kekulé Structures in Conjugated Hydrocarbons. Computers & Chemistry 6: 121

Džonova-Jerman-Blažič B, Trinajstić N (1982b) Application of the Reduced Graph Model to the Enumeration of Kekulé Structures and Conjugated Circuits of Benzenoid Hydrocarbons. Croat Chem Acta 55: 347

Eilfeld P, Schmidt W (1981) Resonance Theoretical Approach to the Calculation of the First IP's of Polycyclic Aromatics. J Electr Spectr Related Phenom 24: 101

El-Basil S (1982) A Novel Topological Method for Counting Kekulé Valence Structures & Conjugated Circuits of Catacondensed Benzenoid Hydrocarbons. Indian J Chem 21B: 561

El-Basil S (1986a) Gutman Trees - Combinatorial-Recursive Relations of Counting Polynomials - Data Reduction Using Chemical Graphs. J Chem Soc Faraday Trans 2 82: 299

El-Basil S (1986b) Combinatorial Clar Sextet Theory - On Valence-Bond Method of Herndon and Hosoya. Theor Chim Acta 70: 53

El-Basil S, Jashari G, Knop JV, Trinajstić N (1984) Note on the Application of the Reduced Graph Model in Conjunction with Search Trees to the Enumeration of Kekulé Structures. Monatsh Chem 115: 1299

El-Basil S, Křivka P, Trinajstić N (1984) Application of the Dualist Model - Generation of Kekulé Structures and Resonant Sextets of Benzenoid Hydrocarbons. Croat Chem Acta 57: 339

El-Basil S, Trinajstić N (1984) Application of the Reduced Graph Model to the Sextet Polynomial. J Mol Struct (Theochem) 110: 1

Erlenmeyer E (1866) Studien über die s.g. aromatischen Säuren. Annal d Chemie u Pharm 137: 327

Gordon M, Davison WHT (1952) Theory of Resonance Topology of Fully Aromatic Hydrocarbons - I. J Chem Phys 20: 428

Graovac A, Babić D, Strunje M (1986) Enumeration of Kekulé Structures in Polymers. Chem Phys Letters 123: 433

Gutman I (1974) Some Topological Properties of Benzenoid Systems. Croat Chem Acta 46: 209

Gutman I (1977) Topological Properties of Benzenoid Systems - An Identity for the Sextet Polynomial. Theor Chim Acta 45: 309

Gutman I (1981a) Topological Properties of Benzenoid Systems - VI - On Kekulé Structure Count. Bull Soc Chim Beograd 46: 411

Gutman I (1981b) Topological Resonance Energy of Very Large Benzenoid Hydrocarbons. Z Naturforsch 36a: 128

Gutman I (1982a) Topological Properties of Benzenoid Molecules. Bull Soc Chim Beograd 47: 453

Gutman I (1982b) The Number of Kekulé Structures in Conjugated Systems Containing a Linear Polyacene Fragment. Croat Chem Acta 55: 371

Gutman I (1982c) On Kekulé Structure Count of Cata-Condensed Benzenoid Hydrocarbons. Match 13: 173

Gutman I (1983) Topological Properties of Benzenoid Systems - XXI - Theorems, Conjectures, Unsolved Problems. Croat Chem Acta 56: 365

Gutman I (1984) Topological Properties of Benzenoid Systems - XXX - An Identity for the Number of Kekulé Structures. Coll Sci Papers Fac Sci Kragujevac 5: 59

Gutman I (1985a) Topological Properties of Benzenoid Systems - XXXIV - Number of Kekulé Structures of Some Peri-Condensed Molecules. Coll Sci Papers Fac Sci Kragujevac 6: 35

Gutman I (1985b) Topological Properties of Benzenoid Systems - XXVIII - Number of Kekulé Structures of Some Benzenoid Hydrocarbons. Match 17: 3

Gutman I (1986) Topological Properties of Benzenoid Systems - XLVIII - Two Contradictory Formulas for Total π-Electron Energy and Their Reconciliation. Match 21: 317

Gutman I, Cioslowski J (1987) Bounds for the Number of Perfect Matchings in Hexagonal Systems. Publ Inst Math (Beograd) (in press)

Gutman I, Cyvin SJ (1986) Recognizing Kekuléan Benzenoid Molecules. J Mol Struct (Theochem) 138: 325

Gutman I, Cyvin SJ (1987a) A New Method for the Enumeration of Kekulé Structures. Chem Phys Letters 136: 137

Gutman I, Cyvin SJ (1987b) Topological Properties of Benzenoid Systems - XXXV - Number of Kekulé Structures of Multiple-Chain Aromatics. Monatsh Chem 118: 541

Gutman I, Cyvin SJ (1988) Hexagonal Systems with Small Number of Perfect Matchings. (To be published)

Gutman I, Hosoya H, Yamaguchi T, Motoyama A, Kuboi N (1977) Topological Properties of Benzenoid Systems - III - Recursion Relation for the Sextet Polynomial. Bull Soc Chim Beograd 42: 503

Gutman I, Petrović S (1981) Topological Properties of Benzenoid Systems VII - Approximate Formulas for Resonance Energy. Bull Soc Chim Beograd 46: 459

Gutman I, Petrović S (1983) On Total π-Electron Energy of Benzenoid Hydrocarbons. Chem Phys Letters 97: 292

Gutman I, Polansky OE (1986) Mathematical Concepts in Organic Chemistry. Springer-Verlag, Berlin

Gutman I, Polansky OE, Zander M (1984) Some Topological Properties of Isomeric Benzenoid Hydrocarbons. Match 15: 145

Gutman I, Randić M (1979) A Correlation Between Kekulé Valence Structures and Conjugated Circuits. Chem Phys 41: 265

Gutman I, Su LX, Cyvin SJ (1987) Identity for the Number of Kekulé Structures of Benzenoid Hydrocarbons. J Serb Chem Soc 52: 263

Hall GG (1973) A Graphical Model of a Class of Molecules. Int J Math Educ Sci Technol 4: 233

Hall GG (1981) Eigenvalues of Molecular Graphs. Publ Inst Math Appl 17: 70

Ham NS (1958) Mobile Bond Orders in the Resonance and Molecular Orbital Theories. J Chem Phys 29: 1229

Harary F (1967) Graphical Enumeration Problems [in] Graph Theory and Theoretical Physics (Harary F, Edit). Academic Press, London; p 1

Harary F (1969) Graph Theory. Addison-Wesley, Reading, Massachusetts

Harary F, Harborth H (1976) Extremal Animals. J Combinat Inf & System Sci 1: 1

Harary F, Read RC (1970) The Enumeration of Tree-Like Polyhexes. Proc Edinburgh Math Soc (Ser II) 17: 1

He WC, He WJ (1985) A Novel Nomenclature of Polycyclic Aromatic Hydrocarbons Without Using Graph Centre. Theor Chim Acta 68: 301

Heilbronner E (1962) Über einen graphentheoretischen Zusammenhang zwischen dem Hückel'schen MO-Verfahren und dem Formalismus der Resonanztheorie. Helv Chim Acta 45: 1722

Herndon WC (1973) Enumeration of Resonance Structures. Tetrahedron 29: 3

Herndon WC (1974a) Resonance Theory - VI - Bond Orders. J Am Chem Soc 96: 7605

Herndon WC (1974b) Resonance Theory and the Enumeration of Kekule Structures. J Chem Educ 51: 10

Herndon WC (1974c) Thermochemical Parameters for Benzenoid Hydrocarbons. Thermochim Acta 8: 225

Herndon WC (1976) Ionization Potentials of π Molecular Hydrocarbons. J Am Chem Soc 98: 887

Herndon WC (1980) Structure-Resonance Theory - A Review of Applications to π-Hydrocarbon Systems. Israel J Chem 20: 270

Herndon WC, Hosoya H (1984) Parametrized Valence Bond Calculations for Benzenoid Hydrocarbons Using Clar Structures. Tetrahedron 40: 3987

Herndon WC, Párkányi C (1976) π-Bond Orders and Bond Lengths. J Chem Educ 53: 689

Hite GE, Metropoulos A, Klein DJ, Schmalz TG, Seitz WG (1986) Extended π-Networks with Multiple Spin-Pairing Phases - Resonance-Theory Calculations on Poly-Polyphenanthrenes. Theor Chim Acta 69: 369

Hosoya H (1973) Topological Index and Fibonacci Numbers with Relation to Chemistry. Fibonacci Quarterly 11: 255

Hosoya H (1986a) Matching and.Symmetry of Graphs. Comp & Maths with Appls 12B: 271

Hosoya H (1986b) How to Design Non-Kekulé Polyhex Graphs? Croat Chem Acta 59: 583

Hosoya H, Ohkami N (1983) Operator Technique for Obtaining the Recursion Formulas of Characteristic and Matching Polynomials as Applied to Polyhex Graphs. J Comput Chem 4: 585

Hosoya H, Yamaguchi T (1975) Sextet Polynomials - A New Enumeration and Proof Technique for the Resonance Theory Applied to the Aromatic Hydrocarbons. Tetrahedron Letters: 4659

Jiang YS (1980) Graph Theory of Molecular Orbitals - Calculation of α_N - the Determinant of Adjacent Matrix and the Stability of Molecules. Scient Sinica 23: 847

John P, Rempel J (1985) Counting Perfect Matchings in Hexagonal Systems. Proc Int Conf on Graph Theory, Eyba 1984. Teubner, Leipzig; p 72

John P, Sachs H (1985a) Wegesysteme und Linearfaktoren in hexagonalen und quadratischen Systemen [in] Graphen in Forschung und Unterricht (Bodendiek R, Schumacher H, Walter G, Edit). Verlag Barbara Franzbecker/Didaktischer Dienst Verlag, Bad Salzdetfurth; p 85

John P, Sachs H (1985b) Calculating the Number of Perfect Matchings and Pauling's Bond Orders in Hexagonal Systems Whose Inner Dual is a Tree. Proc Int Conf on Graph Theory, Eyba 1984. Teubner, Leipzig; p 80

Kekulé A (1865) Sur la constitution des substances aromatiques. Bull Soc Chim Paris, Nouv Ser 3: 98

Kekulé A (1866) Untersuchungen über aromatische Verbindungen. Annal d Chem u Pharm 137: 129

Kiang YS (1980a) Calculation of the Determinant of the Adjacency Matrix and the Stability of Conjugated Molecules. Internat J Quant Chem 18: 331

Kiang YS (1980b) Determinant of Adjacency Matrix and Kekulé Structures. Internat J Quant Chem Symposium 14: 541

Klein DJ, Alexander SA, Seitz WA, Schmalz TG, Hite GE (1986) Wavefunction Comparisons for the Valence-Bond Model for Conjugated π-Networks. Theor Chim Acta 69: 393

Klein DJ, Hite GE, Schmalz TG (1986) Transfer-Matrix Method for Subgraph Enumeration - Applications to Polypyrene Fusenes. J Comput Chem 7: 443

Klein DJ, Hite GE, Seitz WA, Schmalz TG (1986) Dimer Coverings and Kekulé Structures on Honeycomb Lattice Strips. Theor Chim Acta 69: 409

Klein DJ, Schmalz TG, Hite GE, Metropoulos A, Seitz WA (1985) The Poly-Polyphenanthrene Family of Multi-Phase π-Network Polymers in a Valence-Bond Picture. Chem Phys Letters 120: 367

Klein DJ, Schmalz TG, Hite GE, Seitz WA (1986) Resonance in C_{60}, Buckminsterfullerene. J Am Chem Soc 108: 1301

Knop JV, Müller WR, Szymanski K, Trinajstić N (1985) Computer Generation of Certain Classes of Molecules. SKTH/Kemija u industriji (Association of Chemists and Technologists of Croatia), Zagreb

Knop JV, Szymanski K, Trinajstić N, Křivka P (1984) Computer Generation of all 1-Factors for a Class of Graphs with all Vertices of Degree Two or Three. Comp & Maths with Appls 10: 369

Knop JV, Trinajstić N (1980) Chemical Graph Theory - II - On the Graph Theoretical Polynomials of Conjugated Structures. Internat J Quant Chem Symposium 14: 503

Kostochka AV [Косточка АВ] (1985) Критерий существования совершенных паросочетаний в шестиугольных системах. Proc 30 Intern Wiss Koll TH Ilmenau 1985, Vortragsreihe F: 49

Křivka P, Nikolić S, Trinajstić N (1986) Application of the Reduced Graph Model - Enumeration of Kekulé Structures for Certain Classes of Large Benzenoid Hydrocarbons. Croat Chem Acta 59: 659

Longuet-Higgins HC (1950) Some Studies in Molecular Orbital Theory - I - Resonance Structures and Molecular Orbitals in Unsaturated Hydrocarbons. J Chem Phys 18: 265

Marckwald W (1894) Ueber die Constitution der Ringsysteme. Liebig Annal d Chem 279: 1

Müller E, Müller-Rodloff I (1935) Magnetochemische Untersuchungen organischer Stoffe - 1. Mitteilung - Zur Frage der Existenz von Biradikalen. Liebig Annal d Chem 517: 134

Neto N, Scrocco M, Califano S (1966) A Simplified Valence Force Field of Aromatic Hydrocarbons - I - Normal Co-ordinate Calculation for C_6H_6, C_6D_6, $C_{10}H_8$, $C_{10}D_8$, $C_{14}H_{10}$ and $C_{14}D_{10}$. Spectrochim Acta 22: 1981

Newman MS, Lednicer D (1956) The Synthesis and Resolution of Hexahelicene. J Am Chem Soc 78: 4765

Ohkami N, Hosoya H (1983) Topological Dependency of the Aromatic Sextets in Polycyclic Benzenoid Hydrocarbons - Recursive Relations of the Sextet Polynomial. Theor Chim Acta 64: 153

Pauling L (1939) The Nature of the Chemical Bond. Cornell University Press, Ithaca, New York

Pauling L (1980) Bond Numbers and Bond Lengths in Tetrabenzo[de,no,st,c_1,d_1]Heptacene and other Condensed Aromatic Hydrocarbons - A Valence-Bond Treatment. Acta Cryst B36: 1898

Pauling L, Brockway LO, Beach JY (1935) The Dependence of Interatomic Distance on Single Bond - Double Bond Resonance. J Am Chem Soc 57: 2705

Polansky OE, Derflinger G (1967) Zur Clar'schen Theorie Lokaler Benzoider Gebiete in Kondensierten Aromaten. Internat J Quant Chem 1: 379

Polansky OE, Gutman I (1980) Graph-Theoretical Treatment of Aromatic Hydrocarbons V - The Number of Kekulé Structures in an All-Benzenoid Aromatic Hydrocarbon. Match 8: 269

Polansky OE, Rouvray DH (1976a) Graph-Theoretical Treatment of Aromatic Hydrocarbons I - The Formal Graph-Theoretical Description. Match 2: 63

Polansky OE, Rouvray DH (1976b) Graph-Theoretical Treatment of Aromatic Hydrocarbons II - The Analysis of All-Benzenoid Systems. Match 2: 91

Pullman A (1946) Sur le calcul du nombre de formules mésomères de chaque degré d'excitation pour les hydrocarbures aromatiques condensés. Compt Rend Acad Sci Paris 222: 736

Ramaraj R, Balasubramanian K (1985) Computer Generation of Matching Polynomials of Chemical Graphs and Lattices. J Comput Chem 6: 122

Randić M (1975) Graph Theoretical Approach to Local and Overall Aromaticity of Benzenoid Hydrocarbons. Tetrahedron 31: 1477

Randić M (1976a) Conjugated Circuits and Resonance Energies of Benzenoid Hydrocarbons. Chem Phys Letters 38: 68

Randić M (1976b) Enumeration of the Kekulé Structures in Conjugated Hydrocarbons. J Chem Soc Faraday Trans 2 72: 232

Randić M (1980) Resonance Energy of Very Large Benzenoid Hydrocarbons. Internat J Quant Chem 17: 549

Riordan J (1968) Combinatorial Identities. Wiley, New York

Robertson JM (1948) Bond-Length Variations in Aromatic Systems. Acta Cryst 1: 101

Robertson JM, White JG (1947) The Crystal Structure of Pyrene - A Quantitative X-Ray Investigation. J Chem Soc: 358

Sachs H (1984) Perfect Matchings in Hexagonal Systems. Combinatorica 4: 89

Sachs H, Zernitz H (1984) Ein O(n.log n)-Algorithmus zur Auffindung eines Linear-faktors in einem hexagonalen System - Teil I. Wyzsza Szkola Inz, Zielona Gora, Zeszyty Naukowe 75: 101

Scherer JR (1962) Modified Urey-Bradley Force Field for Condensed Aromatic Rings. J Chem Phys 36: 3308

Schmalz TG, Seitz WA, Klein DJ, Hite GE (1986) C_{60} Carbon Cages. Chem Phys Letters 130: 203

Schmidt W (1977) Photoelectron Spectra of Polynuclear Aromatics - V - Correlations with Ultraviolet Absorption Spectra in the Catacondensed Series. J Chem Phys 66: 828

Seitz WA, Klein DJ, Schmalz TG, García-Bach MA (1985) The Poly-Polyacene Family of Multi-Phase π-Network Polymers in a Valence-Bond Picture. Chem Phys Letters 115: 139

Smith FT (1961) Capacitive Energy and the Ionization of Aromatic Hydrocarbons. J Chem Phys 34: 793

Spiegel MR (1971) Finite Differences and Difference Equations. McGraw-Hill, New York

Stein SE, Brown RL (1985) Chemical Theory of Graphite-Like Molecules. Carbon 23: 105

Streitwieser A (1961) Molecular Orbital Theory for Organic Chemists. Wiley, New York

Su LX (1986) Linear Recursion Relations and the Enumeration of Kekulé Structures of Benzenoid Hydrocarbons. Match 20: 229

Swinborne-Sheldrake R, Herndon WC, Gutman I (1975) Kekulé Structures and Resonance Energies of Benzenoid Hydrocarbons. Tetrahedron Letters: 755

Trinajstić N (1983) Chemical Graph Theory, Vol I, II, CRC Press, Boca Raton, Florida

Trinajstić N, Křivka P (1986) On the Reduced Graph Model [in] Mathematics and Computational Concepts in Chemistry (Trinajstić N, Edit). Ellis Horwood Limited, Chichester; p 328

Wheland GW (1935) The Number of Canonical Structures of Each Degree of Excitation for an Unsaturated or Aromatic Hydrocarbon. J Chem Phys 3: 356

Yen TF (1971) Resonance Topology of Polynuclear Aromatic Hydrocarbons. Theor Chim Acta 20: 399

Zander M (1978a) Eine PMO-theoretische Untersuchung der benzogenen Diels-Alder-Reaktion polycyclischer aromatischer Kohlenwasserstoffe. Z Naturforsch 33a: 1395

Zander M (1978b) Empirische Beziehungen der Energie der para-Bande und der Topologie von kata-annellierten aromatischen Kohlenwasserstoffen. Z Naturforsch 33a: 1398

Zander M (1979) Korrelationen zwischen PMO-Lokalisierungsenergien und PO-MO-Charakterordnungen benzoider Teilsysteme in polycyclischen aromatischen Kohlenwasserstoffen. Z Naturforsch 34a: 521

Zhang FJ, Chen RS (1986a) On Kekulé Structure Count of Hexagonal Systems. J Xinjiang Univ 3(3): 10

Zhang FJ, Chen RS (1986b) A Theorem Concerning Polyhex Graphs. Match 19: 179

Zhang FJ, Chen RS, Guo XF (1985) Perfect Matchings in Hexagonal Systems. Graphs and Combinatorics 1: 383

Zhang FJ, Chen RS, Guo XF, Gutman I (1986) An Invariant of the Kekulé Structures of Benzenoid Hydrocarbons. J Serb Chem Soc 51: 537

Zhang FJ, Chen RS, Gutman I (1985) Some Topological Properties of Generalized S,T-Isomers. Match 18: 101

Živković TP, Trinajstić N (1987) On the Number of Kekulé Structures of Unbranched Cata-Condensed Benzenoid Chains. Chem Phys Letters 136: 141

SUBJECT INDEX

I. Gutman, O. E. Polansky

Mathematical Concepts in Organic Chemistry

1986. 28 figures. X, 212 pages.
ISBN 3-540-16235-6

Contents: Introduction. – Chemistry and Topology: Topological Aspects in Chemistry. Molecular Topology. – Chemistry and Graph Theory: Chemical Graphs. Fundamentals of Graph Theory. Graph Theory and Molecular Orbitals. Special Molecular Graphs. – Chemistry and Group Theory: Fundamentals of Group Theory. Symmetry Groups. Automorphism Groups. Some Interrelations between Symmetry and Automorphism Groups. – Special Topics: Topological Indices. Thermodynamic Stability of Conjugated Molecules. Topological Effect on Molecular Orbitals.– Appendices 1–6. – Literature. – Subject Index.

The present book is an attempt to outline some, certainly not all, mathematical aspects of modern organic chemistry. The mathematical concepts of organic chemistry as graph theory, group theory, and topology as well as their numerous interrelations are presented in a concise and simple, but mathematically rigorous fashion. Thus, this book emphasizes for the first time the concepts of mathematical chemistry – new branch of science that is being formed in our own times.

Springer-Verlag
Berlin Heidelberg New York
London Paris Tokyo

Distribution rights for the socialist countries: Akademie-Verlag, Berlin

Lecture Notes in Chemistry